DATE DUE

Unless Recalled Earlier

DEMCO 38-297

Compendium of Organic
Synthetic Methods

Compendium of Organic Synthetic Methods

Volume 7

MICHAEL B. SMITH

DEPARTMENT OF CHEMISTRY
THE UNIVERSITY OF CONNECTICUT
STORRS, CONNECTICUT

A Wiley-Interscience Publication

JOHN WILEY & SONS, INC.
New York • Chichester • Brisbane • Toronto • Singapore

In recognition of the importance of preserving what has been
written, it is a policy of John Wiley & Sons, Inc., to have books
of enduring value published in the United States printed on
acid-free paper, and we exert our best efforts to that end.

Library of Congress Catalog Card Number: 71-162800

ISBN 0-471-60713-4

Printed and bound in the United States of America
by Braun-Brumfield, Inc.

10 9 8 7 6 5 4 3

PREFACE

It has now been about twenty years since Ian and Shuyen Harrison first published the *Compendium of Organic Synthetic Methods*. Its goal was to facilitate the search for functional group transformations in the original literature of organic chemistry. In Volume 2, difunctional compounds were added and this compilation was continued by Louis Hegedus and Leroy Wade for Volume 3 of the series. Wade became the author for Volume 4 and continued with Volume 5. I edited the series beginning with Volume 6 in which I introduced an author index for the first time and added a new chapter (Chapter 15, "Oxides"). Even in this day of rapid dissemination of information by computers, the *Compendium* is a handy desktop reference that hopefully remains a valuable tool to the working organic chemist. The body of organic literature is so large that a "comprehensive one-volume listing of synthetic methods . . ." is impractical at a reasonable price. The *Compendium* is, therefore, a focused and highly representative review of the literature and is offered in that context.

Compendium of Organic Synthetic Methods, Volume 7 presents the functional group transformations, as well as many carbon bond forming reactions, for the literature appearing in the years 1987, 1988, and 1989. The classification schemes of all previous volumes has been used, including Chapter 15, which continues from Volume 6. Difunctional compound appear in Chapter 16. It was noted that many preparations of difunctional compounds include oxides of sulfur, nitrogen or phosphorous. Since the classification scheme made these functional groups "invisible," new sections have been added to the end of Chapter 16. Beginning with Section 378 ("Oxides–Alkynes") and ending with Section 389 ("Oxides–Oxides"), these important difunctional compounds can assume their rightful place in the *Compendium*. I hope these few sections will prove useful to the organic chemistry community. The experienced user of the *Compendium* will require no special instructions for the use of the new sections of Volume 7.

Author citations and the Author Index have been continued from Volume 6. Alphabetized heading have been added and the format changed slightly and it is hoped this addition to the series is useful and will facilitate it use.

The manuscript for Volume 7 was prepared on a MacIntoch-Plus® PC using Microsoft Word® (version 4.0) for word processing. All structures were prepared using ChemDraw™ (version 2.1.3, license #6021) and the manuscript was printed with a LaserWriter® II printer.

Finally, I want to thank my wife Sarah and my son Steven who have shown unfailing patience and devotion during this work. Such support is essential for the completion of any work of this nature and it was never lacking. I also wish to thank Ted Hoffman who has shown unflinching interest and dedication to this series and is truly responsible for its continued life.

MICHAEL B. SMITH

Storrs, Connecticut
July 1991

CONTENTS

ABBREVIATIONS ix

INDEX, MONOFUNCTIONAL COMPOUNDS xiii

INDEX, DIFUNCTIONAL COMPOUNDS xiv

INTRODUCTION xv

1 PREPARATION OF ALKYNES 1

2 PREPARATION OF ACID DERIVATIVES AND ANHYDRIDES 6

3 PREPARATION OF ALCOHOLS 15

4 PREPARATION OF ALDEHYDES 59

5 PREPARATION OF ALKYLS, METHYLENES AND ARYLS 72

6 PREPARATION OF AMIDES 121

7 PREPARATION OF AMINES 141

8 PREPARATION OF ESTERS 165

9 PREPARATION OF ETHERS, EPOXIDES AND THIOETHERS 188

10 PREPARATION OF HALIDES AND SULFONATES 206

11 PREPARATION OF HYDRIDES 219

12 PREPARATION OF KETONES 230

13 PREPARATION OF NITRILES 258

14 PREPARATION OF ALKENES 264

15 PREPARATION OF OXIDES 290

16 PREPARATION OF DIFUNCTIONAL COMPOUNDS 298

AUTHOR INDEX 487

ABBREVIATIONS

Ac   Acetyl,
acac   Acetylacetonate
AIBN   *Axo-bis*-isobutyronitrile
aq.   Aqueous

9-BBN   9-Borabicyclo[3.31]nonane
BINAP   2R,3S-2,2'-bis-(diphenylphosphino)-1,1'-binapthyl
Bn   Benzyl
Bz   Benzoyl

BOC   t-Butoxycarbonyl,
bpy (Bipy)   2,2'-Bipyridyl
Bu   n-Butyl, $-CH_2CH_2CH_2CH_3$
CAM   Carboxamidomethyl
CAN   Ceric ammonium nitrate, $(NH_4)_2Ce(NO_3)_6$
c-   Cyclo-
cat.   Catalytic

Cbz   Carbobenzyloxy,
Chirald   2S,3R-(+)-4-dimethylamino-1,2-diphenyl-3-methylbutan-2-ol
COD   1,5-Cyclooctadienyl
COT   1,3,5-Cyclooctatrienyl
Cp   Cyclopentadienyl
CSA   Camphorsulfonic acid
CTAB   Cetyltrimethylammonium bromide, $C_{16}H_{33}NMe_3{}^+Br$

Cy (c-C_6H_{11})   Cyclohexyl,

°C   Temperature in degrees Centigrade
DABCO   1,4-Diazobicyclo[2.2.2]octane
dba   Dibenzylidene acetone
DBE   1,2-Dibromoethane, $BrCH_2CH_2Br$
DBN   1,8-Diazabicyclo[5.4.0]undec-7-ene
DBU   1,5-Diazabicyclo[4.3.0]non-5-ene

DCC	1,3-Dicyclohexylcarbodiimide, $c-C_6H_{13}-N=C=N-c-C_6H_{13}$
DCE	1,2-Dichloroethane, $ClCH_2CH_2Cl$
DDQ	2,3-Dichloro-5,6-dicyano-1,4-benzoquinone
%de	% Diasteromeric excess
DEA	Diethylamine $HN(CH_2CH_3)_2$
DEAD	Diethylazodicarboxylate $EtO_2C-N=N-CO_2Et$
Dibal-H	Diisobutylaluminum hydride $(Me_2CHCH_2)_2AlH$
Diphos (dppe)	1,2-bis-(Diphenylphosphino)ethane $Ph_2PCH_2CH_2PPh_2$
Diphos-4 (dppb)	1,4-bis-(Diphenylphosphino)butane $Ph_2P(CH_2)_4PPh_2$
DMAP	4-Dimethylaminopyridine
DME	Dimethoxyethane $MeOCH_2CH_2OMe$
DMF	N,N'-Dimethylformamide $H\overset{\displaystyle O}{\underset{}{\overset{\|}{C}}}N(CH_3)_2$
dppf	bis-(Diphenylphosphino)ferrocene
dppp	1,3-bis-(Diphenylphosphino)propane $Ph_2P(CH_2)_3PPh_2$
dvb	Divinylbenzene
e^-	Electrolysis
%ee	% Enantiomeric excess
EE	1-Ethoxyethyl $EtO(Me)HCO-$
Et	Ethyl $-CH_2CH_3$
EDA	Ethylenediamine $H_2NCH_2CH_2NH_2$
EDTA	Ethylenediaminetetraacetic acid
FMN	Flavin mononucleotide
fod	tris-(6,6,7,7,8,8,8)-Heptafluoro-2,2-dimethyl-3,5-octanedionate
Fp	Cyclopentadieny-bis-carbonyl iron
FVP	Flash vacuum pyrolysis
h	hour (hours)
hν	Irradiation with light
1,5-HD	1,5-Hexadienyl
HMPA	Hexamethylphosphoramide $(Me_3N)_3P=O$
HMPT	Hexamethylphorous triamide $(Me_3N)_3P$
iPr	Isopropyl $-CH(CH_3)_2$
LICA (LIPCA)	Lithium cyclohexylisopropylamide
LDA	Lithium diisopropylamide $LiN(iPr)_2$
LHMDS	Lithium hexamethyl disilazide $LiN(SiMe_3)_2$
LTMP	Lithium 2,2,6,6-tetramethylpiperidide
mCPBA	meta-Chloroperoxybenzoic acid
Me	Methyl $-CH_3$
MEM	β-Methoxyethoxymethyl $MeOCH_2CH_2OCH_2-$
Mes	Mesityl $2,4,6$-tri-Me$-C_6H_2$
MOM	Methoxymethyl $MeOCH_2-$
Ms	Methanesulfonyl CH_3SO_2-

MS	Molecular sieves (3Å or 4Å)
MTM	Methylthiomethyl CH_3SCH_2-
NAD	Nicotinamide adenine dinucleotide
NADP	Sodium triphosphopyridine nucleotide
Napth	Napthyl ($C_{10}H_8$)
NBD	Norbornadiene
NBS	N-Bromosuccinimide
NCS	N-Chlorosuccinimide
NIS	N-Iodosuccinimide
Ni(R)	Raney nickel
Oxone	$2\ KHSO_5 \cdot KHSO_4 \cdot K_2SO_4$
(P)	Polymeric backbone
PCC	Pyridinium chlorochromate
PDC	Pyridinium dichromate
PEG	Polyethylene glycol
Ph	Phenyl
PhH	Benzene
PhMe	Toluene
Phth	Phthaloyl
Pip	Piperidine
Pr	n-Propyl $-CH_2CH_2CH_3$
Py	Pyridine
quant.	Quantitative yield
Red-Al	$[(MeOCH_2CH_2O)_2AlH_2]Na$
sBu	sec-Butyl $CH_3CH_2CH(CH_3)$
sBuLi	sec-Butyllithium $CH_3CH_2CH(Li)CH_3$
Siamyl	Diisoamyl $(CH_3)_2CHCH(CH_3)-$
TASF	tris-(Diethylamino)sulfonium difluorotrimethyl silicate
TBAF	Tetrabutylammonium fluoride n-$Bu_4N^+F^-$
TBDMS	t-Butyldimethylsilyl t-$BuMe_2Si$
TBHP (t-BuOOH)	t-Butylhydroperoxide $Me_3C-COOH$
t-Bu	tert-Butyl $-C(CH_3)_3$
TEBA	Triethylbenzylammonium $Bn(CH_3)_3N^+$
TEMPO	Tetramethylpiperdinyloxy free radical
TFA	Trifluoroacetic acid CF_3COOH
TFAA	Trifluoroacetic anhydride $(CF_3CO)_2O$
Tf (OTf)	Triflate $-SO_2CF_3(-OSO_2CF_3)$
THF	Tetrahydrofuran
THP	Tetrahydropyran
TMEDA	Tetramethylethylenediamine $Me_2NCH_2CH_2NMe_2$
TMS	Trimethylsilyl $-Si(CH_3)_3$

TMP	2,2,6,6-Tetramethylpiperidine
Tol	Tolyl $4-C_6H_4CH_3$
Tr	Trityl $-CPh_3$
TRIS	Triisopropylphenylsulfonyl
Ts(Tos)	Tosyl = p-Toluenesulfonyl $4-MeC_6H_4$
)))	Sonication
X_c	Chiral auxiliary

INDEX, MONOFUNCTIONAL COMPOUNDS

Sections—heavy type
Pages—light type

PREPARATION OF →
FROM ↓

FROM \ PREPARATION OF	Alkynes	Carboxylic acids, acid halides, anhydrides	Alcohols, phenols	Aldehydes	Alkyls, methylenes, aryls	Amides	Amines	Esters	Ethers, epoxides	Halides, sulfonates	Hydrides (RH)	Ketones	Nitriles	Alkenes	Oxides
Alkynes	**1** 1	**16** 6	**31** 15	**46** 59	**61** 72			**106** 165	**121** 188			**166** 230		**196** 264	**211** 290
Carboxylic acids, acid halides, anhydrides		**17** 7		**47** 59	**62** 73	**77** 121		**107** 166	**122** 189	**137** 206		**167** 232		**197** 267	**212** 290
Alcohols, phenols		**18** 7	**33** 16	**48** 60	**63** 73	**78** 123	**93** 141	**108** 169	**123** 189	**138** 207	**153** 219	**168** 235	**183** 258	**198** 268	**213** 291
Aldehydes	**4** 2	**19** 8	**34** 17	**49** 62	**64** 74	**79** 124	**94** 142	**109** 171	**124** 192	**139** 210		**169** 241	**184** 254	**199** 270	
Alkyls, methylenes, aryls			**35** 30	**50** 63	**65** 75		**95** 143			**140** 211	**155** 221			**200** 275	**215** 291
Amides		**21** 9	**36** 31	**51** 64		**81** 125	**96** 143	**111** 173	**126** 194			**171** 242	**186** 260		
Amines			**37** 31	**52** 64	**67** 78	**82** 131	**97** 145			**142** 213		**172** 243	**187** 260	**202** 276	**217** 292
Esters		**23** 10	**38** 31	**53** 65	**68** 79	**83** 134	**98** 154	**113** 174	**128** 194	**143** 213	**158** 221	**173** 245	**188** 261	**203** 277	
Ethers, epoxides			**39** 33	**54** 65	**69** 80	**84** 135		**114** 178	**129** 196	**144** 214	**159** 222			**204** 278	**219** 293
Halides, sulfonates, sulfates	**10** 3	**25** 11	**40** 36	**55** 66		**85** 135	**100** 155	**115** 179	**130** 197	**145** 214	**160** 223	**175** 246	**190** 261	**205** 279	**220** 295
Hydrides (RH)		**26** 11	**41** 37	**56** 67	**71** 86		**101** 156	**116** 181	**131** 198	**146** 216		**176** 247			**221** 296
Ketones		**27** 12	**42** 38	**57** 67	**72** 87	**87** 136	**102** 156	**117** 182	**132** 199	**147** 217	**162** 227	**177** 248	**192** 262	**207** 282	
Nitriles		**28** 12		**58** 68	**73** 87	**88** 137	**103** 157	**118** 185			**163** 228			**208**	
Alkenes	**14** 4	**29** 12	**44** 48	**59** 68	**74** 88	**89** 137	**104** 157	**119** 186	**134** 201	**149** 217		**179** 252	**194** 263	**209** 286	
Miscellaneous compounds	**15** 4	**30** 13	**45** 51	**60** 69	**75** 118	**90** 138	**105** 158	**120** 187	**135** 204	**150** 218	**165** 229	**180** 253		**210** 288	**225** 297

PROTECTION

	Sect.	Pg.
Carboxylic acids	30A	13
Alcohols, phenols	45A	51
Aldehydes	60A	70
Amines	105A	163
Ketones	180A	255

Blanks in the table correspond to sections for which no additional examples were found in the literature.

INDEX, DIFUNCTIONAL COMPOUNDS

Sections—heavy type
Pages—light type

	Alkyne	Carboxylic acid	Alcohol	Aldehyde	Amide	Amine	Ester	Ether, epoxide	Halide	Ketone	Nitrile	Alkene	Oxides
Alkyne	**300** 298												
Carboxylic acid		**312** 306											
Alcohol	**302** 299	**313** 307	**323** 315										
Aldehyde			**324** 321										
Amide	**304** 300	**315** 308	**325** 322	**334** 354	**342** 358								
Amine	**305** 301	**316** 308	**326** 325	**335** 354	**343** 358	**350** 372							
Ester	**306** 302	**317** 310	**327** 329	**336** 354	**344** 359	**351** 374	**357** 389						
Ether, epoxide	**307** 302		**328** 334	**337** 354	**345** 361	**352** 378	**358** 391	**363** 411					
Halide	**308** 302	**319** 312	**329** 337		**346** 362	**353** 379	**359** 393	**364** 414	**368** 436				
Ketone	**309** 303	**320** 312	**330** 339	**339** 355	**347** 363	**354** 380	**360** 395	**365** 418	**369** 437	**372** 443			
Nitrile	**310** 304				**348** 366	**355** 383		**366** 427		**373** 447			
Alkene	**311** 304	**322** 314	**332** 347	**341** 356	**349** 366	**356** 384	**362** 401	**367** 428	**371** 439	**374** 448	**376** 461	**377** 462	
Oxides	**378** 473	**379** 473	**380** 473	**381** 474	**382** 474	**383** 476	**384** 477	**385** 478	**386** 479	**387** 480	**388** 481	**389** 485	

Blanks in the table correspond to sections for which no additional examples were found in the literature.

INTRODUCTION

Relationship between Volume 7 and Previous Volumes. *Compendium of Organic Synthetic Methods, Volume 7* presents about 1250 examples of published methods for the preparation of monofunctional compounds, updating the 8100 in Volumes 1–6. In addition, Volume 7 contains about 850 examples of preparations of difunctional compounds with various functional groups, updating the sections introduced in Volume 2. Reviews have long been a feature of this series and Volume 7 adds almost 100 pertinent reviews in the various sections. Chapters 1–14 continue as in Volumes 1–6 and Chapter 15, introduced in Volume 6 continues. Difunctional compounds appear in Chapter 15 in Volumes 1–5 but in Chapter 16 in Volumes 6 and 7. In order to identify oxides of sulfur, nitrogen, and phosphorous that are a key functional group when present with alkynes through alkenes (as the *Compendium* is organized), new sections have been added to the end of Chapter 16. The new sections begin with Section 378 (Oxides–Alkynes), progress to Section 388 (Oxides–Alkenes), and end with Section 389 (Oxides–Oxides). These oxides are integral functional groups throughout organic chemistry and the new sections allow proper attention to be drawn to their preparation in the presence of other functionalities. As in Volume 6, all references to alcohols include thiols, ethers, and thioethers.

The authors for each citation appear below the reaction. The principle author is indicated by underlining (i.e., Zezza, C. A.; Smith, M. B.) rather than the asterisk (*) that was used in Volume 6. Following Chapter 16 is a complete alphabetical listing of all authors (last name, initials).

Classification and Organization of Reactions Forming Monofunctional Compounds. Chemical transformations are classified according to the reacting functional group of the starting material and the functional group formed. Those reactions that give products with the same functional group form a chapter. The reactions in each chapter are further classified into sections on the basis of the functional group of the starting material. Within each section, reactions are loosely arranged in ascending order of year cited (1987–1989), although an effort has been made to put similar reactions together when possible. Review articles are collected at the end of each appropriate section.

The classification is unaffected by allylic, vinylic, or acetylenic unsaturation appearing in both starting material and product, or by increases or decreases in the length of carbon chains; for example, the reactions t-BuOH → t-BuCOOH, PhCH$_2$OH → PhCOOH, and PhCH=CHCH$_2$OH → PhCH=CHCOOH would all be considered as preparations of carboxylic acids from alcohols. Conjugate reduction and alkylation of unsaturated ketones, aldehydes, esters, acids, and nitriles have been placed in category 74 (alkyls from olefins).

The terms hydrides, alkyls, and aryls classify compounds containing reacting hydrogens, alkyl groups, and aryl groups, respectively; for example, RCH$_2$—H → RCH$_2$COOH (carboxylic acids from hydrides), RMe → RCOOH (carboxylic acids from alkyls), RPh → RCOOH (carboxylic acids from aryls). Note the distinction between R$_2$CO → R$_2$CH$_2$ (methylenes from ketones) and RCOR' → RH (hydrides from ketones). Alkylations involving additions across double bonds are found in Section 74 (alkyls, methylenes, and aryls from olefins).

The following examples illustrate the classification of some potentially confusing cases:

RCH=CHCOOH → RCH=CH$_2$	Hydrides from carboxylic acids
RCH=CH$_2$ → RCH=CHCOOH	Carboxylic acids from hydrides
ArH → ArCOOH	Carboxylic acids from hydrides
ArH → ArOAc	Esters from hydrides
RCHO → RH	Hydrides from aldehydes
RCH=CHCHO → RCH=CH$_2$	Hydrides from aldehydes
RCHO → RCH$_3$	Alkyls from aldehydes
R$_2$CH$_2$ → R$_2$CO	Ketones from methylenes
RCH$_2$COR → R$_2$CHCOR	Ketones from ketones
RCH=CH$_2$ → RCH$_2$CH$_3$	Alkyls from olefins
RBr + CHCH → RC≡CR	Acetylenes from halides; also acetylenes from acetylenes
ROH + RCOOH → RCOOR	Esters from alcohols; also esters from carboxylic acids
RCH=CHCHO → RCH$_2$CH$_2$CHO	Alkyls from olefins
RCH=CHCN → RCH$_2$CH$_2$CN	Alkyls from olefins

How to Use the Book to Locate Examples of the Preparation of Protection of Monofunctional Compounds. Examples of the preparation of one functional group from another are located via the monofunctional index on p. xiii, which lists the corresponding section and page. Thus Section 1 contains examples of the preparation of acetylenes from other acetylenes; Section 2, acetylenes from carboxylic acids; and so forth.

Sections that contain examples of the reactions of a functional group are found in the horizontal rows of the index. Thus Section 1 gives examples of the reactions of acetylenes that form acetylenes; Section 16, reactions of acetylenes that form carboxylic acids; and Section 31, reactions of acetylenes that form alcohols.

Examples of alkylation, dealkylation, homologation, isomerization, and transposition are found in Sections 1, 17, 33, and so on, lying close to a diagonal of the index. These sections correspond to such topics as the preparation of acetylenes from acetylenes; carboxylic acids from carboxylic acids; and alcohols, thiols, and phenols from alcohols, thiols, and phenols. Alkylations that involve conjugate additions across a double bond are found in Section 74 (alkyls, methylenes, and aryls from olefins).

Examples of name reactions can be found by first considering the nature of the starting material and product. The Wittig reaction, for instance is in Section 199 (olefins from aldehydes) and Section 207 (olefins from ketones). The aldol condensation can be found in the chapters on difunctional compounds in Section 324 (alcohol, thiol–aldehyde) and in Section 330 (alcohol, thiol-ketone).

Examples of the protection of acetylenes, carboxylic acids, alcohols, phenols, aldehydes, amides, amines, esters, ketones, and olefins are also indexed on p. xiii.

The pairs of functional groups alcohol, ester; carboxylic acid, ester; amine, amide; and carboxylic acid, amide can be interconverted by simple reactions. When a member of these groups is the desired product or starting material, the other member should, of course, also be consulted in the text.

The original literature must be used to determine the generality of reactions, although this is occasionally stated in the citation. This is only done in cases where such generality is stated clearly in the original citation. A reaction given in this book for a primary aliphatic substrate may also be applicable to tertiary or aromatic compounds. This book provides very limited experimental conditions or precautions and the reader is referred to the original literature before attempting a reaction. In no instance should the citation be taken as a complete experimental procedure. Not to refer to the original literature could be hazardous. The original papers usually yield a further set of references to previous work. Subsequent publications can be found by consulting the *Science Citation Index*.

Classification and Organization of Reactions Forming Difunctional Compounds. This chapter considers all possible difunctional compounds formed from the groups acetylene, carboxylic acid, alcohol, thiol, aldehyde, amide, amine, ester, ether, epoxide, thioether, halide, ketone, nitrile, and olefin. Reactions that form difunctional compounds are classified into sections on the basis of the two functional groups of the product. The relative positions of the groups do not affect the classification. Thus preparations of 1,2-aminoalcohols, 1,3-aminoalcohols, and 1,4-aminoalcohols are included in a single section. Difunctional compounds that have an oxide as the second group are found in the monofunctional sections for the nonoxide functional group. Therefore, the nitroketone product of oxidation of a nitroalcohol is found in Section 168 (ketones from alcohols and thiols). Conversion of an oxide to another functional group is generally found in the "Miscellaneous" section, so conversion of a nitroalkane to an amine is found in Section 105 (amines from miscellaneous compounds). The following examples illustrate the application of this classification system:

Difunctional Product	Section Title
$RC{\equiv}C{-}C{\equiv}CR$	Acetylene–acetylene
$RCH(OH)COOH$	Carboxylic acid–alcohol
$RCH{=}CHOMe$	Ether–olefin
$RCHF_2$	Halide–halide
$RCH(Br)CH_2F$	Halide–halide
$RCH(OAc)CH_2OH$	Alcohol–ester
$RCH(OH)CO_2Me$	Alcohol–ester
$RCH{=}CHCH_2CO_2Me$	Ester–olefin
$RCH{=}CHOAc$	Ester–olefin
$RCH(OMe)CH_2SO_2CH_2CH_2OH$	Alcohol–ether
$RSO_2CH_2CH_2OH$	Oxides–alcohol

How to Use the Book to Locate Examples of the Preparation of Difunctional Compounds. The difunctional index on p. xiv gives the section and page corresponding to each difunctional product. Thus Section 327 (alcohol, thiol-ester) contains examples of the preparation of hydroxyesters; Section 323 (alcohol, thiol–alcohol, thiol) contains examples of the preparation of diols.

Some preparations of olefinic and acetylenic compounds from olefinic and acetylenic starting materials can, in principle, be classified in either the monofunctional or difunctional sections; for example, $RCH{=}CHBr \rightarrow RCH{=}CHCOOH$, carboxylic acids from halides (Section 25, monofunctional compounds) or carboxylic acid-olefin (Section 322, difunctional compounds). In such cases both sections should be consulted.

Reactions applicable to both aldehyde and ketone starting materials are in many cases illustrated by an example that uses only one of them.

Many literature preparations of difunctional compounds are extensions of the methods applicable to monofunctional compounds. Thus the reaction RCl → ROH can be extended to the preparation of diols by using the corresponding dichloro compound as a starting material. Such methods are not fully covered in the difunctional sections.

The user should bear in mind that the pairs of functional groups alcohol, ester; carboxylic acids, ester; amine, amide; and carboxylic acid, amide can be interconverted by simple reactions. Compounds of the type RCH(OAc)CH$_2$OAc (ester–ester) would thus be of interest to anyone preparing the diol RCH(OH)CH$_2$OH (alcohol–alcohol).

CHAPTER 1
PREPARATION OF ALKYNES

SECTION 1: ALKYNES FROM ALKYNES

Yeh, M.C.P.; Knochel, P. Tetrahedron Lett., **1989**, 30, 4799.

Baudin, J.-B.; Julia, S.A.; Wang, Y. Tetrahedron Lett., **1989**, 30, 4965.

SECTION 2: ALKYNES FROM ACID DERIVATIVES

NO ADDITIONAL EXAMPLES

SECTION 3: ALKYNES FROM ALCOHOLS AND THIOLS

NO ADDITIONAL EXAMPLES

SECTION 4: ALKYNES FROM ALDEHYDES

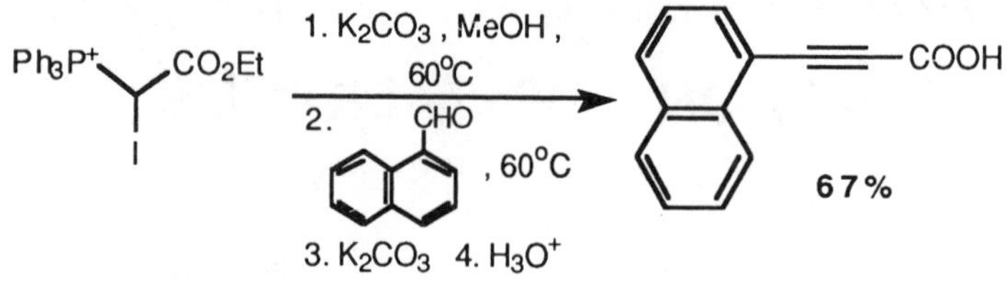

Ph_3P^+ CO_2Et , I

1. K_2CO_3 , MeOH , 60°C

2. CHO (naphthalene) , 60°C

3. K_2CO_3 4. H_3O^+

→ (naphthalene)—$C\equiv C$—COOH   **67%**

Chenault, J.; Dupin, J.-F.E. *Synthesis,* ***1987***, 498.

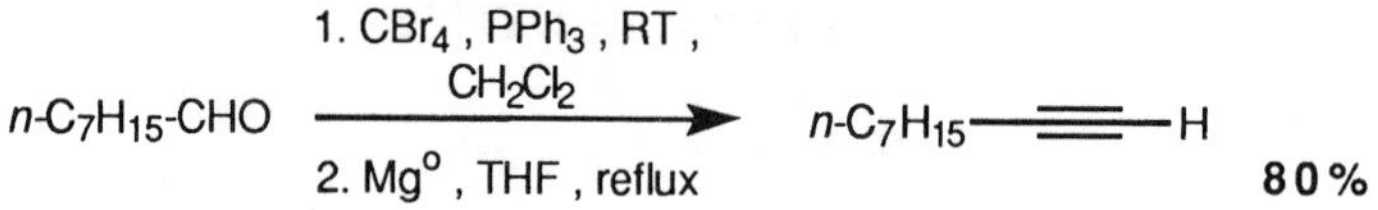

n-C_7H_{15}-CHO

1. CBr_4 , PPh_3 , RT , CH_2Cl_2

2. $Mg°$, THF , reflux

→ n-C_7H_{15}—$C\equiv C$—H   **80%**

van Hijfte, L.; Kolb, M.; Witz, P. *Tetrahedron Lett.,* ***1989***, *30*, 3655.

SECTION 5: ALKYNES FROM ALKYLS, METHYLENES AND ARYLS

NO ADDITIONAL EXAMPLES

SECTION 6: ALKYNES FROM AMIDES

NO ADDITIONAL EXAMPLES

SECTION 7: ALKYNES FROM AMINES

NO ADDITIONAL EXAMPLES

SECTION 8: ALKYNES FROM ESTERS

NO ADDITIONAL EXAMPLES

SECTION 9: ALKYNES FROM ETHERS, EPOXIDES AND THIOETHERS

NO ADDITIONAL EXAMPLES

SECTION 10: ALKYNES FROM HALIDES AND SULFONATES

Ph-I
$\xrightarrow[\substack{3h}]{5\% \text{ Pd(PPh}_3)_4 \text{ , THF , } 22^\circ C}}$

ClZn—≡—⟨

Ph—≡—⟨   90%

Negishi, E.; Akiyoshi, K.; Takahashi, T. *J. Chem. Soc., Chem. Commun.*, **1987**, 477.

$\xrightarrow[\substack{ZnCl_2 \\ CCl_4 \\ C_6H_{13}\text{-}C\equiv C}]{Bu_3SnC\equiv CC_6H_{13}}$ 49%

Zhai, D.; Zhai, W.; Williams, R.M. *J. Am. Chem. Soc.*, **1988**, *110*, 2501.

$\xrightarrow{4 \text{ eq. KOH , Aliquat 336}}$ Bu-C≡C-H   98%

Vinczer, P.; Kovacs, T.; Novak, L.; Szantay, C. *Org. Prep. Proceed. Int.*, **1989**, *21*, 232.

SECTION 11: ALKYNES FROM HYDRIDES

For examples of the reaction RC≡CH → RC≡C-C≡CR[1], see section 300 (Alkyne-Alkyne).

NO ADDITIONAL EXAMPLES

SECTION 12: ALKYNES FROM KETONES

Shioiri, T.; Iwamoto, Y.; Aoyama, T. *Heterocycles*, *1987*, *25*, 1467.

Engler, T.A.; Combrink, K.D.; Ray, J.E. *Synth. Commun.*, *1989*, *19*, 1735.

SECTION 13: ALKYNES FROM NITRILES

NO ADDITIONAL EXAMPLES

SECTION 14: ALKYNES FROM ALKENES

$$(76 \quad : \quad 24) \quad 98\%$$

Caporusso, A.M.; Polizzi, C.; Lardicci, R. *J. Org. Chem.*, *1987*, *52*, 3920

SECTION 15: ALKYNES FROM MISCELLANEOUS COMPOUNDS

Chen, Q.-Y.; He, Y.-B. *Tetrahedron Lett.*, *1987*, *28*, 2387.

SECTION 15A: PROTECTION OF ALKYNES

Iwamura. M.; Ishikawa, T.; Koyama, Y.; Sakuma, K.; Iwamura, H. *Tetrahedron Lett., 1987, 28*, 679.

CHAPTER 2

PREPARATION OF ACID DERIVATIVES

SECTION 16: ACID DERIVATIVES FROM ALKYNES

Moriarty, R.M.; Vaid, R.K.; Duncan, M.P.; Vaid, B.K. *Tetrahedron Lett., 1987, 28*, 2845.

HEPT = hexaethyl phosphoric triamide

Ballistreri, F.P.; Failla, S.; Tomaselli, G.A. *J. Org. Chem., 1988, 53*, 830.

Moiriarty, R.M.; Penmasta, R.; Awasthi, A.K. *J. Org. Chem., 1988, 53*, 6124.

SECTION 17: ACID DERIVATIVES FROM ACID DERIVATIVES

Ph
CO₂H

1. iPr, H, OMe , BuLi, THF
   iPrHN   -95°C
2. EtI , -95°C , 2h
3. H₂O

Ph, H
CO₂H

42%

24% ee , S

Ando, A.; Shioiri, T. *J. Chem. Soc., Chem. Commun.*, *1987*, 656.

SECTION 18: ACID DERIVATIVES FROM ALCOHOLS AND THIOLS

Me
SH
Me

2 eq. SO₂Cl₂ , AcOH
-40→ +30°C

Me O
    ‖
    S
Me Cl

95%

Youn, J.-H.; Herrmann, R. *Synthesis*, *1987*, 72.

$C_7H_{15}CH_2OH$

KMnO₄ - CuSO₄ •5 H₂O
CH₂Cl₂

$C_7H_{15}COOH$

90%

Jefford, C.W.; Wang, Y. *J. Chem. Soc., Chem. Commun.*, *1988*, 634.

OH
OH

NiOOH - anode ,
H₂O - K₂CO₃ , 20°C
e⁻, aq. NaHCO₃ ,
K₂CO₃ , 24h

HOOC   COOH

72%

Ruholl, H.; Schäfer, H.-J. *Synthesis*, *1988*, 54.

OH
'''OH

MoO₂(acac) - t-BuOOH
PhCl , 60°C , 24h

COOH
COOH

79%

Kameda, K.; Morimoto, K.; Imanaka, T. *Chem. Lett.*, *1988*, 1295.

$CH_3(CH_2)_8CH_2OH$

NaBrO₃ •HBr , RT
CCl₄ , t-BuOH
1h

$CH_3(CH_2)_8COOH$

78%

Veeraiah, T.; Periasamy, M. *Synth. Commun.*, *1989*, 19, 2151.

SECTION 19: ACID DERIVATIVES FROM ALDEHYDES

$$\text{(ArCHO)} \xrightarrow[\substack{Mn^{II} \text{ stearate} \\ \text{decane}}]{O_2 \text{ (120 psi)}} \text{(ArCOOH)}$$

75%

Riley, D.P.; Gelman, D.P.;Beck, G.R.; Heintz, R.M. *J. Org. Chem.*, *1987*, *52*, 287.

$$PhCH_2Br \xrightarrow[55^\circ C \text{ , 3h}]{CoCl_2 \text{ , NaBH}_4 \text{ , 5N NaOH , CO}} PhCH_2COOH$$

88%

Satayanarayana, N.; Perisamy, M. *Tetrahedron Lett.*, *1987*, *28*, 2633.

$$Cl-\text{C}_6\text{H}_4-CHO \xrightarrow[\substack{5\% \text{ PhSeO}_2H \text{ , 5h} \\ \text{reflux}}]{30\% \text{ H}_2O_2 \text{ , THF}} Cl-\text{C}_6\text{H}_4-COOH$$

94%

Choi, J.-K.; Chang, Y.-K.; Hong, S.Y. *Tetrahedron Lett.*, *1988*, *29*, 1967.

$$C_8H_{17}OH \xrightarrow[(C_{12}H_{25})_2NMe_2Br \text{ , 90 min}]{RuCl_3 \cdot 3 H_2O \text{ , 80}^\circ C \text{ , CH}_2Cl_2} C_7H_{15}COOH$$

85% conversion   **68%**

Barak, G.; Dakka, J.; Sasson, Y. *J. Org. Chem.*, *1988*, *53*, 3553.

$$O_2N-\text{C}_6\text{H}_4-CHO \xrightarrow[CH_2Cl_2 \text{ , 14h}]{BuMnO_4 \text{ , 36}^\circ C} O_2N-\text{C}_6\text{H}_4-COOH$$

results are poor with aliphatic aldehydes

Srivastava, R.G.; Venkataramani, P.S. *Synth. Commun.*, *1988*, *18*, 2193.

$$Ph\overset{O}{\underset{}{\text{C}}}H \xrightarrow[]{CsSO_4F \text{ , MeCN , 35}^\circ C} Ph\overset{O}{\underset{}{\text{C}}}F$$

>80%

Stavber, S.; Planinšek, Z.; Zupan, M. *Tetrahedron Lett.*, *1989*, *30*, 6095.

$$\text{Ph-CHO} \xrightarrow[\text{steam bath}]{\text{NaBO}_3 \cdot \text{H}_2\text{O} , \text{AcOH} , 2\text{h}} \text{Ph-COOH}$$

90%

Bannerjee, A.; Hazra, B.; Bhattacharya, A.; Bannerjee, S.; Bannerjee, G.C.; Sengupta, S. *Synthesis,* *1989*, 765.

SECTION 20: ACID DERIVATIVES FROM ALKYLS, METHYLENES AND ARYLS

NO ADDITIONAL EXAMPLES

SECTION 21: ACID DERIVATIVES FROM AMIDES

Evans, D.A.; Britton, T.C.; Ellman, J.A. *Tetrahedron Lett.,* *1987,* *28,* 6141.

Eaton, J.T.; Rounds, W.D.; Urbanowicz, J.H.; Gribble, G.W. *Tetrahedron Lett.,* *1988,* *29,* 6553.

SECTION 22: ACID DERIVATIVES FROM AMINES

NO ADDITIONAL EXAMPLES

SECTION 23: ACID DERIVATIVES FROM ESTERS

$$Ph\diagdown CO_2Me \xrightarrow[\text{reflux}]{H_2O \text{ , Dowex-50}} Ph\diagdown COOH$$

$$88\%$$

Basu, M.K.; Sarkar, D.C.; Ranu. B.C. *Synth. Commun., 1989, 19*, 627.

$$\xrightarrow[-30 \to 0^{\circ}C]{Bu_2CuLi \text{ , THF}} C_6H_{13}CH_2COOH \qquad 83\%$$

Kawashima, M.; Sato, T.; Fujisawa. T. *Tetrahedron, 1989, 45*, 403.

Other reactions useful for the hydrolysis of esters may be found in Section 30A (Protection of Carboxylic Acids).

SECTION 24: ACID DERIVATIVES FROM ETHERS, EPOXIDES AND THIOETHERS

NO ADDITIONAL EXAMPLES

SECTION 25: ACID DERIVATIVES FROM HALIDES AND SULFONATES

$$Ph\text{-}I \xrightarrow[\substack{3 \text{ atm CO , } 120^{\circ}C \text{ , DMF} \\ 20h}]{HCO_2^-Ca^+ \text{ , } PdCl_2 \text{ , } PPh_3} Ph\text{-}COOH \qquad 85\%$$

Pri-Bar, I.; Buchman, O. *J. Org. Chem., 1988, 53*, 624.

$$Me\text{—}\bigcirc\text{—}Cl \xrightarrow[Bu_4NBr]{e^- \text{ , } CO_2 \text{ , DMF}} Me\text{—}\bigcirc\text{—}COOH \qquad 78\%$$

Heintz, M.; Sock, O.; Saboureau, C.; Périchon. J. *Tetrahedron, 1988, 44*, 1631.

Kiji, J.; Okano, T.; Nishiumi, W.; Konishi, H. *Chem. Lett., 1988*, 957.

Amer, I.; Alper, H. *J. Org. Chem., 1988, 53*, 5147.

Amer, I.; Alper, H. *J. Am. Chem. Soc., 1989, 111*, 927.
Alper, H.; Amer, I.; Vasapollo, G. *Tetrahedron Lett., 1989, 30*, 2617.

Miura, M.; Okura, K.; Hattori, A.; Nomura, M. *J. Chem. Soc., Perkin Trans. I, 1989*, 73.

SECTION 26: ACID DERIVATIVES FROM HYDRIDES

Elsheimer, S.; Slattery, D.K.; Michael, M.; Weeks, J.; Topoleski, K. *J. Org. Chem., 1989, 54*, 3992.

SECTION 27: ACID DERIVATIVES FROM KETONES

Piccolo, O.; Spreafico, F.; Visentii, G. *J. Org. Chem., 1987, 52,* 10.

SECTION 28: ACID DERIVATIVES FROM NITRILES

Vo-Quang, Y.; Marais, D.; Vo-Quang, L.; LeGoffic, F.; Thiéry, A.; Maestracci, M.; Arnaud, A.; Galzy, P. *Tetrahedron Lett., 1987, 28,* 4057.

Rounds, W.D.; Eaton, J.T.; Urbanowicz, J.H.; Gribble, G.W. *Tetrahedron Lett., 1988, 29,* 6557.

SECTION 29: ACID DERIVATIVES FROM ALKENES

Urata, H.; Fujita, A.; Fuchikami, T. *Tetrahedron Lett., 1988, 29,* 4435.

Oguchi, T.; Ura, T.; Ishii, Y.; Ogawa, M. *Chem. Lett., 1989*, 857.

SECTION 30: ACID DERIVATIVES FROM MISCELLANEOUS COMPOUNDS

Ali, S.M.; Tanimoto, S. *J. Chem. Soc., Chem. Commun., 1988*, 1465.

REVIEW:

"Selenium and Tellurium Isologues of Carboxylic Acid Derivatives"

Kato, S.; Murai, T.; Ishida, M. *Org. Prep. Proceed. Int., 1986, 18,* 369.

SECTION 30A: PROTECTION OF CARBOXYLIC ACID DERIVATIVES

Chen, J.; Zhou, X.-J. *Synth. Commun., 1987, 17,* 161.

94%

Jaouadi, M.; Martinez, J.; Castro, B.; Barcelo, G.; Sennyey, G.; Senet, J.-P. *J. Org. Chem.*, *1987*, *52*, 2364.

Misra, P.K.; Hashmi, S.A.N.; Haq, W.; Katti, S.B. *Tetrahedron Lett.*, *1989*, *30*, 3569.

Barrett, A.G.M.; Lebold, S.A.; Zhang, X. *Tetrahedron Lett.*, *1989*, *30*, 7317.

83%

Perni, R.B. *Synth. Commun.*, *1989*, *19*, 2883.

Other reactions useful for the protection of carboxylic acids are included in Section 107 (Esters from Carboxylic Acids and Acid Halides) and Section 23 (Carboxylic Acids from Esters).

CHAPTER 3

PREPARATION OF ALCOHOLS

SECTION 31: ALCOHOLS AND THIOLS FROM ALKYNES

syn:anti = 94:6

Zweifel, G.; Hahn, G.R.; Shoup, T.M. *J. Org. Chem.*, **1987**, *52*, 5484.

Wulff, W.D.; Xu, Y.-C. *Tetrahedron Lett.*, **1988**, *29*, 415.

SECTION 32: ALCOHOLS AND THIOLS FROM ACID DERIVATIVES

NO ADDITIONAL EXAMPLES

SECTION 33: ALCOHOLS AND THIOLS FROM ALCOHOL AND THIOLS

Boger, D.L.; Coleman, R.S. *Tetrahedron Lett.*, **1987**, *28*, 1027.

trans:cis = 94:6)

Feghouli, G.; Vanderesse, R.; Fort, Y.; Caubere, P. *Tetrahedron Lett.*, **1988**, *29*, 1383.

Ceković, Z.; Ilijiv, D. *Tetrahedron Lett.*, **1988**, *29*, 1441.

Nishio, T. *J. Chem. Soc., Chem. Commun.*, **1989**, 205.

SECTION 34: ALCOHOLS AND THIOLS FROM ALDEHYDES

The following reaction types are included in this section:
A. Reductions of Aldehydes to Alcohols
B. Alkylation of Aldehydes, forming Alcohols.

Coupling of Aldehydes to form Diols is found in Section 323 (Alcohol-Alcohol).

88%

Syper, L. *Synthesis*, *1989*, 167.

SECTION 34A: REDUCTIONS OF ALDEHYDES TO ALCOHOLS

$$PhCHO \xrightarrow[\text{aq. DMF}]{HOCH_2SO_2Na \text{ , } 100°C} PhCH_2OH \quad 75\%$$

Harris, A.R.; Mason, T.J. *Synth. Commun.*, *1989*, *19*, 529.

70%

Shibata, I.; Yoshida, T.; Baba, A.; Matsuda, H. *Chem. Lett.*, *1989*, 619.

SECTION 34B: ALKYLATION OF ALDEHYDES, FORMING ALCOHOLS

ASYMMETRIC ALKYLATIONS

1. (+)-DPMPM , hexane

reflux , 20 min

PhCHO

2. Et$_2$Zn , 0°C , 4h

Ph, H

HO Et   92% ee , S

98%

1. (-)-erythro PMDM

2. Et$_2$Zn

Ph, H

71% ee , R

HO Et

quant.

R^1 = R^2 = Ph = (+)-DPMDM

R^1 = Ph, R^2 = H = (-)-erytho PMDM

Soai. K.; Ookawa, A.; Ogawa, K.; Kaba, T. *J. Chem. Soc., Chem. Commun.,* ***1987,*** 467.

Ph Me

Me

HO N N OH

Me (CH$_2$)$_3$

catalyst =   Me Ph   59% , 85% ee , R

Soai. K.; Nishi, M.; Ito, Y. *Chem. Lett.,* ***1987,*** 2405.

Li-N N-Li + PhCHO

Et$_2$Zn , PhMe

Ph Et

HO H   68%

92% ee , R

Soai. K.; Niwa, S.; Yamada, Y.; Inoue, H. *Tetrahedron Lett.,* ***1987,*** *28,* 4841.

$+$CH$_2$CH$+$$_x$

Me Ph

N OH

catalyst =   Me   67%   (61% ee , S)

Soai. K.; Niwa, S.; Watanabe, M. *J. Org. Chem.,* ***1988,*** *53,* 927.
Soai. K.;; Ookawa, A.; Kaba, T.; Ogawa, K. *J. Am. Chem. Soc.,* ***1987,*** *109,* 7111.

Minowa, N.; Mukaiyama, T. *Bull. Chem. Soc., Jpn., 1987, 60,* 6397.

Casiraghi, G.; Cornia, M.; Casnati, G.; Fava, G.G.; Belicchi, M.F.; Zetta, L. *J. Chem. Soc., Chem. Commun., 1987,* 794.

Tomioka, K.; Nakajima, M.; Koga, K. *Tetrahedron Lett., 1987, 28,* 1291.

Boldrini, G.P.; Lodi, L.; Tagliavini, E.; Tavasco, C.; Trombini, C.; Umani-Ronchi, A. *J. Org. Chem., 1987, 52,* 5447.

Krämer, T.; Hoppe, D. *Tetrahedron Lett., 1987, 28,* 5149.

Corey, E.J.; Hannon, F.J. *Tetrahedron Lett., 1987, 28,* 5233.

Muchow, G.; Vannoorenberghe, Y.; Buono, G. *Tetrahedron Lett., 1987, 28,* 6163.

Roush, W.R.; Ando, K.; Powers,.D.B.; Halterman, R.L.; Palkowitz, A.D. *Tetrahedron Lett., 1988, 29,* 5579.

$$\text{(cyclohexane-NHSO}_2\text{CF}_3\text{, NHSO}_2\text{CF}_3) \xrightarrow[\begin{subarray}{c}\text{2. Et}_2\text{Zn}\\ \text{3. PhCHO}\end{subarray}]{\text{1. Ti(OiPr)}_4} \text{Ph}\underset{\text{OH}}{\overset{}{\diagup}}\text{Et} \quad 99\%$$

98% ee , S

Yoshioka, M.; Kawakita, T.; Ohno, M. *Tetrahedron Lett.*, *1989*, *30*, 1657.
Takahashi, H.; Kawakita, T.; Yoshioka, M.; Kobayashi, S.; Ohno, M. *Tetrahedron Lett.*, *1989*, *30*, 7095.

$$\text{C}_6\text{F}_5\text{CHO} \longrightarrow \text{C}_6\text{F}_5\underset{\text{OH}}{\diagup}\diagdown\text{SPh} \quad 90\%$$

88% ee

Maruoka, K.; Hoshino, Y.; Shirasaka, T.; Yamamoto, H. *Tetrahedron Lett.*, *1989*, *30*, 3967.

$$\xrightarrow[-50^\circ\text{C , 2h}]{n\text{-C}_8\text{H}_{17}\text{CHO:TiCl}_4} \text{C}_8\text{H}_{17}\underset{\text{OH}}{\diagup}\diagdown \quad 61\%$$

48% ee , S

Chan, T.H.; Wang, D. *Tetrahedron Lett.*, *1989*, *30*, 3041.

$$\text{Ph}\diagdown\diagup\text{CHO} \xrightarrow[\begin{subarray}{c}\text{CrCl}_2\text{, THF , DMF}\\ 25^\circ\text{C , 2h}\end{subarray}]{\text{MeCH=CHCHCl}_2} \text{Ph}\diagdown\diagup\diagdown\text{(OH)}\diagdown\diagup\text{Cl} \quad 95\%$$

E:Z   =   96:4
anti:syn = 93:7

Takai, K.; Kataoka, Y.; Utimoto, K. *Tetrahedron Lett.*, *1989*, *30*, 4389.

Ph-CHO $\xrightarrow{\text{, PhMe , 0}^{\circ}\text{C , 24h}}$

88%

78% ee , R

Joshi, N.N.; <u>Srebnik, M.</u>; Brown, H.C. *Tetrahedron Lett.*, *1989*, *30*, 5581.

Ph-CHO $\xrightarrow[\text{Et}_2\text{Zn , PhMe}]{5\%}$

83%

97% ee , R

<u>Tanaka, K.</u>; Ushio, H.; Suzuki, H. *J. Chem. Soc., Chem. Commun.*, *1989*, 1700.

$\xrightarrow[\substack{3.\ \text{PhCH}_2\text{CHO} \\ 4.\ \text{HCl , O}_2}]{\substack{1.\ n\text{-PrMgBr} \\ 2.\ \text{THF}-\text{MgCl} \\ 25^{\circ}\text{C}}}$

(2.5   :   1)   80%

<u>Collins, S.</u>; Kuntz, B.A.; Hong, Y. *J. Org. Chem.*, *1989*, *54*, 4154.

PhCHO $\xrightarrow[\substack{5\%\ \text{NiBr}_2\text{(bpy) , Bu}_4\text{NBr} \\ \text{Zn anode}}]{\text{, e}^-\text{, DMF , RT}}$

85%

<u>Durandetti, S.</u>; Sibille, S.; Périchon, J. *J. Org. Chem.*, *1989*, *54*, 2198.

REVIEW:

"Asymmetric Reductions with Organoborane Reagents"

Midland, M.M. *Chem. Rev.*, *1989*, 89, 1553.

NON-ASYMMETRIC ALKYLATIONS

Guo, B.-S.; Doubleday, W.; Cohen, T. *J. Am. Chem. Soc.*, *1987*, 109, 4710.

Smaardijk, Ab.A.; Wynberg, H. *J. Org. Chem.*, *1987*, 52, 135.

Hosomi, A.; Kohra, S.; Tominaga, Y. *J. Chem. Soc., Chem. Commun.*, *1987*, 1517.

Coppi, L.; Mordini, A.; Taddei, M. *Tetrahedron Lett.*, *1987*, 28, 965.

C_6H_{13} ⌒ CHO $\xrightarrow[\text{15h}, \; \diagup\diagup\diagdown I]{\text{Sb}^0, \text{THF-HMPA, reflux}}$

OH
⌇ C_6H_{13}

80%

Butsugan, Y.; Ito, H.; Araki, S. *Tetrahedron Lett.*, **1987**, 28, 3707.

Me ⌒⌒ SnBu₃ (Me) $\xrightarrow{\text{PhCHO}, \text{150}^\circ\text{C}, \text{18h}}$

OH
Ph ⌇ ⌇ Me
Me   Me

72%

Hull, C.; Mortlock, S.V.; Thomas, E.J. *Tetrahedron Lett.*, **1987**, 28, 5343.

Me ⟩ OMe / Me ⟩ OSiMe₃ $\xrightarrow[\text{CH}_2\text{Cl}_2, \text{RT}, 2\text{h}]{\text{PhCHO}, \text{SmCl}_3}$

Me Me
Ph ⌇ CO_2Me
OSiMe₃

66%

+

Me Me
Ph ⌇ CO_2Me
OH

28%

Vougioukas, A.E.; Kagan, H.B. *Tetrahedron Lett.*, **1987**, 28, 5513.

Ph ⌒ Te ⟋ Bu $\xrightarrow[\text{2. PhCHO}]{\text{1. BuLi, THF}}$

Ph ⌒ ⌒ Ph
OH

74%

Hiiro, T.; Kambe, N.; Ogawa, A.; Miyoshi, N.; Murai, S.; Sonoda, N. *Angew. Chem. Int. Ed., Engl*, **1987**, 26, 1187.

⟨furan⟩ CHO $\xrightarrow[\text{2. 5% HCl}]{\text{1. PbBr}_4 \cdot \text{Al}^0, \text{DMF} \quad \diagup\diagdown\text{Br}, 2\text{h}}$

⟨furan⟩ ⌇ ⌒⌒
OH

85%

Tanaka, H.; Yamashita, S.; Hamatani, T.; Ikemoto, Y.; Torii, S. *Synth. Commun.*, **1987**, 17, 789.

Masuyama. Y.; Takahara, J.P.; Kurusu, Y. *J. Am. Chem. Soc., 1988, 110*, 4473.

Araki. S.; Butsugan, Y. *Chem. Lett., 1988,* 457.

Yeh, M.C.P.; Knochel. P.; Santa, L.E. *Tetrahedron Lett., 1988, 29*, 3887.

syn:anti = 11:1

Lipshutz. B.H.; Ellsworth, E.L.; Behling, J.R. *Tetrahedron Lett., 1988, 29*, 893.

Minato, M.; Tsuji. T. *Chem. Lett., 1988,* 2049.

n-C$_7$H$_{15}$-CHO

1. SbEt$_3$, =/⟍-Br
85°C , 6h
2. aq. EtOH , 30 min

n-C$_7$H$_{15}$ ⟍ OH ⟍=

91%

Chen, C.; Shen, Y.; Huang, Y.-Z. *Tetrahedron Lett.*, *1988, 29*, 1395.

1. C$_9$H$_{19}$CHO , 200°C
6h
2. H$_3$O$^+$

(pyridine)—CO$_2$SiMe$_3$ → (pyridine)—CH(OH)—C$_9$H$_{19}$

73%   C$_9$H$_{19}$

Effenberger, F.; König, J. *Tetrahedron, 1988, 44*, 3281.

CH$_3$CO—(CH$_2$)$_7$—CHO

3 =/⟍-OAc , PdCl$_2$(PhCN)$_2$
SnCl$_2$, DMI , 50°C , 20h

CH$_3$CO—(CH$_2$)$_7$—CH(OH)—CH$_2$—CH=CH$_2$

1,3-dimethylimidazolidin-2-one

Masuyama, Y.; Hayashi, R.; Otake, K.; Kurusu, Y. *J. Chem. Soc., Chem. Commun.*, *1988*, 44.

=⟍—CH(OBn)—OBn

PhCHO , CrCl$_2$, THF
Me$_3$SiI , -30°C , 3h

=⟍—CH(OBn)—CH(OH)—Ph + =⟍—CH(OBn)—CH(OH)—Ph

(88   :   12) **98%**

Takai, K.; Nitta, K.; Utimoto, K. *Tetrahedron Lett.*, *1988, 29*, 5263.

Ph—C(OSiMe$_3$)=CH—Me

1. BiCl$_3$, PhCHO , CH$_2$Cl$_2$
RT , 50 min
2. MeOH , 1N HCl
3. H$_2$O

Ph—CO—CH(Me)—CH(OH)—Ph **95%**

erythro:threo = 56:44

Ohki, H.; Wada, M.; Akiba, K. *Tetrahedron Lett.*, *1988, 29*, 4719.

$$\text{(propyl)}\!-\!ZnI \quad \xrightarrow[\text{2. H}_3\text{O}^+]{\substack{\text{1. C}_5\text{H}_{11}\text{CO , Me}_3\text{SiCl} \\ \text{PhMe , DMA , 60}^\circ\text{C , 4h}}} \quad \text{product}$$

OH

C₅H₁₁

66%

Tamaru, Y.; Nakamura, T.; Sakaguchi, M.; Ochiai, H.; Yoshida, Z. *J. Chem. Soc., Chem. Commun.,* **1988**, 610.

$$\xrightarrow[\substack{\text{3. PhCHO ,} \\ 30 \to +20^\circ\text{C}}]{\substack{\text{1. } n\text{-BuLi , -50} \to \text{-70}^\circ\text{C} \\ \text{2. } t\text{-BuLi}}}$$

HO   Ph

73%

Barluenga, J.; Fañanás, J.; Foubelo, F.; Yus, M. *J. Chem. Soc., Chem. Commun.,* **1988**, 1135.

$$\xrightarrow[\text{2. H}_3\text{O}^+]{\substack{\text{1. 6\% TASF , DMPU , RT} \\ \text{PhCHO}}}$$

Ph

OH

93%

DMPU = 1,3-dimethyl-3,4,5,6-tetrahydro-2(1H)-pyrimidinone

Fujita, M.; Obayashi, M.; Hiyama, T. *Tetrahedron,* **1988**, *44*, 4135.

$$\text{PhCHO} \quad \xrightarrow[\text{2. H}_3\text{O}^+]{\text{1. I-CH}_2\text{CO}_2\text{Et , THF , In}^\circ} \quad$$

Ph

CO₂Et

OH   **90%**

Araki, S.; Ito, H.; Butsugan, Y. *Synth. Commun.,* **1988**, *18*, 453.

$$\text{CHO} \quad \xrightarrow{\text{SmI}_2 \text{ , THF , 0}^\circ\text{C , MeOH}} \quad$$

Ph

—Ph   —Ph

IOH   +   IOH

(2.5   :   1)   73%

Enholm, E.J.; Trivellas, A. *Tetrahedron Lett.,* **1989**, *30*, 1063.

PhCHO $\xrightarrow[\text{CsF , 0°C , 1h}]{\text{SiF}_3}$

OH
Ph
Me

+

OH
Ph
Me

(99   :   1)   92%

Kira, M.; Hino, T.; Sakurai, H. *Tetrahedron Lett.*, *1989*, *30*, 1099.

1. OMgBr (2-*tert*-butylphenol magnesium bromide), ether
2. CH$_2$Cl$_2$, 3h

OH OH, BOC-pyrrolidine, *tert*-butyl phenol

+

OH OH, BOC-pyrrolidine, *tert*-butyl phenol

(>100   :   1)   65%

Bigi, F.; Casnati, G.; Sartori, G.; Araldi, G.; Bocelli, G. *Tetrahedron Lett.*, *1989*, *30*, 1121.

SnMe$_3$ / CHO spiro compound $\xrightarrow{\text{EtAlCl}_2 , \text{CH}_2\text{Cl}_2 , 0°C}$ spiro-OH compound

92%

Macdonald, T.L.; Delahunty, C.M.; Mead, K.; O'Dell, D.E. *Tetrahedron Lett.*, *1989*, *30*, 1473.

Bu / Bu$_3$Sn / OBOM $\xrightarrow[\text{BF}_3 \cdot \text{OEt}_2]{\text{C}_6\text{H}_{13}\text{CHO}}$

OH
Bu   C$_6$H$_{13}$
OBOM

+

OH
Bu   C$_6$H$_{13}$
OBOM

(87   :   13)   58%

20% OH   12% OH

Marshall, J.A.; Gung, W.Y. *Tetrahedron Lett.*, *1989*, *30*, 2183.

Me / SnBu$_3$ $\xrightarrow[\text{MeCN , RT , 12h}]{n\text{-C}_6\text{H}_{13}\text{CHO , CoCl}_2}$

Me   C$_6$H$_{13}$
OH

70%

Iqbal, J.; Joseph, S.P. *Tetrahedron Lett.*, *1989*, *30*, 2421.

Boeckman Jr., R.K.; O'Connor, K.J. *Tetrahedron Lett.*, *1989*, *30*, 3271.

Masuyama, Y.; Takahara, J.P.; Kurusu, Y. *Tetrahedron Lett.*, *1989*, *30*, 3437.
Masuyama, Y.; Otake, K.; Kurusu, Y. *Tetrahedron Lett.*, *1988*, *29*, 3563.

Cossy, J.; Pete, J.P.; Portella, C. *Tetrahedron Lett.*, *1989*, *30*, 7361.

Qiu, W.; Wang, Z. *J. Chem. Soc., Chem. Commun.*, *1989*, 356.

Sibille, S.; Mcharek, S.; Perichon, J. *Tetrahedron*, *1989*, *45*, 1423.

REVIEWS:

"Organoaluminums in Organic Synthesis"

Maruoka, K.; Yamamoto, H. *Tetrahedron, 1988, 44*, 5001.

"Acyclic Stereocontrol via Allylic Organometallic Compounds"

Yamamoto, Y. *Acc. Chem. Res., 1987, 20*, 243.

SECTION 35: ALCOHOLS AND THIOLS FROM ALKYLS, METHYLENES AND ARYLS

No examples of the reaction $RR^1 \rightarrow ROH$ (R^1 = alkyl, aryl, etc.) occur in the literature. For reactions of the type $RH \rightarrow ROH$ (R = alkyl or aryl) see Section 41 (Alcohols and Phenols from Hydrides).

Kira, M.; Kobayashi, M.; Sakurai, H. *Tetrahedron Lett., 1987, 28*, 4081.

Fleming, I.; Sanderson, P.E.J. *Tetrahedron Lett., 1987, 28*, 4229.

Sato, K.; Kira, M.; Sakurai, H. *Tetrahedron Lett., 1989, 30*, 4375.

Tamao, K.; Hayashi, T.; Ito, Y. *Tetrahedron Lett., 1989, 30*, 6533.

76%

Lohray, B.B.; Enders, D. Helv. Chim. Acta, **1989**, 72, 980.

SECTION 36: ALCOHOLS AND THIOLS FROM AMIDES

92%

Fu, J.-M.; Sharp, M.J.; Snieckus, V. Tetrahedron Lett., **1988**, 29, 5459.

SECTION 37: ALCOHOLS AND THIOLS FROM AMINES

78%

Brook, M.A.; Jahangir Synth. Commun., **1988**, 18, 893.

SECTION 38: ALCOHOLS AND THIOLS FROM ESTERS

$C_8H_{17}OH$ 88%

Zahalka, H.A.; Alper, H. Tetrahedron Lett., **1987**, 28, 2215.

Burke, S.D.; Deaton, D.N.; Olsen, R.J.; Armistead, D.M.A.; Blough, B.E. *Tetrahedron Lett.*, **1987**, *28*, 3905.

Bianco, A.; Passacantilli, P.; Righi, G. *Synth. Commun.*, **1988**, *18*, 1765.

Yadav, V.; Fallis, A.G. *Tetrahedron Lett.*, **1989**, *30*, 3283.

Jenner, G.; Nahmed, E.M.; Leismann, H. *Tetrahedron Lett.*, **1989**, *30*, 6501.

Liu, H.-J.; Luo, W. *Synth. Commun.*, **1989**, *19*, 387.

SECTION 39: ALCOHOLS AND THIOLS FROM ETHERS, EPOXIDES AND THIOETHERS

SnCl$_4$, CH$_2$Cl$_2$, 4h

reflux

96%

<u>Taylor, S.K.</u>; Davisson, M.E.; Hissom Jr., B.R.; Brown, S.L.; Pristach, H.A.; Schramm, S.B.; Harvey, S.M. *J. Org. Chem.*, *1987*, *52*, 425.

1.5 Na(PEG)$_2$BH$_2$, THF, 80°C

5 h

95%

PEG 400 = polyethylene glycol

<u>Santaniello, E.</u>; Ferraboschi, P.; Fiecchi, A.; Grisenti, P.; Manzocchi, A. *J. Org. Chem.*, *1987*, *52*, 671.

NaOEt, EtOH

25°C

77%   14%

<u>Benedetti, F.</u>; Fabrissin, S.; Gianferrara, T.; Risaliti, A. *J. Chem. Soc., Chem. Commun.*, *1987*, 406.

1. 1.5 eq BuLi, -78°C
   THF/hexane

2. -20°C, 3 min

30%

syn:anti = 10:1

<u>Schreiber, S.L.</u>; Goulet, M.T. *Tetrahedron Lett.*, *1987*, *28*, 1043.

n-C$_{10}$H$_{21}$ ～ OH

Me$_2$Cu(CN)Li$_2$, -20°C

THF:DMEU
2h

DMEU = 1,3-dimethyl-2-imidazolidinone

n-C$_{10}$H$_{21}$ 〜 OH
         Me
         OH

84

+

         Me
n-C$_{10}$H$_{21}$ 〜 OH
         OH

16

:

Chong, J.M.; Cyr, D.R.; Mar, E.K. *Tetrahedron Lett.*, *1987*, *28*, 5009.

BF$_3$•OEt$_2$, -78°C

CH$_2$Cl$_2$

86%

Shimazaki, M.; Hara, H.; Suzuki, K.; Tsuchihashi, G. *Tetrahedron Lett.*, *1987*, *28*, 5891.

Bu$_3$SnPh , DMF , H$_2$O
PdCl$_2$(MeCN)$_2$

85%

Echavarren, A.M.; Tueting, D.R.; Stille, J.K. *J. Am. Chem. Soc.*, *1988*, *110*, 4039.

1. n-Bu$_3$SnH , AIBN
   DME , NaI , 70°C , 1h
2. SiO$_2$, MeOH

77%

Bonini, C.; DiFabio, R. *Tetrahedron Lett.*, *1988*, *29*, 819.

OSiMe$_2$$t$-Bu

Me

BuLi , THF , -78°C

t-BuMe$_2$SiO   Me

OH

74%

Z:E = 70:30

anti:syn = 91:9

Nakai, E.; Nakai, T. *Tetrahedron Lett.*, *1988*, *29*, 5409.

1. sBuLi , THF
TMEDA , -50°C
2.
3. H_3O^+

(2   :   1) 72%

Schaumann, E.; Kirschning, A. *Tetrahedron Lett.*, *1988*, *29*, 4281.

PBu$_3$, THF

[1-napthyl lithium , THF
CuI/PPh$_3$, 0°C]

(1   :   6) 56%

Wu, T.-C.; Rieke, R.D. *Tetrahedron Lett.*, *1988*, *29*, 6753.

1. $\left[\begin{array}{c} \text{H-N} \quad \text{N-Me} \\ \text{BuLi , PhMe} \end{array}\right]$

2. <15°C , PhMe

3. BuLi , 70°C

56%

Gilles, B.; Loft, M.S. *Synth. Commun.*, *1988*, *18*, 191.

Na [PhSeB(OEt)$_3$]

EtOH

89%

Miyashita, M.; Suzuki, T.; Yoshikoshi, A. *Tetrahedron Lett.*, *1989*, *30*, 1819.
Miyashita, M.; Hoshino, M.; Suzuki, T.; Yoshikoshi, A. *Chem. Lett.*, *1988*, 507.

[Bu$_2$Cu(CN)Li$_2$, Bu$_3$SnH]

THF , -78°C

81%

Lipshutz, B.H.; Ellsworth, E.L.; Dimock, S.H.; Reuter, D.C. *Tetrahedron Lett.*, *1989*, *30*, 2065.

REVIEW:

"S_N2' Additions of Organocopper Reagents to Vinyloxiranes"

Marshall, J.A. Chem. Rev., 1989, 89, 1503.

Additional examples of ether cleavages may be found in Section 45A (Protection of Alcohols and Thiols).

SECTION 40:   ALCOHOLS AND THIOLS FROM HALIDES AND SULFONATES

(4   :   96)   **77%**

Canonne, P.; Bernatchez, M. J. Org. Chem., 1987, 52, 4025.

74%

Boyer, J.H.; Natesh, A. Synthesis, 1988, 980.

74%

>99.5% ee

Chan, T.H.; Pellon, P. J. Am. Chem. Soc., 1989, 111, 8737.

84%

Gingras, M.; Chan, T.H. Tetrahedron Lett., 1989, 30, 279.

SECTION 41: ALCOHOLS AND THIOLS FROM HYDRIDES

Mn(TPP)Cl , imidazole

MeCO₃H , e⁻, aq. MeCN

LiClO₄

TPP = tetraphenylporphyrin

(85   :   15) 75%

Leduc, P.; Battioni, P.; Bartoli, J.F.; Mansuy, D. *Tetrahedron Lett.*, *1988*, *29*, 205.

1. Li⁰ , BuCl
   20°C
2. B(OBu)₃ •H₂O

35%

Einhorn, J.; Luche, J.-L.; Demerseman, P. *J. Chem. Soc., Chem. Commun.*, *1988*, 1350.

Beauveria sulfurescens
ATCC 7159

63%

Archelas, A.; Fourneron, J.-D.; Furstoss, R. *J. Org. Chem.*, *1988*, *53*, 1797.

70% aq. *t*-BuOO*t*-Bu
AIBN , O₂ , 60°C , 15h

66%

Sabol, M.R.; Wiglesworth, C.; Watt, D.S. *Synth. Commun.*, *1988*, *18*, 1.

1. DMSO , NaS$_2$O$_8$
Cu(NO$_3$)$_2$, 50°C
NNO$_3$

80% conversion

72%

Cort, A.D.; Mandolini, L.; Panaioli, S. *Synth. Commun.*, *1988*, *18*, 613.

RuCl$_3$, 4 NaIO$_4$, 55°C

CCl$_4$:MeCN:H$_2$O
20h

90%

Me
(endo:exo = 3:1)

Me

Tenaglia, A.; Terranova, E.; Waegell, B. *Tetrahedron Lett.*, *1989*, *30*, 5271, 5275.

SECTION 42: ALCOHOLS AND THIOLS FROM KETONES

The following reaction types are included in this section:
A. Reductions of Ketones to Alcohols
B. Alkylations of Ketones, forming Alcohols

Coupling of ketones to give diols is found in Section 323 (Alcohol → Alcohol).

SECTION 42A: REDUCTION OF KETONES TO ALCOHOLS

ASYMMETRIC REDUCTION

Aspergillus niger (ATCC 9142)

H$_2$O , glucose , yeast ,
soya meal , NaCl, K$_2$HPO$_4$
pH 6.5 , 24 h

80%

98:2 R:S

Belan, A.; Bolte, J.; Fauve, A.; Gourcy, J.G.; Veschambre, H. *J. Org. Chem.*, *1987*, *52*, 256.

R = Me   50   :   50
                >95%ee   >95%ee

R = Et   5   :   95
                >95%ee   90%ee

Buisson, D.; Sanner, C.; Larcheveque, M.; Azerad, R. *Tetrahedron Lett., 1987, 28,* 3939.

98%

44% ee , R

Brown, H.C.; Cho, B.T.; Park, W.S. *J. Org. Chem., 1987, 52,* 4020.

quant

87% ee , R

Takahashi, H.; Morimoto, T.; Achiwa, K. *Chem. Lett., 1987,* 855.

62%

98% ee , S

Brown, H.C.; Chandrasekharan, J.; Ramachandran, P.V. *J. Am. Chem. Soc., 1988, 110,* 1539.

Youn, I.K.; Lee, S.W.; Pak, C.S. *Tetrahedron Lett.*, *1988*, *29*, 4453.

Vannoorenberghe, Y.; Buono, G. *Tetrahedron Lett.*, *1988*, *29*, 3235.

Jung, M.E.; Hogan, K.T. *Tetrahedron Lett.*, *1988*, *29*, 6199.

Brown, H.C.; Cho, B.T.; Park, W.S. *J. Org. Chem.*, *1988*, *53*, 1231.

(66   :   34)   88%

Itoh, T.; Takagi, Y.; Fujisawa, T. Tetrahedron Lett., **1989**, 30, 3811.

80% ee , R   quant

Balavoine, G.; Clinet, J.C.; Lellouche, I. Tetrahedron Lett., **1989**, 30, 5141.

Chirald = 2S,3R-(+)-4-dimethylamino-
1,2-diphenyl-3-methylbutan-2-ol

46% ee   98%

Feghouli, A.; Vanderesse, R.; Fort, Y.; Caubére, P. J. Chem. Soc., Chem. Commun.,
1989, 224.

REVIEW:

"Asymmetric Reductions with Organoborane Reagents"

Midland, M.M. Chem. Rev., **1989**, 89, 1553.

NON-ASYMMETRIC REDUCTION

95%

Kira, M.; Sato, K.; Sakurai, H. *J. Org. Chem.*, *1987, 52*, 948.

99%

Komiya, S.; Tsutsumi, O. *Bull. Chem. Soc., Jpn.*, *1987, 60*, 3423.

53%

also obtained 47% of condensation products

Sanchez, R.; Scott, W. *Tetrahedron Lett.*, *1988, 29*, 139.

>97%   8%

Ward, D.E.; Rhee, C.K.; Zoghaib, W.M. *Tetrahedron Lett.*, *1988, 29*, 517.

71%

aliphatic ketones give no reaction; reduces only cyclic ketones

Hulce, M.; LaVaute, T. *Tetrahedron Lett.*, *1988, 29*, 525.

(96   :   4) 90%

Bloch, R.; Gilbert, L.; Virard, C. *Tetrahedron Lett.*, **1988**, 29, 1021.

erythro:threo = 54:46

Shibata, I.; Suzuki, T.; Baba, A.; Matsuda, H. *J. Chem. Soc., Chem. Commun.*, **1988**, 882.

Kaspar, J.; Trovarelli, A.; Lenarda, M.; Graziani, M. *Tetrahedron Lett.*, **1989**, 30, 2705.

63%

Toda, F.; Kiyoshige, K.; Yagi, M. *Angew. Chem. Int. Ed., Engl.*, **1989**, 28, 320.

81%

Khai, B.T.; Arcelli, A. *J. Org. Chem.*, **1989**, 54, 949.

96%

Sarkar, A.; Rao, B.R.; Konar, M.M. *Synth. Commun.*, **1989**, 19, 2313.

89%

Toda, F.; Tanaka, K.; Tange, H. *J. Chem. Soc., Perkin Trans. I*, *1989*, 1555.

90%

Radhakrishna, A.S.; Prasad Rao, K.R.K.; Nigam, S.C.; Bakthavatchalam, R.; Singh, B.B. *Org. Prep. Proceed. Int.*, *1989*, *21*, 373.

SECTION 42B: ALKYLATION OF KETONES, FORMING ALCOHOLS

Aldol reactions are listed in Section 330 (Ketone-Alcohol)

80%

Sibille, S.; d'Incon, E.; Leport, L.; Massebiau, M.-C.; Perichon, J. *Tetrahedron Lett.*, *1987*, *28*, 55.

86%

96% ee , S

Yamamoto, K.; Ando, H.; Chikamatsu, H. *J. Chem. Soc., Chem. Commun.*, *1987*, 334.

Bakers Yeast , D-glucose

air , H₂O , 35°C

(90   :   10) 75%

Brooks, D.W.; Mazdiyasni, H.; Grothaus, P.G. *J. Org. Chem., 1987, 52*, 3233.

[activated Zn°/Me₃SiCl]

ether , 0°C   79%

Picotin, G.; Miginiac, Ph. *Tetrahedron Lett., 1987, 28*, 4551.

3 eq. TBAF , THF , 25°C

12h

n-C₆H₁₃—SiMe₃ n-C₆H₁₃   Me   68%

Bulman Page, P.C.; Rosenthal, S.; Williams, R.V. *Tetrahedron Lett., 1987, 28*, 4455.

HSi(OMe)₃ , ether , 0°C , 20h

S,S-   61%

58% ee

Kohra, S.; Hayashida, H.; Tominaga, Y.; Hosomi, A. *Tetrahedron Lett., 1988, 29*, 89.

PhMe , 95°C , 18h

sealed tube   98%

Danheiser, R.L.; Nishida, A.; Savariar, S.; Trova, M.P. *Tetrahedron Lett., 1988, 29*, 4917.

Ph—C(=O)—Ph

2 eq. PhSCH$_2$SiMe$_3$, 5°C
────────────────────────────
TBAF , THF

PhS—CH$_2$—C(Ph)(Ph)—OSiMe$_3$

96%

Kitteringham, J.; Michell, M.B. *Tetrahedron Lett.*, *1988*, *29*, 3319.

Me—C(=O)—C$_6$H$_{13}$

CH$_2$=CHCH$_2$CH$_2$I , In° , DMF
────────────────────────────
RT , 1h

Me, OH, CH$_2$=CH—CH$_2$—C—C$_6$H$_{13}$

89%

Araki, S.; Ito, H.; Butsugan, Y. *J. Org. Chem.*, *1988*, *53*, 1831.

cyclohexanone

Ph—O—CH$_2$—C(=O)—Cl , SmI$_2$
────────────────────────────
THF , RT , 1 min

cyclohexane with OH and CH$_2$—O—CH$_2$—Ph

61%

Sasaki, M.; Collin, J.; Kagan, H.B. *Tetrahedron Lett.*, *1988*, *29*, 4847.

cycloheptanone

1. [Bu$_3$SnCH$_2$OMOM]
   [THF , BuLi , -78°C]
   -78°C , 5 min
2. satd NH$_4$Cl

cycloheptane with OH and CH$_2$—OMOM

93%

Johnson, C.R.; Medich, J.R. *J. Org. Chem.*, *1988*, *53*, 4131.

Et$_3$C—C(=O)—Me

MeMgBr , CeCl$_3$, THF ,0°C
────────────────────────────

Et$_3$C—C(Me)(OH)—Me

95%

Imamoto, T.; Takiyama, N.; Nakamura, K.; Hatajima, T.; Kamiya, Y. *J. Am. Chem. Soc.*, *1989*, *111*, 4392.

(25:1)

Molander, G.A.; Kenny, C. *J. Am. Chem. Soc.*, **1989**, *111*, 8236.
Molander, G.A.; Kenny, C. *Tetrahedron Lett.*, **1987**, *28*, 4367.

Matsuzawa, S.; Isaka, M.; Nakamura, E.; Kuwajima, I. *Tetrahedron Lett.*, **1989**, *30*, 1975.

Ujikawa, O.; Inanaga, J.; Yamaguchi, M. *Tetrahedron Lett.*, **1989**, *30*, 2837.

(94   :   6)   70%

Fan, R.; Hudlicky, T. *Tetrahedron Lett.*, **1989**, *30*, 5533.

Cahiez, G.; Chavant, P.-Y. *Tetrahedron Lett.*, **1989**, *30*, 7373.

Kariv-Miller, E.; Maeda, H.; Lombardo, F. *J. Org. Chem.*, *1989*, *54*, 4022.

Shono, T.; Kashimura, S.; Mori, Y.; Hayashi, T.; Soejima, T.; Yanaguchi,.Y. *J. Org. Chem.*, *1989*, *54*, 6001.

Review:

"Use of Activation Methods for Organo-Zinc Reagents"

Erdik, E. *Tetrahedron*, *1987*, *43*, 2203.

SECTION 43: ALCOHOLS AND THIOLS FROM NITRILES

NO ADDITIONAL EXAMPLES

SECTION 44: ALCOHOLS AND THIOLS FROM ALKENES

1. *n*-BuLi , THF-hexane
   -78°C , 20 min

2. PhCHO , -78°C

cis:trans = 95:5

60%

Krief, A.; Barbeaux, P. *J. Chem. Soc., Chem. Commun.*, *1987*, 1214.

1. MeBH$_2$, THF , 0°C , 5 min

2. [alkene chain]

0°C , 1h

3. CO   4. H$_2$O$_2$, NaOH

Me

OH

71%

Srebnik, M.; Cole, T.E.; <u>Brown, H.C.</u> *Tetrahedron Lett.*, *1987*, *28*, 3771.

Cl$_2$AlH , O$_2$, THF

0.2 BEt$_3$

78% OH

endo:exo = 1:3

Maruoka, K.; Sano, H.; Shinoda, K.; <u>Yamamoto, H.</u> *Chem. Lett.*, *1987*, 73.

OSiMe$_2$*t*-Bu

3 eq catecholborane

3% Rh(PPh$_3$)$_3$Cl

THF , 25°C

OSiMe$_2$*t*-Bu

"$_{\prime\prime}$OH

68%

+

OSiMe$_2$*t*-Bu

OH

9%

<u>Evans, D.A.</u>; Fu, G.C.; Hoveyda, A.H. *J. Am. Chem. Soc.*, *1988*, *110*, 6917.

H-B,   Me

1. THF,

0°C

Me   Me

, 12h

2

2. H$_2$O$_2$, NaOH

HO   H

63%

93% ee , S

<u>Brown, H.C.</u>; Vara Prasad, J.V.N.; Zaidlewicz, M. *J. Org. Chem.*, *1988*, *53*, 2911.

1. [Rh(COD)Cl]$_2$
   2 BINAP
   THF , -25°C
   ─────────────
   catecholborane

2. H$_2$O$_2$, NaOH

99%

64% ee

Burgess, K.; Ohlmeyer, M.J. *J. Org. Chem.*, **1988**, *53*, 5178.

1. [benzene-fused dioxaborole] B-H , DME
   [Rh(COD)$_2$]BF$_4$, -78°C
   ─────────────────
   (+)-BINAP , 2h

2. NaOH , H$_2$O$_2$

Ph⟍Me
HO H
91%

96% ee , R

Hayashi, T.; Matsumoto, Y.; Ito, Y. *J. Am. Chem. Soc.*, **1989**, *111*, 3426.

1. BH$_3$, ether
   ─────────────────
2. 4 eq. NaBO$_3$ •4 H$_2$O
   H$_2$O , 2h

OH

94%

Kabalka, G.W.; Shoup, T.M.; Goudgaon, N.M. *Tetrahedron Lett.*, **1989**, *30*, 1483.

1. BnEt$_3$N$^+$BH$_4^-$, Me$_3$SiCl
   CH$_2$Cl$_2$, 0°C
   ─────────────────
2. aq. K$_2$CO$_3$

Ph   Ph
     +
"OH   OH

(3   :   1)   84%

Baskaran, S.; Gupta, V.; Chidambaram, N.; Chandrasekaran, S. *J. Chem. Soc., Chem. Commun.*, **1989**, 903.

Co(TFA)$_2$, O$_2$
C$_8$H$_{17}$CH=CH$_2$ ───────────────── C$_8$H$_{17}$──Me + C$_8$H$_{17}$──Me
Et$_3$SiH , 4h
n-PrOH , 75°C

OH
73%

O
10%

Isayama, S.; Mukaiyama, T. *Chem. Lett.*, **1989**, 569.

SECTION 45: ALCOHOLS AND THIOLS FROM MISCELLANEOUSCOMPOUNDS

$$\text{Cl}--\text{S}-\text{S}--\text{Cl} \xrightarrow[\text{MeOH}]{\text{NH}_2\text{NH}_2,\text{THF}} 2 \ \text{Cl}--\text{SH}$$

89%

Maiti, S.N.; Spevak, P.; Singh, M.P.; Micetich, R.G.; Reddy, A.V.N. *Synth. Commun.,* **1988**, *18*, 575.

86%

Herndon, J.W.; Wu, C. *Tetrahedron Lett.,* **1989**, *30*, 6461.

67%

Boivin, J.; El Kaim, L.; Kervagoret, J.; Zard, S.Z. *J. Chem. Soc., Chem. Commun.,* **1989**, 1006.

REVIEW:

"Boronic Esters in Stereodirected Synthesis"

Matteson, D.S. *Tetrahedron,* **1989**, *45*, 1859.

SECTION 45A: PROTECTION OF ALCOHOLS AND THIOLS

72%

Kim, S.; Park, J.H. *Tetrahedron Lett.,* **1987**, *28*, 439.

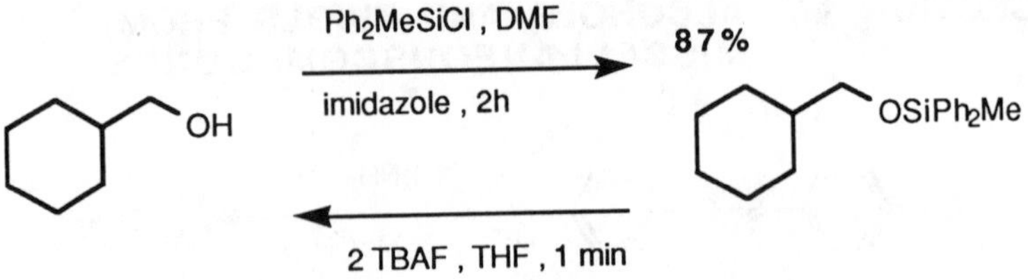

Ph2MeSiCl , DMF
87%
imidazole , 2h
OH
OSiPh2Me
2 TBAF , THF , 1 min

Denmark, S.E.; Hammer, R.P.; Weber, E.J.; Habermas, K.L. *J. Org. Chem.*, ***1987***, *52*, 165.

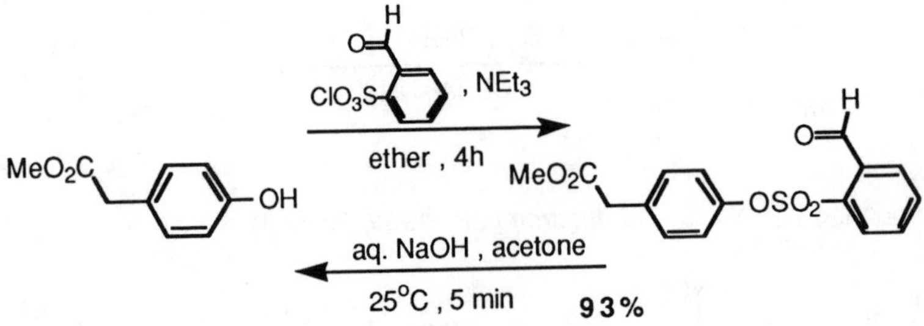

H
O
ClO3S
, NEt3
ether , 4h
MeO2C
OH
MeO2C
OSO2
H
O
aq. NaOH , acetone
25°C , 5 min
93%

Shashidhar, M.S.; Bhatt, M.V. *J. Chem. Soc., Chem. Commun.*, ***1987***, 654.

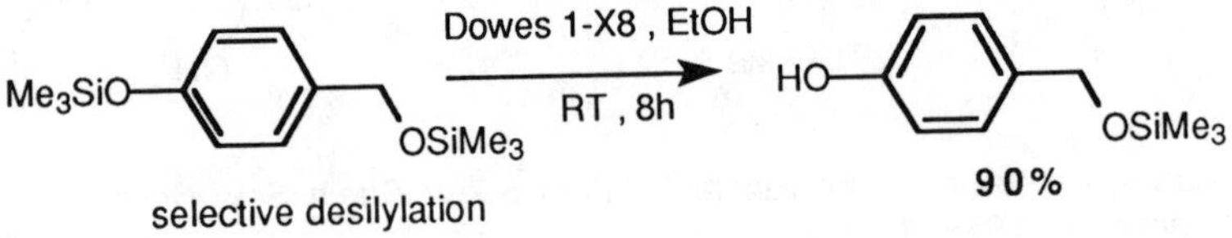

Dowes 1-X8 , EtOH
RT , 8h
Me3SiO
OSiMe3
HO
OSiMe3
90%
selective desilylation

Kawazoe, Y.; Nomura, M.; Kondo, Y.; Kohda, K *Tetrahedron Lett.*, ***1987***, *28*, 4307.

1. 1.2 eq. (Me$_3$Si)$_2$NH ; 0.1 eq. Me$_3$SiCl
2. 1.3 eq. (Me$_3$Si)$_2$NH ; 1.2 eq. Me$_3$SiCl
3. 2.5 eq. (Me$_3$Si)$_2$NH ; 2.5 eq. Me$_3$SiCl ; cat. DMAP

Cossy, J.; Pole, P. *Tetrahedron Lett., 1987, 28,* 6039.

Hanamoto, T.; Hayama, T.; <u>Katsuki, T.</u>; Yamaguchi, M. *Tetrahedron Lett., 1987, 28,* 6329.

<u>Gras, J.-L.</u>; Nouguier, R.; Mchich, M. *Tetrahedron Lett., 1987, 28,* 6601.

<u>Kozikowski, A.P.</u>; Wu, J.-P. *Tetrahedron Lett., 1987, 28,* 5125.

CTAN = ceric triethylammonium nitrate

Maione, A.M.; Romeo, A. Synthesis, **1987**, 250.

Widmer, U. Synthesis, **1987**, 568.

Jansson, K.; Frejid, T.; Kihlberg, J.; Magnusson, G. Tetrahedron Lett., **1988**, 29, 361.

B = pyrimidine or purine

Markiewicz, W.T.; Nowakowska, B.; Adrych, K. Tetrahedron Lett., **1988**, 29, 1561.

1. $(MeO)_3CH$, CH_2Cl_2 , RT

cat. camphorsulfonic acid

2. Dibal-H , hexane/CH_2Cl_2
   -78 - 0°C

3. aq. NaOH

Takasu, M.; Naruse, Y.; Yamamoto, H. *Tetrahedron Lett.*, **1988**, *29*, 1947.

Me_2S , $(PhCO_2)_2$

MeCN , 0°C , 4h

88%

Medina, J.C.; Salomon, M.; Kyler, K.S. *Tetrahedron Lett.*, **1988**, *29*, 3773.

Me_2t-BuSiO

1. 48% HBr , KF
   t-BuBr , 24h , 25°C

2. 2N aq. HCl

95%

Sinhababu, A.K.; Kawase, M.; Borchardt, R. *Synthesis*, **1988**, 710.

TBAF , HMPA

MS 4Å , RT , 1h

94%

Kan, T.; Hashimoto, M.; Yanagiya, M.; Shirahama, H. *Tetrahedron Lett.*, **1988**, *29*, 5417.

NH

Cl_3C⟋OMPM , p-TsOH

CH_2Cl_2 , 5h

77%

DDQ

MPM = 4-methoxyphenyl

Nakajima, N.; Horita, K.; Abe, R.; Yonemitsu, O. *Tetrahedron Lett.*, **1988**, *29*, 4139.

Otera, J.; Niibo, Y.; Chikada, S.; Nozaki, H. *Synthesis, 1988*, 328.

Saunders, D.G. *Synthesis, 1988*, 377.

Johnston, R.D.; Marston, C.R.; Krieger, P.E.; Goe, G.L. *Synthesis, 1988*, 393.

Bolitt, V.; Mioskowski, C.; Shin, D.-S.; Falck, J.R. *Tetrahedron Lett., 1988, 29*, 4583.

Shekhani, M.S.; Khan, K.M.; Mahmood, K. *Tetrahedron Lett., 1988, 29*, 6161.

Ph$_2$Si(Cl)Ot-Bu

$$C_{11}H_{23}CH_2OH \xrightarrow[\text{NEt}_3 , \text{Ch}_2\text{Cl}_2]{} 91\% \quad C_{11}H_{23}CH_2OSiPh_2Ot\text{-Bu}$$

$$\xleftarrow{\text{TBAF}}$$

Gillard, J.W.; Fortin, R.; Morton, H.E.; Yoakim, C.; Quesnelle, C.A.; Daignault, S.; Guindon, Y. *J. Org. Chem.*, *1988*, *53*, 2602.

t-BuPh$_2$SiO⟋OSiMe$_2$$t$-Bu $\xrightarrow[20^{\circ}C]{\text{PPTS , 2h}}$ t-BuPh$_2$SiO⟋OH

(CH$_2$)$_5$ (CH$_2$)$_5$ 88%

PPTS = pyridinium p-toluenesulfonate

Prakash, C.; Saleh, S.; Blair, I.A. *Tetrahedron Lett.*, *1989*, *30*, 19.

⟋OH $\xrightarrow{\text{5 eq. NaH , THF , 16h}}$ ⟋OSiPh$_2$$t$-Bu

SiPh$_2$$t$-Bu   91%

Spinazzé, P.G.; Keay, B.A. *Tetrahedron Lett.*, *1989*, *30*, 1765.

MeO⟋OH $\xrightarrow[\text{hv (350 nm)}]{}$ 90%

Church, G.; Ferland, J.-M.; Gauthier, J. *Tetrahedron Lett.*, *1989*, *30*, 1901.

C$_6$H$_{13}$ ⟋ OMe⟋O⟋O⟋SiMe$_3$ $\xrightarrow[80^{\circ}C , 5h]{\text{TBAF , MS 4Å , DMPU}}$ C$_6$H$_{13}$

Me   95% Me OH

DMPU = N,N'-dimethylpropylene urea

Lipshutz, B.H.; Miller, T.A. *Tetrahedron Lett.*, *1989*, *30*, 7149.

Wagner, A.; Heitz, M.-P.; Mioskowski, C. *J. Chem. Soc., Chem. Commun.*, **1989**, 1619.

Bakos, T.; Vincze, I. *Synth. Commun.*, **1989**, *19*, 523.

Poss, A.J.; Smyth, M.S. *Synth. Commun.*, **1989**, *19*, 3363.

Park, J.H.; Kim, S. *Chem. Lett.*, **1989**, 629.

CHAPTER 4
PREPARATION OF ALDEHYDES

SECTION 46: ALDEHYDES FROM ALKYNES

$n\text{-}C_6H_{13}$ ═══ H

1. $Cp_2Zr\cdot HCl$, THF, 25°C
   H_2
2. n-Bu-NC, 0 - 45°C
3. 50% AcOH, -78°C - RT

⟶ (product) CHO

93% , E

Negishi, E.; Swanson, D.R.; Miller, S.R. *Tetrahedron Lett.,* **1988,** 29, 1631.

SECTION 47: ALDEHYDES FROM ACID DERIVATIVES

(naphthalene)-COOH

$Me_2CHCMe_2BHCl\cdot SMe_2$

CH_2Cl_2, RT, 24h

⟶ (naphthalene)-CHO

56%

Brown, H.C.; Cha, J.S.; Yoon, N.M.; Nazer, B. *J. Org. Chem.,* **1987,** 52, 5400.

$C_6H_{11}COOH$

Kugelrohr (110-140°C)

(silane structure: Ph, Si, H, H, NMe$_2$, naphthalene)

⟶ $C_6H_{11}CHO$

68%

Corriu, R.J.P.; Lanneau, G.F.; Perrot, M. *Tetrahedron Lett.,* **1987,** 28, 3941.

Corriu, R.J.P.; Lanneau, G.F.; Perrot, M. *Tetrahedron Lett.*, *1988*, 29, 1271.

Cha, J.S.; Oh, S.Y.; Lee, K.W.; Yoon, M.S.; Lee, J.C. *Heterocycles*, *1988*, 27, 595.
Cha, J.S.; Kim, J.E.; Oh, S.Y.; Kim, J.D. *Tetrahedron Lett.*, *1987*, 28, 4575.
Cha, J.S.; Kim, J.E.; Yoon, M.S.; Kim, Y.S. *Tetrahedron Lett.*, *1987*, 28, 6231.

Meyers, A.I.; Lutomski, K.A.; Laucher, D. *Tetrahedron*, *1988*, 44, 3107.

REVIEW:

"Recent Developments in the Synthesis of Aldehydes by Reduction of Carboxylic Acids and Their Derivatives with Metal Hydrides. A Review."

Cha, J.S. *Org. Prep. Proceed. Int.*, *1989*, 21, 451.

SECTION 48: ALDEHYDES FROM ALCOHOLS AND THIOLS

CH$_2$Cl$_2$, pH 8.6 , 0°C

0.1M aq. HBr

3 min

two phase

also oxidizes 2° alcohols

(2-octanol - 2-octanone, 99%)

98%

Anelli, P.L.; Biffi, C.; Montanari, F.; Quici, S. *J. Org. Chem.*, *1987*, *52*, 2559.

KMnO$_4$ / TDA-1 / CH$_2$Cl$_2$

PhCH$_2$OH ————————————→ PhCHO

TDA-1 = tris[2-(2-methoxyethoxy)ethyl]amine   82%

McKillop, A.; Mills, L.S. *Synth. Commun.*, *1987*, *17*, 647.

HCrO$_4$ • Al silicate , 48h

C$_5$H$_{11}$OH ————————————→ C$_4$H$_9$CHO   74%

pet ether

Lou, J.-D.; Wu, Y.-Y. *Synth. Commun.*, *1987*, *17*, 1717.

1. LDA ; MeI
2. LDA ; EtBr
3. KOH
4. e⁻ , MeOH , NEt$_3$,
Pt electrodes , 0°C

EtO$_2$C / OH / CO$_2$Et ————————————→ Me / EtO$_2$C / CHO

42%

Renaud, P.; Hürzeler, M.; Seebach, D. *Helv. Chim. Acta*, *1987*, *70*, 292.

1. Cl$_3$CO$_2$CCl , DMSO
CH$_2$Cl$_2$, -70°C
2. NEt$_3$, -70°C → RT

Ph ⁄＼⁄ OH ————————————→ Ph ⁄＼⁄ CHO

89%

Takano, S.; Inomata, K.; Tomita, S. *Tetrahedron Lett.*, *1988*, *29*, 6619.

HCrO$_4$⁻ (Kieselguhr)

BuCH$_2$OH ————————————→ Bu-CHO

pet ether   83%

Lou, J.-D. *Synth. Commun.*, *1989*, *19*, 1841.

97%

Daumas, M.; Vo-Quang, Y.; Vo-Quang, L.; LeGoffic, F. *Synthesis, 1989*, 64.

92%

Yamazaki, S.; Yamazaki, Y. *Chem. Lett., 1989*, 1361.

Related Methods: Ketones from Alcohols and Phenols (Section 168)

SECTION 49: ALDEHYDES FROM ALKYNES

Conjugate reductions and Michael Alkylations of conjugated aldehydes are listed in Section 74 (Alkyls from Alkenes).

1. THF , -20°C

2. 4M HCl , 20°C
   30 min, pH 1-2

47%

Tay, M.K.; Aboryaoude, E.E.; Collingnon, N.; Savignac, Ph. *Tetrahedron Lett., 1987, 28*, 1263.

Ph-CHO

1. MeSCH₂SO₂Ph , K₂CO₃ , iPrOH
2. NaTeH , EtOH , RT , 3h
3. TiCl₄ , MeCN
4. H₃O⁺   Z:E = 74:26

PhCH₂-CHO

49%

Huang, X.; Zhang, H.-Z. *Synthesis, 1989*, 42.

Related Methods: Aldehydes from Ketones (Section 57)
                          Ketones from Ketones (Section 177)
                          Also via: Alkenyl aldehydes (Section 341)

SECTION 50: ALDEHYDES FROM ALKYLS, METHYLENES AND ARYLS

Marx, J.N.; Bih, Q.-R. *J. Org. Chem.*, *1987*, *52*, 336.

Sakakura, T.; Tanaka, M. *J. Chem. Soc., Chem. Commun.*, *1987*, 758.

Kreh, R.P.; Spotnitz, R.M.; Lundquist, J.T. *Tetrahedron Lett.*, *1987*, *28*, 1067.

Cambanis, A.; Bäuml, E.; Mayr, H. *Synthesis*, *1989*, 128.

Li, W.-S.; Liu, L.K. *Synthesis*, *1989*, 293.

SECTION 51: ALDEHYDES FROM AMIDES

$$\text{Ph-C(=O)-NHNH}_2 \xrightarrow[\text{THF , MeOH}]{\text{CuCl}_2 \cdot 2\,\text{H}_2\text{O , NaBH}_4} \text{PhCHO}$$

82%

<u>Attanasi, O.A.</u>; Serra-Zanetti, F.; Tosi, G. *Org. Prep. Proceed. Int.*, *1988*, 20, 405.

SECTION 52: ALDEHYDES FROM AMINES

$$\text{Me-C}_6\text{H}_4\text{-CH=N-OH} \xrightarrow[\text{Fe(CO)}_5\text{/TONSIL}]{\text{PhMe , reflux , 3.2h}} \text{Me-C}_6\text{H}_4\text{-CHO}$$

TONSIL = bentonitic earth   **96%**

<u>Alvarez, C.</u>; Cano, A.C.; Rivera, V.; Márquez, C. *Synth. Commun.*, *1987*, 17, 279.

1. $\overset{\text{Me}}{\underset{\text{Ph}}{\text{C}}}\text{CHO}$, PhMe , reflux , 3h

2. TiCl$_4$, PhMe , 110°C

3. 2M HCl , 10 min

56%

72% de (81% ee)

<u>Bailey, P.D.</u>; Harrison, M.J. *Tetrahedron Lett.*, *1989*, 30, 5341.

PhCh_2NH_2

1. Cl$_2$P(=O)OPh , DMSO ,
   CH$_2$Cl$_2$, CO , 15 min

2. NEt$_3$, -10 → +20°C
   45 min

3. aq. (COOH)$_2$, 20°C
   30 min

Ph-CHO

79%

<u>Liu, H.-J.</u>; Nyangullu, J.M. *Synth. Commun.*, *1989*, 19, 3407.

Related Methods: Ketones from Amines (Section 172)

SECTION 53: ALDEHYDES FROM ESTERS

[LiAlH$_4$, Et$_2$NH , pentane]

$$CH_3(CH_2)_{10}CO_2Me \xrightarrow{\qquad RT , 6h \qquad} CH_3(CH_2)_{10}CHO$$

93%

Cha. J.S.; Kwon, S.S. *J. Org. Chem.*, *1987*, *52*, 5486.

1. ⟋OAc , Pd(PPh$_3$)$_4$
   ──────────────────────
   THF , PPh$_3$

2. heat , 10% aq. HCl

14% ee

53%

Hiroi. K.; Abe, J.; Suya, K.; Sato, S. *Tetrahedron Lett.*, *1989*, *30*, 1543.

SECTION 54: ALDEHYDES FROM ETHERS, EPOXIDES AND THIOETHERS

CH$_2$Cl$_2$, -78°C

E:Z = 7:93

64%

Maruoka, K.; Nonoshita, K.; Banno, H.; Yamamoto. H. *J. Am. Chem. Soc.*, *1988*, *110*, 8922.

CH$_2$Cl$_2$, -78°C , 30 min

200 mol%

94%

Maruoka, K.; Nagahara, S.; Ooi, T.; Yamamoto. H. *Tetrahedron Lett.*, *1989*, *30*, 5607.

$$PhCH_2OMe \xrightarrow[\text{isooctane}]{Cu(NO_3)_2 - SiO_2 , 1h} Ph\text{-}CHO$$

91%

Nishiguchi, T.; Bougauchi, M. *J. Org. Chem.*, **1989**, *54*, 3001.

REVIEW:

"The Thermal Aliphatic Claisen Rearrangement"

Ziegler, F.E. *Chem. Rev.*, **1988**, *88*, 1423.

Related Methods: Ketones from Ethers and Epoxides (Section 174)

SECTION 55: ALDEHYDES FROM HALIDES AND SULFONATES

$$C_7H_{15}CH_2I \xrightarrow[\text{2. PhMe , reflux , 12h}]{\text{1. [... , sBuLi , THF, -78°C], -78°C}} \text{product}$$

70%

Funk, R.L.; Bolton, G.L. *J. Am. Chem. Soc.*, **1988**, *110*, 1290.

$$PhCH_2Br \xrightarrow[\substack{\text{2. PMA   3. TFAA} \\ \text{4. 20\% } K_2CO_3}]{\substack{\text{1.} \quad \text{(pyrazine-SH)} \\ \text{Na}_2\text{CO}_3 , \text{DME}}} Ph\text{-}CHO$$

94%

PMA = permalic acid

Shimazaki, M.; Nakanishi, T.; Mochizuki, M.; Ohta, A. *Heterocycles*, **1988**, *27*, 1643.

e⁻ , DMF , Bu₄NBr

$$Ph\text{-}MgBr \xrightarrow[\substack{\text{Mg anode , stainless} \\ \text{steel cathode}}]{e^- , DMF , Bu_4NBr} Ph\text{-}CHO$$

75%

Saboureua, C.; Troupel, M.; Sibille, S.; d'Incan, E.; Périchon, J. *J. Chem. Soc., Chem. Commun., **1989**, 895.*

SECTION 56: ALDEHYDES FROM HYDRIDES

Brown, S.H.; Crabtree, R.H. *J. Chem. Soc., Chem. Commun., **1987**, 970.*

SECTION 57: ALDEHYDES FROM KETONES

89%

Moskal, J.; van Leusen, A.M. *Recl. Trav. Chim., Pays-Bas, **1987**, 106, 137.*

62%

E:Z = 95:5

Nuzillard, J.-M.; Boumendjel, A.; Massiot, G. *Tetrahedron Lett., **1989**, 30, 3779.*

SECTION 58: ALDEHYDES FROM NITRILES

Ph-CN $\xrightarrow{\text{THF , RT , 24h}}$ Ph-CHO

98%

$$\left[\underset{\overset{|}{H}}{B} \right]^{-} K^{+}$$

Cha, J.S.; Yoon, M.S. *Tetrahedron Lett.,* ***1989****, 30,* 3677.

SECTION 59: ALDEHYDES FROM ALKENES

$$\xrightarrow[\text{Rh}_2(\mu St\text{-Bu})_2(\text{CO})_2\text{L}_2]{\text{CO/H}_2\text{O , 80}^\circ\text{C}}$$

∼∼∼CHO + ∼∼∼CHO

$L = P\left(\text{} \right)_3 \quad$ pH 6 , 15 h   (18   :   1) 53%

SO₃Na

Escaffre, P.; Thorez, A.; Kalck, P. *J. Chem. Soc., Chem. Commun.,* ***1987****,* 146.

$$\xrightarrow[\substack{\text{H}_2\text{ , CO , SnCl}_2 \\ \text{2650 psi , 60}^\circ\text{C , 8h}}]{}$$

CO₂t-Bu

Ph₂P—N

Cl—Pt—PPh₂

Cl

65% conversion

A/B = 0.60

CHO   CHO

OMe   OMe

A   B

(73% ee)

Parrinello, G.; Stille, J.K. *J. Am. Chem. Soc.,* ***1987****, 109,* 7122.

$$\xrightarrow[\substack{\text{1 atm O}_2\text{ , Na}_2\text{HPO}_4 \\ \text{DMF , 50}^\circ\text{C}}]{\text{HO} \quad \text{OH} \quad\text{, PdCl}_2\text{ , CuCl}}$$

75%

Hosokawa , T.; Ohta, T.; Kanayama, S.; Murahashi, S. *J. Org. Chem.,* ***1987****, 52,* 1758.

E:Z = 2:1

SnCl$_4$, MeCN ,0°C

5.5h

42%

trans:cis = 18:1

Simmons, D.P.; Reichlin, D.; Skuy, D. *Helv. Chim. Acta,* **1988,** *71,* 1000.

400 psi H$_2$, CO

RhH(CO)(PPh$_3$)$_3$

PPh$_3$, 14h , 80°C

(96   :   4)   80%

Doyle, M.M.; Jackson, W.R.; Perlmutter, P. *Tetrahedron Lett.,* **1989,** *30,* 233.

Related Methods: Ketones from Alkenes (Section 179)

SECTION 60: ALDEHYDES FROM MISCELLANEOUS COMPOUNDS

1. NaOH , EtOH

2. aq. H$_2$SO$_4$ '
   pentane

66%

can also generate ketones

Chikashita, H.; Morita, Y.; Itoh, K. *Synth. Commun.,* **1987,** *17,* 677.

1. 1.3 N NaOH , RT , 3h

2. H$_2$SO$_4$

68%

extension of the Nef reaction

Lou, J.-D.; Lou, W.-X. *Synthesis,* **1987,** 179.

SECTION 60A: PROTECTION OF ALDEHYDES

PhCHO

$\xrightarrow[\text{1. MeI , 10h}]{p\text{-TsOH , PhH , reflux}}$

$\xleftarrow{\text{2. K}_2\text{CO}_3 \text{ , 60°C}}$

Katritzky, A.R.; Fan, W.-Q.; Li, Q.-L. *Tetrahedron Lett.*, *1987*, *28*, 1195.

1.

$\xrightarrow[\text{3. SiO}_2 \text{ , H}_2\text{O}]{\text{2. LDA , THF ; PhCH}_2\text{I}}$

80%

>65% ee

Agami, C.; Couty, F. *Tetrahedron Lett.*, *1987*, *28*, 5659.

$\xrightarrow[\text{2. Me}_2\text{S , MeOH}]{\begin{array}{c}\text{1. 70% H}_2\text{O}_2 \\ \text{Cl}_3\text{CO}_2\text{H ,} \\ \text{CH}_2\text{Cl}_2 \text{ , } t\text{-BuOH}\end{array}}$

80% 4

Myers, A.G.; Fundy, M.A.M.; Lindstrom Jr., P.A. *Tetrahedron Lett.*, *1988*, *29*, 5609.

$\xrightarrow[\text{aq. MeOH}]{(\text{CF}_3\text{CO}_2)_2\text{IPh , 1 min}}$

92%

Stork, G.; Zhao, K. *Tetrahedron Lett.*, *1989*, *30*, 287.

Ph—CH=CH—CHO $\xrightarrow[\text{CH(OMe)}_3]{\text{RhCl}_3\text{(triphos)}, \text{MeOH}}$ Ph—CH=CH—CH(OMe)$_2$

85%

Ott, J.; Ramos Tombo, G.M.; Schmid, B.; Venanzi, L.M.; Wang, G.; Ward, T.R. *Tetrahedron Lett.*, *1989*, *30*, 6151.

CHAPTER 5

PREPARATION OF ALKYLS, METHYLENES AND ARYLS

This chapter lists the conversion of functional groups into methyl, ethyl, propyl, etc. as well as methylene (CH$_2$), phenyl, etc.

SECTION 61: ALKYLS, METHYLENES AND ARYLS FROM ALKYNES

Et—≡—Et → Me$_3$SiCl, Pd/C, THF / reflux, 3h → hexaethylbenzene **quant.**

Morris Jr., P.E.; Kiely, D.E. *J. Org. Chem.*, *1987*, *52*, 1149.

Et—≡—Et → Me$_3$SiCl, Pd/C, THF / reflux, 3h → hexaethylbenzene **quant.**

Jhingan, A.K.; Maier, W.F. *J. Org. Chem.*, *1987*, *52*, 1161.

EtOH, 12h, 25°C / 2% RhCl(PPh$_3$)$_3$

Neeson, S.J.; Stevenson, P.J. *Tetrahedron Lett.*, *1988*, *29*, 813.

Grigg, R.; Scott, R.; Stevenson, P. *J. Chem. Soc., Perkin Trans. I,* **1988**, 1357.

SECTION 62: ALKYLS, METHYLENES AND ARYLS FROM ACID DERIVATIVES

Gilday, J.P.; Paquette, L.A. *Tetrahedron Lett.,* **1988**, *29*, 4505.

Rigo, B.; Fasseur, D.; Cherepy, N.; Couturier, D. *Tetrahedron Lett.,* **1989**, *30*, 7057.

SECTION 63: ALKYLS, METHYLENES AND ARYLS FROM ALCOHOLS AND THIOLS

Rao, M.S.C.; Rao, G.S.K. *Synthesis,* **1987**, 231.

S:R = 96:4   89%

Harding, K.E.; Davis, C.S. *Tetrahedron Lett.*, *1988*, *29*, 1891.

32%

Nye, S.A.; Potts, K.T. *Synthesis*, *1988*, 375.

97%

Olah, G.A.; Wu, A.-h.; Farooq, O. *J. Org. Chem.*, *1989*, *54*, 1452.

PhOH $\xrightarrow[\text{CH}_2\text{Cl}_2 \text{ , 3h}]{\text{AlEt}_3 \text{ , Tf}_2\text{O , NEt}_3 \text{ , -20}^{\circ}\text{C}}$ PhCH$_2$CH$_3$ 94%

Hirotaq, K.; Isobe, Y.; Maki, Y. *J. Chem. Soc., Perkin Trans. I*, *1989*, 2513.

SECTION 64: ALKYLS, METHYLENES AND ARYLS FROM ALDEHYDES

83%

Reetz, M.T.; Kyung, S.-H. *Chem. Ber.*, *1987*, *120*, 123.

63%

DiRaddo, P.; Harvey. R.G. Tetrahedron Lett., *1988, 29*, 3885.

75%

Nakano, T.; Shirai, H.; Tamagawa,H.; Ishii. Y.; Ogawa,M. *J. Org. Chem., 1988, 53*, 5181.

Related Methods: Alkyls, Methylenes and Aryls from Ketones (Section 72)

SECTION 65: ALKYLS, METHYLENES AND ARYLS FROM ALKYLS, METHYLENES AND ARYLS

(72   :   28) 63%

Comins. D.L.; Mantlo, N.B. *Tetrahedron Lett., 1987, 28*, 759.

96%

Olah. G.A.; Surya Prakash, G.K.; Iyer, P.S.; Tashiro, M.; Yamamoto, T. *J. Org. Chem., 1987, 52*, 1881.

Meyers, A.I.; Brown, J.D.; Laucher, D. *Tetrahedron Lett.*, *1987*, *28*, 5279, 5283.

Sharp, M.J.; Cheng, W.; Snieckus, V. *Tetrahedron Lett.*, *1987*, *28*, 5093, 5097.

Matsuzaka, H.; Hiroe, Y.; Iwasaki, M.; Ishii, Y.; Koyasu, Y.; Hidai, M. *J. Org. Chem.*, *1988*, *53*, 3832.

Pansegrau, P.D.; Rieker, W.F.; Meyers, A.I. *J. Am. Chem. Soc.*, *1988*, *110*, 7178.

30%   52%   18%

Olah, G.A.; Farooq, O.; Farina, S.M.F.; Olah, J.A. *J. Am. Chem. Soc.*, *1988*, *110*, 2560.

quant.

Larock, R.C.; Baker, B.E. *Tetrahedron Lett.*, *1988*, *29*, 905.

51%

Kondo, T.; Tantayanon, S.; Tsuji, Y.; Watanabe, Y. *Tetrahedron Lett.*, *1989*, *30*, 4137.

46%   14%

Petrillo, G.; Novi, M.; Oell;Erba,C. *Tetrahedron Lett.*, *1989*, *30*, 6911.

Toda, F.; Tanaka, K.; Iwata, S. *J. Org. Chem., 1989, 54*, 3007.

SECTION 66: ALKYLS, METHYLENES AND ARYLS FROM AMIDES

NO ADDITIONAL EXAMPLES

SECTION 67: ALKYLS, METHYLENES AND ARYLS FROM AMINES

Takemura, H.; Shinmyozu, T.; Inazu, T. *Tetrahedron Lett., 1988, 29*, 1031.

Hosomi, A.; Hoashi, K.; Tominaga, Y.; Otaka, K.; Sakurai, H. *J. Org. Chem., 1987, 52*, 2947.

Wenkert, E.; Han, A.-L.; Jenny, C.-J. *J. Chem. Soc., Chem. Commun., 1988*, 975.

SECTION 68: ALKYLS, METHYLENES AND ARYLS FROM ESTERS

$$2\% \, Cu \, , \, -30^{\circ}C \quad (96 \quad : \quad 4)$$
$$5\% \, Cu \, , \quad 0^{\circ}C \quad (14 \quad : \quad 86)$$

Bäckvall, J.-E.; Sellén, M. J. Chem. Soc., Chem. Commun., **1987**, 827.

$$(5.5 \quad : \quad 1) \quad 82\%$$

Trost, B.M.; Lautens, M. J. Am. Chem. Soc., **1987**, 109, 1469.

71%

Amri, H.; Villieras, J. Tetrahedron Lett., **1987**, 28, 5521.

95%

Bolitt, V.; Mioskowski, C.; Falck, J.R. Tetrahedron Lett., **1989**, 30, 6027.

SECTION 69: ALKYLS, METHYLENES AND ARYLS FROM ETHERS, EPOXIDES AND THIOETHERS

The conversion ROR → RR' (R' = alkyl, aryl) is included in this section.

Johnstone, R.A.W.; McLean, W.N. *Tetrahedron Lett.*, *1988*, 29, 5553.

Menicagli, R.; Malanga, C.; Finato, B.; Lardicci, L. *Tetrahedron Lett.*, *1988*, 29, 3373.

Barton, D.H.R.; Ozbalik, N.; Ramesh, M. *Tetrahedron Lett.*, *1988*, 29, 3533.

Yamada, J.; Satô, H.; Yamamoto, Y. *Tetrahedron Lett.*, *1989*, 30, 5611.

Citterio, A.; Santi, R.; Fiorani, T.; Strologo, S. *J. Org. Chem.*, *1989*, 54, 2703.

REVIEW:

"Synthesis of Novel Benzenoid Molecules by Low-Valent-Titanium Deoxygenation"

Wong, H.N.C. Acc. Chem. Res., **1989**, 22, 145

SECTION 70: ALKYLS, METHYLENES AND ARYLS FROM HALIDES AND SULFONATES

The replacement of halogen by alkyl or aryl groups is included in this section. For the conversion of RX → RH (X = halogen) see Section 160 (Hydrides from Halides and Sulfonates).

(1 : 3) 81%

Echavarren, A.M.; Stille, J.K. J. Am. Chem. Soc., **1987**, 109, 5478.

94%

Rathke, M.W.; Vogiazoglou, D. J. Org. Chem., **1987**, 52, 3697.

75%

Scott, W.J. J. Chem. Soc., Chem. Commun., **1987**, 1755.

Br(CH$_2$)$_7$ — CH(OAc)—CH=CH$_2$

$\xrightarrow[\substack{\text{Mo(CO)}_6, \\ \text{PhMe , reflux}}]{\text{NaCH(CO}_2\text{Me)}_2}$

Br(CH$_2$)$_7$—CH(CH=CH$_2$)—CH(CO$_2$Me)(CO$_2$Me)

+ Br(CH$_2$)$_7$—CH=CH—CH$_2$—CH(CO$_2$Me)(CO$_2$Me)

(5.5   :   1) 82%

Minato, A.; Suzuki, K.; Tamao, K. *J. Am. Chem. Soc.*, **1987**, *109*, 1257.

$\xrightarrow[\text{3\% NiCl}_2\text{(dppp)}]{\text{PhMgBr , PhH}}$

67%

Comins, D.L.; Herrick, J.J. *Heterocycles*, **1987**, *26*, 2159.

$\xrightarrow[\text{-78 - +22}^\circ\text{C}]{\text{BuLi , ether}}$

86%   OSiMe$_2$t-Bu

Negishi, E.; Zhang, Y.; Bagheri, V. *Tetrahedron Lett.*, **1987**, *28*, 5793.

OP(=O)(OEt)$_2$

$\xrightarrow[\text{Pd(PPh}_3)_4 \text{ , 0}^\circ\text{C}]{\text{Bu}_3\text{MnLi , THF}}$

50%   +   39%

Fugami, K.; Oshima, K.; Utimoto, K. *Chem. Lett.*, **1987**, 2203.

$\xrightarrow[\text{Cu}_2\text{Br}_2 \text{ , -78}^\circ\text{C} \rightarrow \text{RT}]{\text{MeMgBr , THF}}$

75%

Bell, T.W.; Hu, L.-Y.; Patel, S.V. *J. Org. Chem.*, **1987**, *52*, 3847.

1. SOCl$_2$, CCl$_4$, 60°C

2. NalO$_4$, RuCl$_3$ •3 H$_2$O

   MeCN , H$_2$O , 25°C

3. BnMgCl , THF

   Li$_2$CuCl$_4$, -78°C   65%

Gao, Y.; Sharpless, K.B. *J. Am. Chem. Soc.*, *1988*, *110*, 7538.

OTf   Me$_3$SiO   5% [PdCl(η^3C$_3$H$_5$)]$_2$

               20% PPh$_3$

               HMPA , 12h

68%

Aoki, S.; Fujimura, T.; Nakamura, E.; Kuwajima, I. *J. Am. Chem. Soc.*, *1988*, *110*, 3296.

THF , TASF , 50°C , 2oh

PdCl/$_2$

81%

TASF = tris-(diethylamino)sulfonium
       difluorotrimethyl silicate

Hatanaka, Y.; Hiyama, T. *Tetrahedron Lett.*, *1988*, *29*, 97.

e$^-$, SCE electrode , pH 6.8

[Ru(trpy)(bpy)(H$_2$O)]$^{+2}$

phosphate buffer , aq BuOH   48%

trpy = tripyridyl

Mudurro, J.M.; Chiericato Jr., G.; DeGiovani, W.F.; Romero, J.R. *Tetrahedron Lett.*, *1988*, *29*, 765.

Br   1. Mg$^\circ$, THF

2. 0.5   Cl

   THF   Cl

   RT , 2h   90%

Cheng, J.-W.; Luo, F.-T. *Tetrahedron Lett.*, *1988*, *29*, 1293.

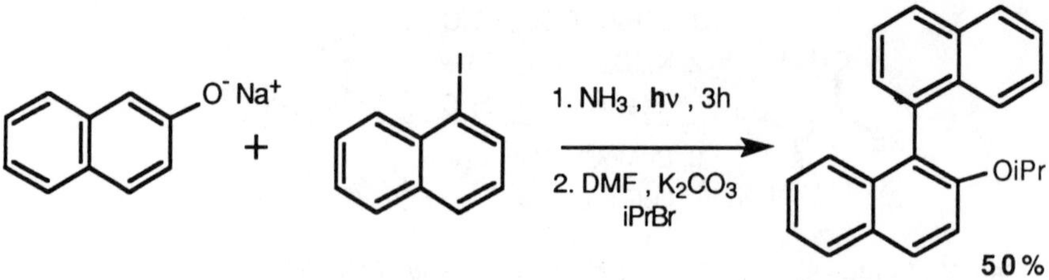

O⁻ Na⁺
+
I
1. NH3 , hν , 3h
2. DMF , K2CO3
iPrBr
OiPr

50%

Beugelmans, R.; Bois-Choussy, M.; Tang, Q. *Tetrahedron Lett.*, ***1988***, *29*, 1705.

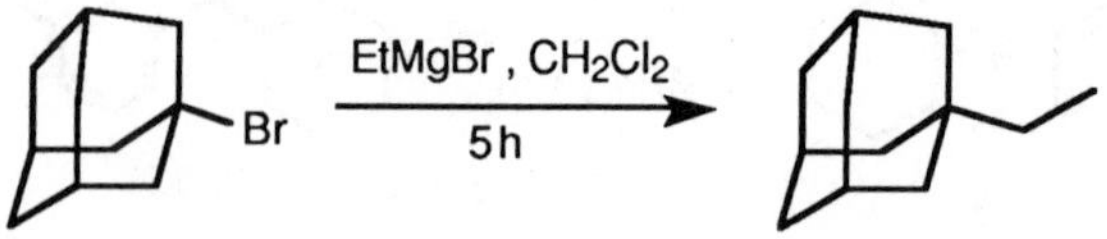

Br
EtMgBr , CH2Cl2
5h

61%

Ohno, M.; Shimizu, K.; Ishizaki, K.; Sasaki, T.; Eguchi, S. *J. Org. Chem.*, ***1988***, *53*, 729.

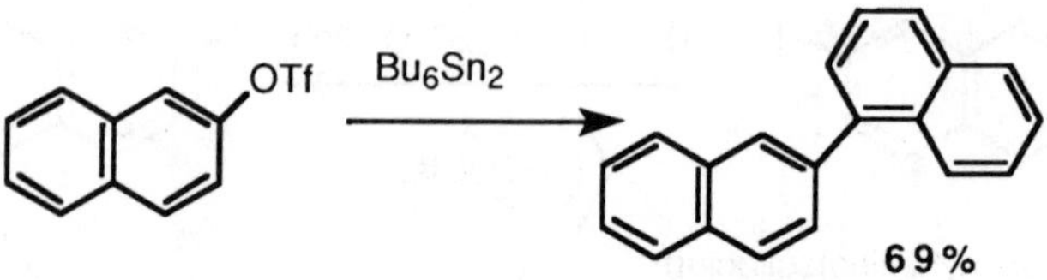

OTf
Bu6Sn2

69%

Crisp, G.T.; Papadopoulos, S. *Aust. J. Chem.*, ***1988***, *41*, 1711.

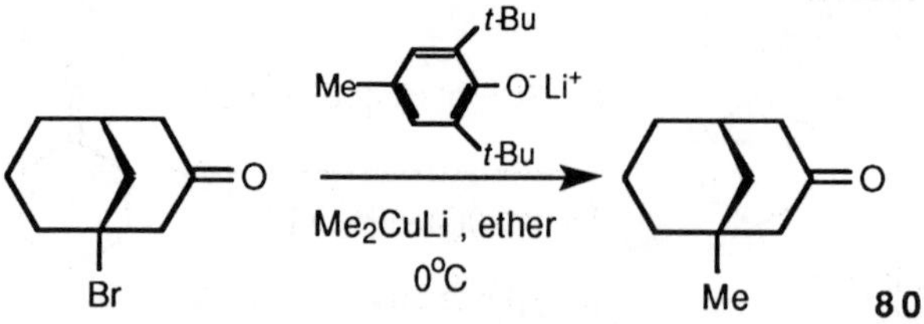

t-Bu
Me
O⁻ Li⁺
t-Bu
O
Br
Me2CuLi , ether
0°C
O
Me

80%

Kraus, G.A.; Yi, P. *Synth. Commun.*, ***1988***, *18*, 473.

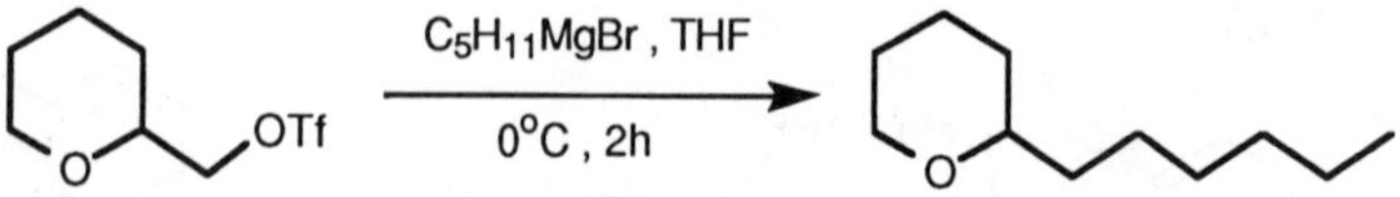

O
OTf
C5H11MgBr , THF
0°C , 2h
O

84%

Kotsuki, H.; Kadota, I.; Ochi, M. *Tetrahedron Lett.*, ***1989***, *30*, 1281.

(100   :   0)   98%

2 eq. ,   45 min   (12   :   1)   91%
0.3 eq. , 2h   (1   :   2)   79%

Matsumoto, T.; Katsuki, M.; Suzuki, K. *Tetrahedron Lett., 1989, 30*, 833.

(4   :   1)   quant

Winkler, J.D.; Sridar, V.; Siegel, M.G. *Tetrahedron Lett., 1989, 30*, 4943.

$$FO_2SCF_2CO_2Me \text{ , CuI}$$

PhCH$_2$Br $\xrightarrow{\qquad\qquad}$ PhCh$_2$CF$_3$

DMF , 70°C , 2h   81%

Chen, Q.-Y.; Wu, S.-W. *J. Chem. Soc., Chem. Commun., 1989*, 705.

MeO$_2$C—〈 〉—I $\xrightarrow[\text{Tl}_2\text{CO}_3 \text{ , THF , 50°C}]{\text{PdCl}_2\text{(dppf) , 16h}}$ MeO$_2$C—〈 〉—C$_8$H$_{17}$ 88%

Sato, M.; Miyaura, N.; Suzuki, A. *Chem. Lett., 1989*, 1405.

I—〈 〉—OEt $\xrightarrow[\text{DMF}]{\substack{\text{PhSiF}_2\text{Et , KF , 70°C} \\ \text{10h , 5\% (}\eta^3\text{-C}_3\text{H}_5\text{PdCl)}_2}}$ Ph—〈 〉—OEt   81%

Hatyanaka, Y.; Fukushima, S.; Hiyama, T. *Chem. Lett., 1989*, 1711.

Miyura, N.; Ishiyama, T.; Sasaki,H.; Ishikawa, M.; Satoh, M.; Suzuki, A. J. Am. Chem. Soc., 1989, 111, 314.

REVIEW:

"Applications of Higher Order Mixed Organocuprates to Organic Synthesis"

Lipshutz, B.H. Synthesis, 1987, 325.

SECTION 71: ALKYLS, METHYLENES AND ARYLS FROM HYDRIDES

This section lists examples of the reaction of RH → RR' (R,R' = alkyl or aryl). For the reaction C=CH → C=C-R (R = alkyl or aryl), see Section 209 (Alkenes from Alkenes). For alkylations of ketones and esters, see Section 177 (Ketones from Ketones) and Section 113 (Esters from Esters).

Lehnert, E.K.; Sawyer, J.S.; Macdonald, T.L. Tetrahedron Lett., 1989, 30, 5215.

REVIEW:

"Arylation Reactions of Organobismuth Reagents"

Finet, J.-P. Chem. Rev., 1989, 89, 1487.

SECTION 72: ALKYLS, METHYLENES AND ARYLS FROM KETONES

The conversions $R_2C=O$ → R-R, R_2CH_2, R_2CHR', etc. are listed in this section.

Bu₃SnH , PhH , AIBN

reflux , 1.5h

$$Bu_3SnH , PhH , AIBN$$

PhCH₃

73%

Schmidt, K.; O'Neal, S.; Chan, T.C.; Alexis, C.P.; Uribe, J.M.; Lossener, K.; <u>Gutierrez, C.G.</u> Tetrahedron Lett., **1989**, 30, 7301.

1. NiCl₂(dppe) , PhH
MeMgl , reflux , overnight

2. H₂O

93%

Yang, P.-F.; Ni, Z.-J.; <u>Luh, T.-Y.</u> J. Org. Chem., **1989**, 54, 2261.

REVIEW:

"Formation of Six-Membered Aromatic Rings by Cyclialkylation of Some Aldehydes and Ketones"

Bradsher, C.K. Chem. Rev., **1987**, 87, 1277.;

SECTION 73 ALKYLS, METHYLENES AND ARYLS FROM NITRILES

variationof Bruylant's reaction   **89%**

<u>Kudzma, L.V.</u>; Spencer, H.K.; Severnak, S.A. Tetrahedron Lett., **1988**, 29, 6827.

SECTION 74: ALKYLS, METHYLENES AND ARYLS FROM ALKENES

The following reaction types are included in this section:

A.   Hydrogenation of Alkenes (and Aryls)
B.   Formation of Aryls

C. Alkylations and Arylations of Alkenes
D. Conjugate Reduction of Conjugated Aldehydes, Ketones, Acids, Esters and Nitriles
E. Conjugate Alkylations
F. Cyclopropanations, including halocyclopropanations

SECTION 74A: Hydrogenation of Alkenes (and Aryls)

Reduction of aryls to dienes are listed in Section 377 (Alkene-Alkene).

Brown, J.M.; James, A.P. *J. Chem. Soc., Chem. Commun.*, *1987*, 181.

Negoro, T.; Ikeda, Y. *Org. Prep. Proceed. Int.*, *1987*, 19, 71.

Ohta, T.; Takaya, H.; Kitamura, M.; Nagai, K.; Noyori, R. *J. Org. Chem.*, *1987*, 52, 3176.
Takaya, H.; Ohta, T.; Sago, N.; Kimobayashi, H.; Akutagawa, S.; Inoue, S.; Kasahara, I.;
Noyori, R. *J. Am. Chem. Soc.*, *1987*, 109, 1596.

Ofosu-Asante, K.; Stock, L.M. *J. Org. Chem.*, **1987**, *52*, 2939.

Cesarotti, E.; Chiesa, A.; Prati, L. *Gazz. Chim. Ital.*, **1987**, *117*, 129.

BINAP = R-(+)-2,2'-bis-(diphenylphosphno) 1,1'-binapthyl

Kawano, H.; Ishii, Y.; Ikariya, T.; Saburi, M.; Yoshikawa, S.; Ucida, Y.; Kumobayashi, H. *Tetrahedron Lett.*, **1987**, *28*, 1905.

Rabideau, P.W.; Karrick, G.L. *Tetrahedron Lett.*, **1987**, *28*, 2481

[RhCl₃ , TPPTS]

$$\xrightarrow{\text{H}_2 \text{ , aq. EtOH , RT}}$$

76%

TPPTS = triphenylphosphine meta trisulfonate

Larpent, C.; Dabard, R.; Patin, H. Tetrahedron Lett., **1987**, *28*, 2507.

[PhI(OAc)₂ , NH₂NH₂ •H₂O]

$$\xrightarrow{\text{CH}_2\text{Cl}_2}$$

PhSEt

85%

new method to generate diimide

Moriarty, R.M.; Vaid, R.K.; Duncan, M.P. *Synth. Commun.*, **1987**, *17*, 703.

[salen • (PdCl₄)⁻ K⁺ , NEt₃]

$$\xrightarrow{\text{H}_2 \text{ , EtOH , 3h}}$$

>99%

Kerr, J.M.; Suckling, C.J. Tetrahedron Lett., **1988**, *29*, 5545.

H₂ , [Rh(COD)Cl]₂ , MeOH

quant

70% ee , S

Ar = NMe₂

Morimoto, T.; Chiba, M.; Achiwa, K. Tetrahedron Lett., **1988**, *29*, 4755.

Me
|
D₃C–C=C–COOH
| |
        Ph

D_3C ... Me, COOH, Ph

50 atm H₂ , MeOH/THF
NEt₃ , 50h

additive , AgBF₄
1% RhCl(NBD)

98.4%ee

= additive

Hayashi, T.; Kawamura, N.; Ito, Y. *Tetrahedron Lett.*, **1988**, *29*, 5969.

70 atm H₂ , 20°C
EtOH/PhH

93%

67% ee , S

Sunjić, V.; Habuš, I.; Comisso, G.; Moimas, F. *Gazz. Chim. Ital.*, **1989**, *119*, 229.

REVIEWS:

"Recent Advances in Catalytic Asymmetric Reactions Promoted by Transition Metal Complexes"

Ojima, I.; Clos, N.; Bastos, C. *Tetrahedron*, **1989**, *45*, 6901.

"Ammonium Formate in Organic Synthesis: A Versatile Agent in Catalytic Hydrogen Transfer Reductions"

Ram, S.; Ehrenkaufer, R.E. *Synthesis*, **1988**, 91.

SECTION 74B: Formation of Aryls

Feigenbaum, A.; Pete, J.-P.; Poquet-Dhimane, A. *Tetrahedron Lett.*, *1988*, *29*, 73.

Suginone, H.; Senboku, H.; Yamada, S. *Tetrahedron Lett.*, *1988*, *29*, 79.

SECTION 74C: Alkylations and Arylations of Alkenes

Amos, P.C.; Whiting, D.A. *J. Chem. Soc., Chem. Commun.*, *1987*, 510.

Uno, M.; Takahashi, T.; Takahashi, S. *J. Chem. Soc., Chem. Commun.*, *1987*, 785.

Negishi, E.; Takahashi, T.; Baba, S.; Van Horn, D.E.; Okukado, N. *J. Am. Chem. Soc.*, *1987*, *109*, 2393.

Bailey, W.F.; Nurmi, T.T.; Patricia, J.J.; Wang, W. *J. Am. Chem. Soc., 1987, 109,* 2442.

Cacchi, S.; Ciattini, P.G.; Morera, E.; Ortar. G. *Tetrahedron Lett., 1987, 28,* 3039.

Hatanaka, Y.; Hiyama. T. *Tetrahedron Lett., 1987, 28,* 4715.

Piers. E.; Jean, M.; Marrs, P.S. *Tetrahedron Lett., 1987, 28,* 5075.

Fiandanese, V.; Marchese, G.; Naso. F.; Ronzini, L. *Synthesis, 1987,* 103.

49%

cis/trans = 1.9

Porter, N.A.; Chang, V.H.-T.; Magnin, D.R.; Wright, B.T. *J. Am. Chem. Soc.*, **1988**, *110*, 3554.

87%

Marino, J.P.; Laborde, E.; Paley, R.S. *J. Am. Chem. Soc.*, **1988**, *110*, 966.

tppe = 1,2-bis-triphenyl-phosphinoethane

Sekiya, K.; Nakamura, E. *Tetrahedron Lett.*, **1988**, *29*, 5155.

86%

(4:1)

Negishi, E.; Zhang, Y.; O'Connor, B. *Tetrahedron Lett.*, **1988**, *29*, 2915.

4% Pd(OAc)$_2$, MeCN
9% PPh$_3$, Ag$_2$CO$_3$
80°C , 3d

72%

Larock, R.C.; Song, H.; Baker, B.E.; Gong, W.H. *Tetrahedron Lett.*, *1988*, 29, 2919.

5% Pd(OAc)$_2$, 2d

HCO$_2$Na , Na$_2$CO$_3$
80°C , DMF , Bu$_4$NCl

83%

Larock, R.C.; Stinn, D.E. *Tetrahedron Lett.*, *1988*, 29, 4687.

Pb(OAc)$_4$, RT
CHCl$_3$, 1 min

65%

Moloney, M.G.; Pinhey, J.T. *J. Chem. Soc., Perkin Trans. I*, *1988*, 2847.

(Bu$_3$Sn)$_2$, AIBN , PhH

hv (sunlamp)

73%

Winkler, J.D.; Sridar, V. *Tetrahedron Lett.*, *1988*, 29, 6219.

Bu$_3$SnH , AIBN

PhH , reflux
4h

96%

α:β = 86:14

DeMesmaeker, A.; Hoffmann, P.; Ernst, B. *Tetrahedron Lett.*, *1988*, 29, 6585.

NO₂

1. AlEt₃ , hexane , 0°C
2. ether , 0.1N HCl
───────────────→
0°C

NO₂   Et

+

O   Et

80%   20%
(cis:trans = 75:25)

Pecunioso, A.; Menicagli, R. *J. Org. Chem.*, *1988, 53*, 45.

Br   Bu₃Sn⌒⌒SO₂Ph
──────────────→
PhH , AIBN , hν , 12h

⌒⌒SO₂Ph

79%

Keck, G.E.; Byers, J.H.; Tafesh, A.M. *J. Org. Chem.*, *1988, 53*, 1127.

BuLi , THF , -78 → 0°C
──────────────────→

C₆H₁₃⌒O⌒SnBu₃

Me

C₆H₁₃   O

54%

cis:trans = 11:1

Broka, C.A.; Lee, W.J.; Shen, T. *J. Org. Chem.*, *1988, 53*, 1336.

PhI , DMF , 80°C
F₂C=CFZnBr ──────────────→ F₂C=CFPh
3% Pd(Ph₃)₄ , 10h   74%

Heinze, P.L.; Burton, D.J. *J. Org. Chem.*, *1988, 53*, 2714.

Me⌒⌒ZnBr
          Me
NiBr₂(PBu₃)₂
──────────────→
CH₂Cl₂ , 25°C
16h

Me⌒⌒⌒O
Me   O   Me
                        Me

50%

Yanagisawa, A.; Habaue, S.; Yamamoto, H. *J. Am. Chem. Soc.*, *1989, 111*, 366.

1. $C_4H_9CH=CH_2$, hν
Bu$_3$SnSnBu$_3$

2. Bu$_3$SnH

Me, CO$_2$Me / I, CO$_2$Me → Me, CO$_2$Me / C$_6$H$_{13}$, CO$_2$Me

66%

Curran, D.P.; Chen, M.-H.; Spletzer, E.; Seong, C.M.; Chang, C.-T. *J. Am. Chem. Soc.*, *1989*, *111*, 8872.

1. *t*-BuLi , pentane/ether
-78°C

2. TMEDA , -78°C→ RT

3. MeOH

84%

Bailey, W.F.; Rossi, K. *J. Am. Chem. Soc.*, *1989*, *111*, 765.

PhI , 2.5% Pd(OAc)$_2$

Bu$_4$NCl , HCO$_2$Na
DMF , RT , 24h

71%

Larock, R.C.; Johnson, P.L. *J. Chem. Soc., Chem. Commun.*, *1989*, 136.

SO$_2$Cl

$\diagup$ CO$_2$Bu , PdCl$_2$(PhCN)$_2$

K$_2$CO$_3$, Bzoct$_3$NCl

CO$_2$Bu

95%   E

Miura, M.; Hashimoto, H.; Itoh, K.; Nomura, M. *Tetrahedron Lett.*, *1989*, *30*, 975.

Br, SPh , 9-BBN

Pd(PPh$_3$)$_4$, NaOH
THF

64%

OMOM   SPh / OMOM

Hoshino, Y.; Ishiyama, T.; Miyama, N.; Suzuki, A. *Tetrahedron Lett.*, *1989*, *30*, 3983.

4 eq. TiCl$_4$, CH$_2$Cl$_2$

-78°C , 3 min

76%

Angle, S.R.; Louie, M.S. *Tetrahedron Lett.*, *1989*, *30*, 5741.

C$_7$H$_{15}$

1. MeLi

2. MeI , -30°C

3. BF$_3$ •OEt$_2$, aq. dioxane
   20°C

C$_7$H$_{15}$ Me

66%

E:Z = 82:18

Baudin , J.-B.; Julia ,S A. *Tetrahedron Lett.*, *1989*, *30*, 1963, 1967.

Si(OMe)$_2$Me

[PdCl(η-C$_3$H$_5$)]$_2$, 5h

P(OEt)$_3$, THF
TBAF , 50°C

Ph

90%

Tamao, K.; Kobayashi, K.; Ito, Y. *Tetrahedron Lett.*, *1989*, *30*, 6051.

REVIEWS:

"Olefin Synthesis via Organometallic Coupling Reactions of Enol Triflates"

Scott, W.J.; McMurry, J.E. *Acc. Chem. Res.*, *1988*, *21*, 48.

"1,2-Additions to Heteroatom Substituted Olefins by Organopalladium
Reagents:

Daves Jr., G.D.; Hallberg, A. *Chem. Rev.*, *1989*, *89*, 1433.

SECTION 74D: Conjugate Reduction of α,β-Unsaturated Aldehydes, Ketones, Acids, Esters and Nitriles

9:1 Zn:NiCl$_2$, »)))

HOCH$_2$CH$_2$OH , H2O
2h   98%

Petrier, C.; Luche, J.-L. Tetrahedron Lett., **1987**, 28, 2347.

MeO$_2$C

TiCl$_4$•LiAlH$_4$, NEt$_3$, 24h
105°C   59%

MeO$_2$C
CO$_2$Me

CO$_2$Me

Hung, C.W.; Wong, H.N.C. Tetrahedron Lett., **1987**, 28, 2393.

OH H
Ph   N
     Me

hv (254 nm) , -40°C
CH$_2$Cl$_2$

OBn
Me

O
OBn   68%
Me   H

44% ee , R

Piva, O.; Henin, F.; Muzart, J.; Pete, J.-P. Tetrahedron Lett., **1987**, 28, 4825.

CO$_2$Me

1. Mg° , MeOH , RT
4h
2. 3N HCl

CO$_2$Me
70%

Hudlicky, T.; Sinai-Zingde, G.; Natchus, M.G. Tetrahedron Lett., **1987**, 28, 5287.

NO$_2$

1. Bu$_3$SnH
2. H$_2$F$_2$, MeOH

NO$_2$

Ph   Me

Ph   Me   50%

Aizpurua, J.M.; Oiarbide, M.; Palomo, C. Tetrahedron Lett., **1987**, 28, 5365.

Nag, A.; Sarkar, A.; Sarkar, S.K.; Palit, S.K. *Synth. Commun.*, *1987*, *17*, 1007.

Camps, F.; Coll, J.; Guitart, J. *Tetrahedron*, *1987*, *43*, 2329.

Gil, G.; Ferre, E.; Barre, M.; LePetit, J. *Tetrahedron Lett.*, *1988*, *29*, 3797.

Brunner, H.; Leitner, W. *Angew. Chem. Int. Ed., Engl*, *1988*, *27*, 1180.

Kim, K.E.; Park, S.B.; Yoon, N.M. *Synth. Commun.*, *1988*, *18*, 89.

Mahoney, W.S.; Brestensky, D.M.; Stryker, J.M. *J. Am. Chem. Soc.*, *1988*, *110*, 291.

Brestensky, D.M.; Stryker, J.M. *Tetrahedron Lett.*, *1989*, *30*, 5677.

Rhodes, R.A.; Boykin, D.W. *Synth. Commun.*, *1988*, *18*, 681.

Hazarika, M.J.; Barua, N.C. *Tetrahedron Lett.*, *1989*, *30*, 6567.

DBNE = (1S,2R)-N,N-dibutylnorephedrine   74% ee

Soai, K.; Hayasaka, T.; Ugajin, S. *J. Chem. Soc., Chem. Commun.*, *1989*, 516.

Leutenegger, U.; Madin, A.; Pfaltz, A. Angew. Chem. Int. Ed., Engl., 1989, 28, 60.

Huang, X.; Zhang, H.-Z. Synth. Commun., 1989, 19, 97.

Goudgaon, N.M.; Wadgaonkar, P.P.; Kabalka, G.W. Synth. Commun., 1989, 19, 805.

de la Cal, M.T.; Cristobal, B.I.; Cuadrado, F.; González, A.M.; Pulido, F.J. Synth. Commun., 1989, 19, 1039.

Ohta, H.; Kobayashi, N.; Ozaki, K. J. Org. Chem., 1989, 54, 1802.

82%

Nishiyama, Y.; Makino, Y.; Hamanaka, S.; Ogawa, A.; Sonoda, N. *Bull. Chem. Soc., Jpn.,* **1989,** *62,* 682.

SECTION 74E: Conjugate Alkylations

96%

Johnson, C.R.; Marren, T.J. *Tetrahedron Lett.,* **1987,**28, 27.

95%

Johnson, C.R.; Dhanoa, D.S. *J. Org. Chem.,* **1987,** *52,* 1885.

(18   :   1) 83%

Bunce, R.A.; Wamsley, E.J.; Pierce, J.D.; Shellhammer Jr., A.J.; Drumwright, R.E. *J. Org. Chem.,* **1987,** *52,* 464.

Cooke Jr., M.P.; Widener, R.K. *J. Org. Chem.*, **1987**, *52*, 1381.

Basavaiah, D.; Gowriswari, V.V.L.; Bharathi, T.K. *Tetrahedron Lett.*, **1987**, *28*, 4591.

Dieter, J.W.; Li, Z.; Nicholas, K.M. *Tetrahedron Lett.*, **1987**, *28*, 5415.

Panek, J.S.; Sparks, M.A. *Tetrahedron Lett.*, **1987**, *28*, 4649.

Keinan, E.; Perez, D. *J. Org. Chem.*, **1987**, *52*, 2576.

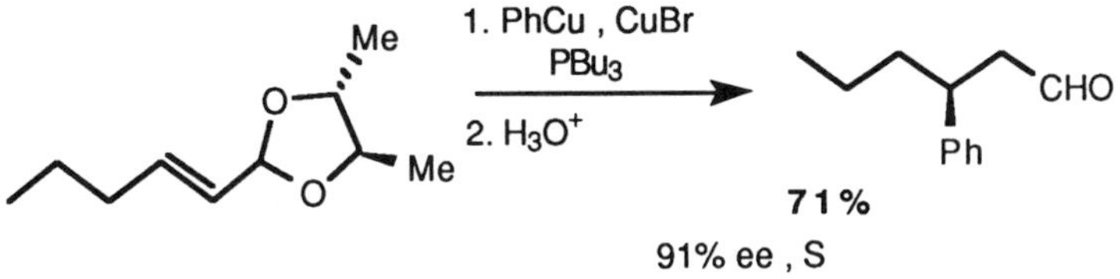

Me
O
O
Me
1. PhCu , CuBr
PBu3
2. H3O+
CHO
Ph
71%
91% ee , S

Mangeney, P.; Alexakis, A.; Normant, J.F. *Tetrahedron Lett.*, ***1987***, *28*, 2363.

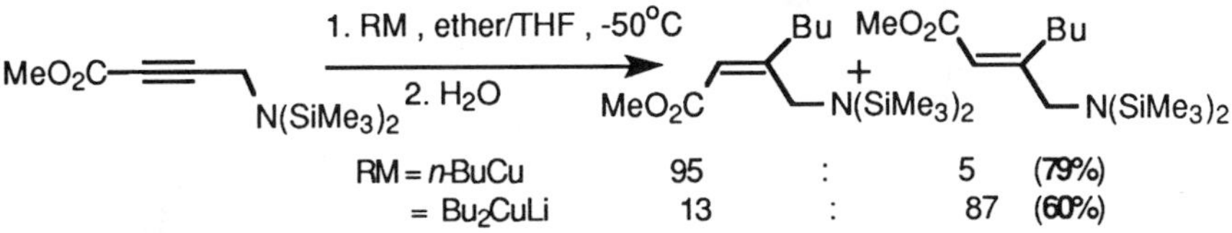

MeO2C
N(SiMe3)2
1. RM , ether/THF , -50°C
2. H2O
Bu
MeO2C
N(SiMe3)2
+
MeO2C
Bu
N(SiMe3)2
RM = n-BuCu 95 : 5 (79%)
= Bu2CuLi 13 : 87 (60%)

Corriu, R.J.P.; Moreau, J.J.E.; Vernhet, C. *Tetrahedron Lett.*, ***1987***, *28*, 2963.

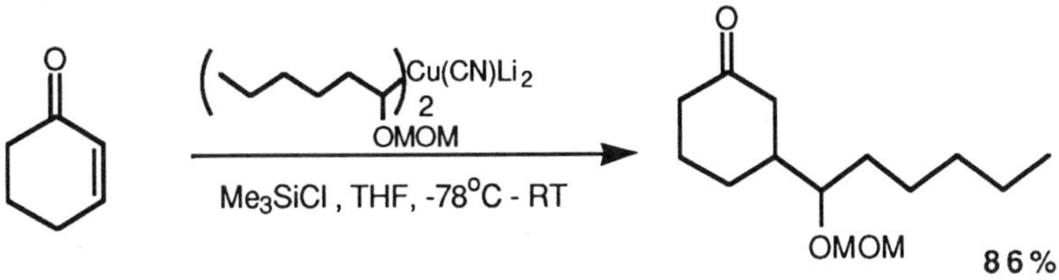

O
Cu(CN)Li2
2
OMOM
Me3SiCl , THF, -78°C - RT
O
OMOM
86%

Linderman, R.J.; Godfrey, A.; Horne, K. *Tetrahedron Lett.*, ***1987***, *28*, 3911.

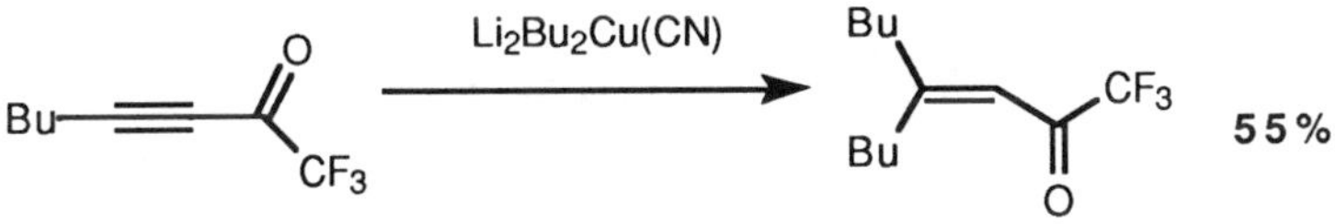

Bu
O
CF3
Li2Bu2Cu(CN)
Bu
Bu
CF3
O
55%

Linderman, R.J.; Lonikar, M.S. *Tetrahedron Lett.*, ***1987***, *28*, 5271.

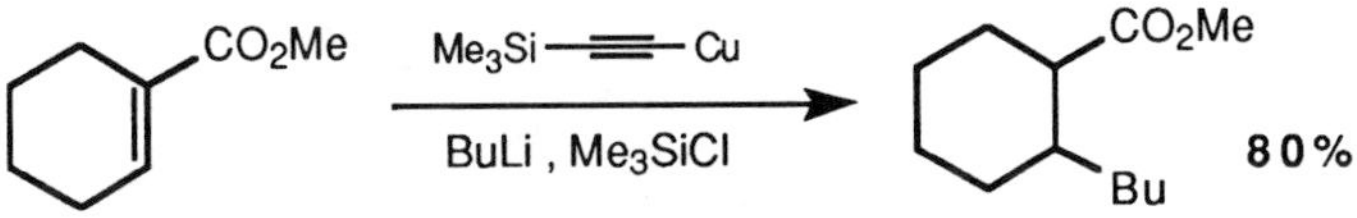

CO2Me
Me3Si Cu
BuLi , Me3SiCl
CO2Me
Bu
80%

Sakata, H.; Kuwajima, I. *Tetrahedron Lett.*, ***1987***, *28*, 5719.

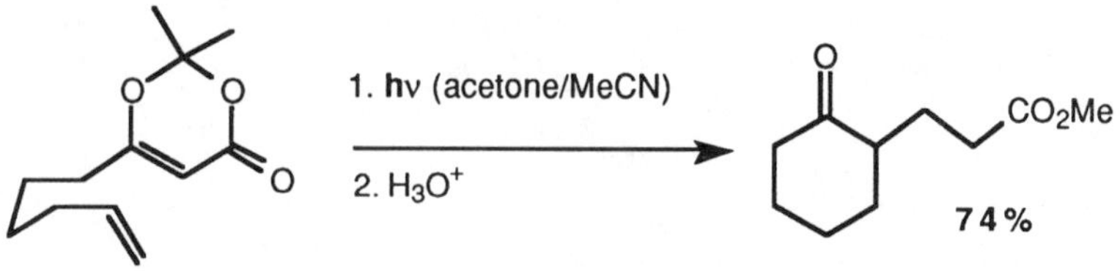

1. hν (acetone/MeCN)
2. H_3O^+
CO_2Me
74%

Winkler, J.D.; Hey, J.P.; Hannon, F.J.; Williard, P.G. *Heterocycles*, ***1987***, *25*, 55.

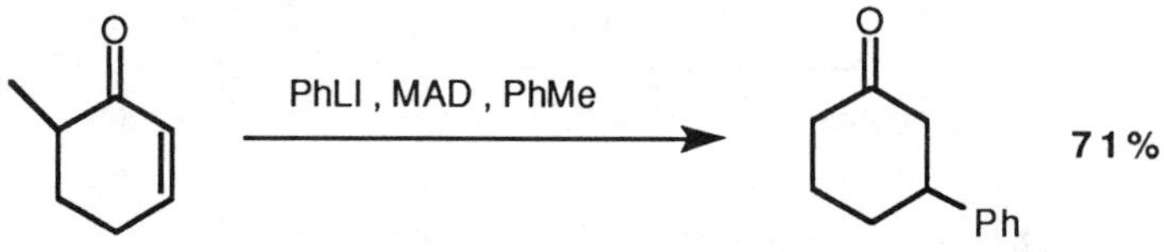

PhLI , MAD , PhMe
Ph
71%

cis:trans = 33:67

MAD = bis(2,6-di*t*-butyl)-4-methyl phenoate MeAl

Maruoka, K.; Nonishita, K.; Yamamoto, H. *Tetrahedron Lett.*, ***1987***, *28*, 5723.

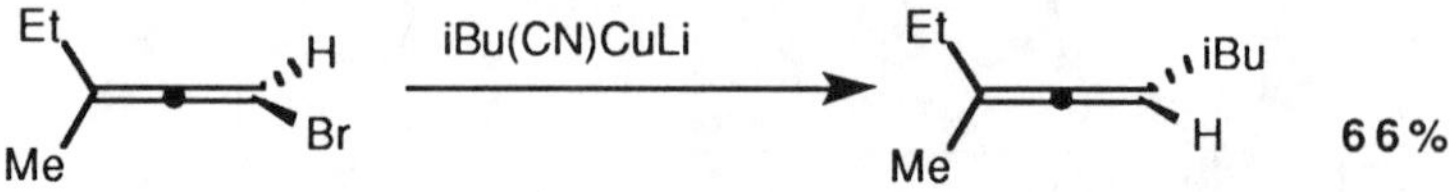

Et
Me
H
Br
iBu(CN)CuLi
iBu
66%

Caporusso, A.M.; Polizzi, C.; Lardicci, L. *Tetrahedron Lett.*, ***1987***, *28*, 6073.

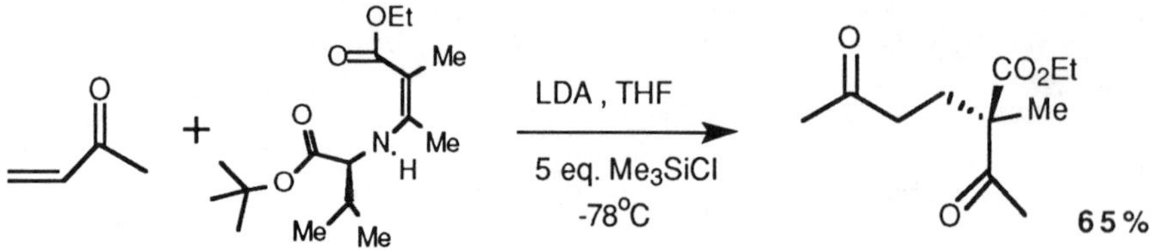

OEt
Me
N
H
LDA , THF
5 eq. Me_3SiCl
-78°C
CO_2Et
65%

73% ee , R

Tomioka, K.; Seo, W.; Ando, K.; Koga, K. *Tetrahedron Lett.*, ***1987***, *28*, 6637.

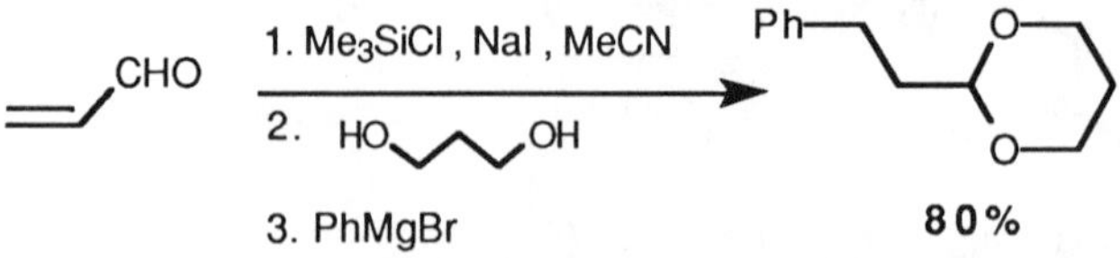

CHO
1. Me_3SiCl , NaI , MeCN
2. HO OH
3. PhMgBr
Ph
80%

Lee, Y.-S.; del Valle, L.; Larson, G.L. *Synth. Commun.*, ***1987***, *17*, 385.

BuCH₂NO₂ $\xrightarrow[\text{Amberlyst-A21}]{\overset{\text{CO}_2\text{Me}}{=\!\!=\!\!\diagup}\text{, RT , 6h}}$ Bu~CH(NO₂)CH₂CH₂CO₂Me

77%

Ballini, R.; Petrini, M.; Rosini, G. *Synthesis*, **1987**, 711.

$$\text{(MeCuL*)Li , ether} \quad -78^{\circ}\text{C}$$

66%

77% ee

L* =

Dieter, R.K.; Tokles, M. *J. Am. Chem. Soc.*, **1987**, *109*, 2040.

$$\xrightarrow[\substack{\text{MeCN , cat. TrClO}_4 \\ 3d}]{\diagup\!\!\diagdown\text{SiMe}_2\text{Ph , 0}^{\circ}\text{C}}$$

68%

Hayashi, M.; Mukaiyama, T. *Chem. Lett.*, **1987**, 289.

LiPhSCu~~~CuSPhLi THF

$$\xrightarrow{-15^{\circ}\text{C , 10 min}}$$

96%

Wender, P.A.; White, A.W. *J. Am. Chem. Soc.*, **1988**, *110*, 2218.

$$\xrightarrow[\text{THF , RT , 1h}]{\overset{\text{SnBu}_3}{=\!\!=\!\!\diagdown}\text{, Me}_2\text{Cu(CN)Li}_2}$$

95%

Behling, J.R.; Babiak, K.A.; Ng, J.S.; Campbell, A.L.; Moretti, R.; Koerner, M.; Lipshutz, B.H. *J. Am. Chem. Soc.*, **1988**, *110*, 2641.

$$2 \ Me_2Cu(CN)Li_2$$
THF , -78°C , 30 min

76%

trans:cis = 56:1

Smith III, A.B.; Dunlap, N.K.; Sulikowski, G.A. *Tetrahedron Lett., 1988, 29*, 443.

BuLi , THF , -78°C
(inverse addition)

77%

Cooke Jr., M.P. *J. Org. Chem., 1987, 52*, 5729.

, vitamin B_{12}
e^- , $h\nu$, DMF , 5.5h
$LiClO_4$, Pt anode
C cathode

70%

Arliński, R.; Stankiewicz, J. *Tetrahedron Lett., 1988, 29*, 1601.

$Zn(Cu)$, , aq. EtOH
RT , ⠀⠀⠀»)))

90%

Luche, J.-L.; Allavena, C. *Tetrahedron Lett., 1988, 29*, 5369.

1. 0.5% $Pd(PPh_3)_4$, PhH , 5h
$PhCl_2SiSiMe_3$, 80°C
2. MeLi , ether
3. H_3O^+

78%

Hayashi, T.; Matsumoto, Y.; Ito, Y. *Tetrahedron Lett., 1988, 29*, 4147.

85%

Ng, J..; Behling, J.R.; Campbell, A.L.; Nguyen, D.; Lipshutz, B. *Tetrahedron Lett.,* *1988, 29,* 3045.

80%

Jansen, J.F.G.A.; Feringa, B.L. *Tetrahedron Lett., 1988, 29,* 3593.

87%

Yeh, M.C.P.; Knochel, P.; Butler, W.M.; Berk, S.C. *Tetrahedron Lett., 1988, 29,* 6693.

75%

Asaoka, M.; Shima, K.; Takei, H. *J. Chem. Soc., Chem. Commun., 1988,* 430.

78%

Bambal, R.; Kemmitt, R.D.W. *J. Chem. Soc., Chem. Commun., 1988,* 734.

87%

Fang, J.-M.; Chang, H.-T.; Lin, C.-C. *J. Chem. Soc., Chem. Commun.*, *1988*, 1385.

trans:cis = 1.8:1

Little, R.D.; Fox, D.P.; Van Hijfte, L.; Dannecker, R.; Sowell, G.; Wolin, R.L.; Moëns, L.; Baizer, M.M. *J. Org. Chem.*, *1988, 53*, 2287.

trans:cis = 97:3

Tietze, L.F.; Beifuss, U.; Ruther, M.; Rühlmann, A.; Antel, J.; Sheldrick, G.M. *Angew. Chem. Int. Ed., Engl, 1988, 27*, 1186.

$$C_{11}H_{23}I \xrightarrow[\text{AIBN , 80°C , MeCN}]{\text{CH}_2\text{=CHCN , Bu}_3\text{GeH , 8h}} C_{11}H_{23}CH_2CH_2C\equiv N$$

71%

Pike, P.; Hershberger, S.; Hershberger,J. *Tetrahedron, 1988, 44*, 6295.

Kjonaas, R.A.; Hoffer, R.K. *J. Org. Chem., 1988, 53*, 4133.

Ph, trans-CH=CH, C(=O), Ph

1. [Bu$_2$N, OH, Me, Ph (amino alcohol), Ni(acac)$_2$, 80°C, PhMe]
RT
2. Me$_2$Zn, -30°C

Ph, CH(OH), CH$_2$, C(=O), Ph   72%

40% ee, R

Soai, K.; Yokoyama, S.; Hayasaka, T.; Ebihara, K. *J. Org. Chem.*, *1988*, *53*, 4149.
Soai, K.; Hayasaka, T.; Ugajin, S.; Yokoyama, S. *Chem. Lett.*, *1988*, 1571.

cyclohexyl-I

CH$_2$=CH-CN, AIBN, PhMe
(Me$_3$Si)$_3$SiH, 90°C, 5h

cyclohexyl-CH$_2$CH$_2$-CN   90%

Giese, B.; Kopping, B.; Chatgilialoglu, C. *Tetrahedron Lett.*, *1989*, *30*, 681.

C$_7$H$_{15}$, Br

CH$_2$=CH-CO$_2$Et, THF
NiCl$_2$·6 H$_2$O, Zn^0
Py, 200h

C$_9$H$_{19}$, CO$_2$Et   69%

Sustmann, R.; Hopp, P.; Hull, P. *Tetrahedron Lett.*, *1989*, *30*, 689.

cyclohex-2-enone

BuMnCl, 1% CuCl
THF, 0°C, 30 min

3-butylcyclohexanone   95%

Cahiez, G.; Alami, M. *Tetrahedron Lett.*, *1989*, *30*, 3541, 3545, 7365.

Me, CO$_2$Et / CO$_2$Et

BuMnCl, THF
10°C

Me, CO$_2$Et; Bu, CO$_2$Et   67%

Cahiez, G.; Alami, M. *Tetrahedron*, *1989*, *45*, 4163.

Me$_3$Sn(Bu)Cu(CN)Li$_2$

THF , -78°C , 30 min

81%

Lipshutz, B.H.; Reuter, D.C. *Tetrahedron Lett.*, **1989**, *30*, 4617.

1. LDA , THF , -78°C

2. -100°C , /\CO$_2$Me

3. H$_3$O$^+$

80%

Casey, M.; Manage, A.C.; Gairns, R.S. *Tetrahedron Lett.*, **1989**, *30*, 6919.
Casey, M.; Manage, A.C.; Nezhat, L. *Tetrahedron Lett.*, **1988**, *29*, 5821.

Me$_2$Cu(CN)Li$_2$, THF , -78°C

83%

Girard, C.; Romain, I.; Ahmar, M.; Bloch, R. *Tetrahedron Lett.*, **1989**, *30*, 7399.

Me_/\CO$_2$Et

ZnI

CuCN , HMPA , 70°C

1.5h

78%

Tamaru, Y.; Tanigawa, H.; Yamamoto, T.; Yoshida, Z. *Angew. Chem. Int. Ed., Engl.*, **1989**, *28*, 351.

1. Me$_3$SnLi

2. MeI

3. Me$_3$SiOTf

33%

Sato, T.; Watanabe, T.; Hayata, T.; Tsukui, T. *Tetrahedron*, **1989**, *45*, 6401.

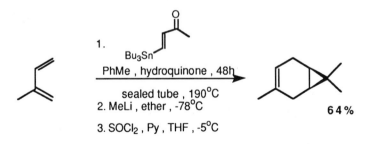

Pecunioso, A.; Menicagl, R. *J. Org. Chem.*, *1989*, *54*, 2391.

REVIEWS:

"Applications of Higher Order Mixed Organocuprates to Organic Synthesis"

Lipshutz, B.H. *Synthesis*, *1987*, 325.

"Asymmetric Synthesis of Carbon-Carbon Bonds Using Sulfinyl Cycloalkenones, Alkenolids and Pyrones"

Posner, G.A. *Acc. Chem. Res.*, *1987*, *20*, 72.

"Michael Addition of Orgnolithium Compounds"

Hunt, D.A. *Org. Prep. Proceed. Int.*, *1989*, *21*, 705.

SECTION 74F: Cyclopropanations, including Halocyclopropanations

Johnson, C.R.; Kadow, J.F. *J. Org. Chem.*, *1987*, *52*, 1493.

MeO₂C   CO₂Me

$$5\% \text{ Pd(dppe)}_2 \text{ , THF}$$

$$25^\circ C \text{ , } 18h$$

MeO₂C   CO₂Me

71%

Bäckvall, J.-E.; Vågberg, J.O.; Zercher, C.; Genêt, J.P.; Denis, A. *J. Org. Chem.*, **1987**, *52*, 5430.

CO₂Et

1. NaH

2. PhI , DMSO , 95°C

4% Pd(dba)₂ , 88h

CO₂Et

Ph   CO₂Et

Me

CO₂Et

43%

Fournet, G.; Balme, G.; Gove, J. *Tetrahedron Lett.*, **1987**, *28*, 4533.

1. LDA , THF-hexane , 1.5h
   -78°C
2. SmI₂ , CH₂I₂ , THF
   RT , 1h
3. aq. HCl

HO

58%

Imamoto, T.; Takiyama, N. *Tetrahedron Lett.*, **1987**, *28*, 1307.

SnMe₃

$$\text{BF}_3 \cdot \text{OEt}_2 \text{ , CH}_2\text{Cl}_2$$

-78°C

Me

Me   OH

62%

cis:trans = 64:36

Sato, T.; Watanabe, M.; Murayama, E. *Synth. Commun.*, **1987**, *17*, 781.

Ph

$$\text{Sm}^\circ \text{ , CH}_2\text{I}_2$$

THF , -78°C → RT

OH

Ph   OH

+

Ph   OH

(>200   :   1)   88%

Molander, G.A.; Etter, J.B. *J. Org. Chem.*, **1987**, *52*, 3942.

1. ClCH$_2$Li , THF , -78°C
2. Li^o , -78→ -30°C
3. aq. HCl

42%

Barluenga, J.; Fernandez-Simon, J.L.; Cancellon, J.M.; Yus, M. *Synthesis, 1987*, 584.

1. $=\overset{O}{\underset{Me}{\overset{\parallel}{S}}}-Me$

2. MeOH , H$^+$

56%

95% de

Meyers, A.I.; Romine, J.L.; Fleming, S.A. *J. Am. Chem. Soc., 1988, 110*, 7245.

t-BuOK , *t*-BuOH
1h

50%

Mori, M.; Kanda, N.; Ban, Y.; Aoe, K. *J. Chem. Soc., Chem. Commun., 1988*, 12.

(CO)$_5$Cr

, neat , 100°C

3.5h

73%

Wienand, A.; Reissig, H.-U. *Tetrahedron Lett., 1988, 29*, 2315.
Buchert, M.; Reissig, H.-U. *Tetrahedron Lett., 1988, 29*, 2319.

2 eq. SmI$_2$, RT
RT , 1 min

51%

Sasaki, M.; Collin, J.; Kagan, H.B. *Tetrahedron Lett., 1988, 29*, 6105.

Ambler, P.W.; Davies. S.G. *Tetrahedron Lett.*, *1988*, 29, 6979, 6983.

Podder, R.K.; Sarkar, T.K.; Sarkar, R.K.; Ray, S.C. *Indian J. Chem.*, *1988*, 27B, 217.

Reddy, D.B.; Reddy, P.S.; Seshamma, T.; Padmaja, A.; Reddy, M.V.R. *Indian J. Chem.*, *1988*, 27B, 658.

Thomsen, M.W.; Handwerker, B.M.; Katz, S.A.; Fisher, S.A. *Synth. Commun.*, *1988*, 18, 1433.

EtO$_2$C $\diagdown$ $\diagup$ CO$_2$Et   Ph$-\!\equiv$

CH$_2$Cl$_2$

Rh(OAc)$_4$, reflux

20 min

85%   +   **6%**

Rh$_2$(OAc)$_2$, 5d

33%

<u>Davies, H.M.L.</u>; Romines, K.R. *Tetrahedron,* **1988,** *44,* 3347.

EtO$_2$C $\diagdown$ $\diagup$ CO$_2$Et $\diagup$ Ph

Rh$_2$(OAc)$_4$,

CH$_2$Cl$_2$, reflux

10 min

(8.3   :   1)   **96%**

<u>Davies, H.M.L.</u>; Clark, T.J.; Church, L.A. *Tetrahedron Lett.,* **1989,** *30,* 5057.

excess Sm(Hg)

CH$_2$I$_2$, THF

-78°C → RT

(>30   :   1)   **64%**

<u>Molander, G.A.</u>; Harring, L.S. *J. Org. Chem.,* **1989,** *54,* 3525.

e$^-$, DMF , Bu$_4$N$^+$ Cl$^-$

Pt anode , Hg$_{pool}$

cathode

58%

<u>Gassman, P.G.</u>; Lee, C. *J. Am. Chem. Soc.,* **1989,** *111,* 739.

PhCO$_2$Et   Sm^0 , CH$_2$I$_2$, THF

50°C , 1.2h   **76%**

<u>Imamoto, T.</u>; Kamiya, Y.; Hatajima, T.; Takahashi, H. *Tetrahedron Lett.,* **1989,** *30,* 5149.

PhS—C(SiMe₃)(SPh) →(with epoxide SiMe₃, THF)→ PhS, Me₃Si cyclopropane SiMe₃ **76%**

Schaumann, E.; Friese, C. *Tetrahedron Lett.*, **1989**, *30*, 7033.

$CH_2=CH$—CO₂Me →(Br₂C(CN)₂, In°, DMF / LiI)→ cyclopropane MeO₂C, CN, CN **94%**

Araki, S.; Butsugan, Y. *J. Chem. Soc., Chem. Commun.*, **1989**, 1286.

(pinene structure) →(2% TiCl₄, Zn-Cu, CH₂I₂ / 10% CuCl, 52°C, 2h)→ (cyclopropanated structure) **61%**

Friedrich, E.C.; Lunetta, S.E.; Lewis, E.J. *J. Org. Chem.*, **1989**, *54*, 2388.

Br₂C(CO₂Et)₂ →($CH_2=CH$—CO₂Et, Bu₂Te / neat, RT, 16h)→ cyclopropane CO₂Et, CO₂Et, CO₂Et **73%**

Matsuki, T.; Hu, N.X.; Aso, Y.; Otsubo,T.; Ogura, F. *Bull. Chem. Soc., Jpn.*, **1989**, *62*, 2105.

REVIEW:

"Cyclopropanes from Reactions of Transition-Metal-Carbene Complexes with Olefins"

Brookhart, M.; Studabaker, W.B. *Chem. Rev.*, **1987,** *87*, 411.

SECTION 75:   ALKYLS, METHYLENES AND ARYLS FROM MISCELLANEOUS COMPOUNDS

Semmelhack, M.F.; Ho, S.; Steigerwald, M.; Lee, M.C. *J. Am. Chem. Soc.*, *1987*, *109*, 4397.

Snider, B.B.; Patricia, J.J. *J. Org. Chem.*, *1989*, *54*, 38.

E:Z = 1:4

Yamamoto, Y.; Hatsuya, S.; Yamada, J. *J. Chem. Soc., Chem. Commun.*, *1988*, 86.

Jacobi, P.A.; Kravitz, J.I. *Tetrahedron Lett.*, *1988*, *29*, 6873.

Tamura, R.; Tamai, S.; Suzuki, H. *Tetrahedron Lett.*, *1989*, *30*, 2413.

REVIEWS:

"Newer Methods of Arylation"

Abramovitch, R.T.; Barton, D.H.R.; Finet, J.-P. *Tetrahedron, 1988, 44*, 3039.

"Strained Organic Compounds" - A series of Reviews

various authors, *Chem. Rev., 1989, 89*, 973-1270.

"Anionic Approaches to the Construction of Cyclopentanoids"

Hudlicky, T.; Price, J.D. *Chem. Rev., 1989, 89*, 1467.

CHAPTER 6
PREPARATION OF AMIDES

SECTION 76: AMIDES FROM ALKYNES

NO ADDITIONAL EXAMPLES

SECTION 77: AMIDES FROM ACID DERIVATIVES

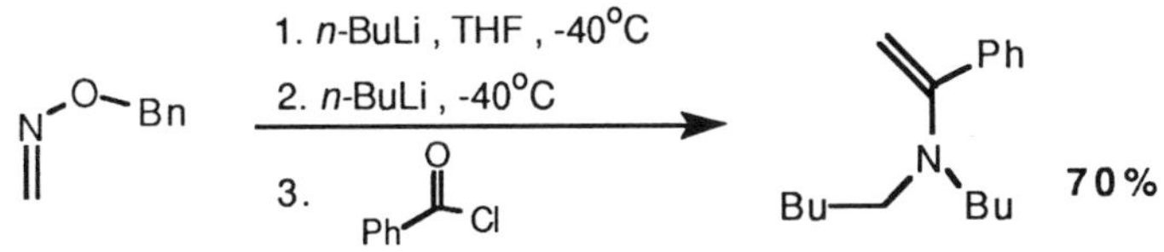

1. n-BuLi , THF , -40°C
2. n-BuLi , -40°C
3.
O
Ph
Cl
N
O
Bn
Ph
N
Bu
Bu
70%

Basha, A.; Brooks, D.W. *J. Chem. Soc., Chem. Commun.*, ***1987***, 305.

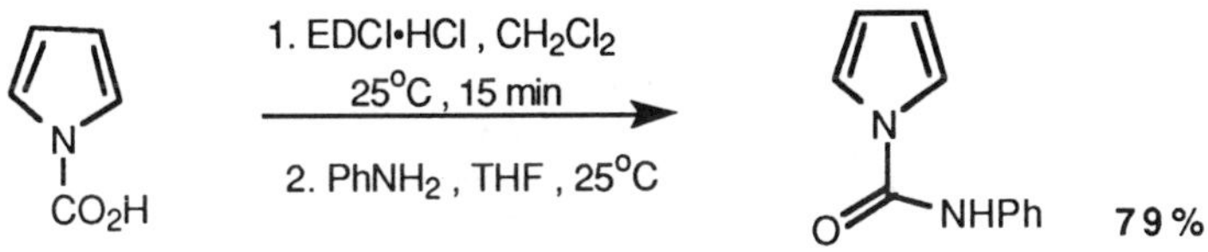

1. EDCI•HCl , CH2Cl2
25°C , 15 min
2. PhNH2 , THF , 25°C
N
CO2H
N
O
NHPh
79%

EDCI = 1-[3-(dimethylamino)propyl]-3-ethyl carbodiimide

Boger, D.L.; Patel, M. *J. Org. Chem.*, ***1987***, *52*, 2319.

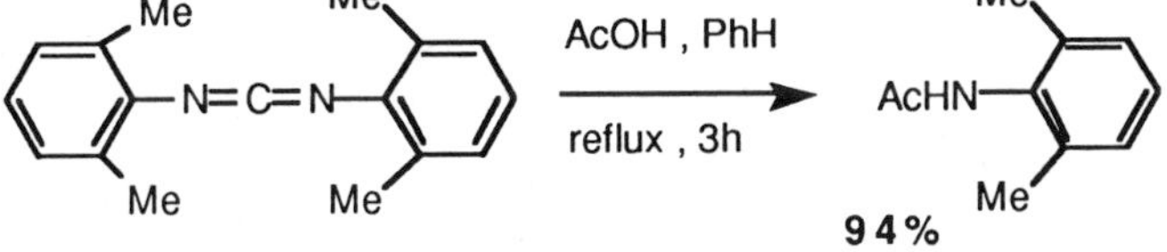

Me
Me
N=C=N
Me
Me
AcOH , PhH
reflux , 3h
Me
AcHN
Me
94%

Vovk, M.V.; Dorohkov, V.I.; Samarai, L.I. *J. Org. Chem., U.S.S.R.*, ***1987***, *23*, 1212.

PhO⌒COOH + [structure: N-imine with Ph, Me, CO₂Me] → [Me₂N·PCl₂ (O), NEt₃, CH₂Cl₂, RT, 24h] → [β-lactam product with PhO, Ph, Me, N-CO₂Me] **60%**

Cossío, F.P.; Ganbua, I.; García, J.M.; Lecea, B.; Palomo, C. *Tetrahedron Lett., 1987, 28*, 1945.

[structure: COOH with NHMe chain] → 1. Ti(OiPr)₄, ClCH₂CH₂Cl 5h / 2. aq. NaOH → [N-methyl pyrrolidinone] **85%**

Mader, M.; Helquist, P. *Tetrahedron Lett., 1988, 29*, 3049.

[structure: HOOC, Me, Me, NHBn] → Ph₂P-Cl (O), MeCN, NEt₃, 0.01M, 80°C, 10h → [β-lactam with Me, Me, N-Bn] **85%**

Kim, S.; Lee, P.H.; Lee, T.A. *J. Chem. Soc., Chem. Commun., 1988*, 1242.
Kim, S.; Lee, P.H.; Lee, T.A. *Synth. Commun., 1988, 18*, 247.

[structure: HOOC, NHTs, OSiPh₂t-Bu chain] → RT, DCC, 4-PPY, CH₂Cl₂ → [β-lactam with OSiPh₂t-Bu, N-Ts] **83%**

4-PPY = 4-pyrrolidino pyridine

Tanner, D.; Somfai, P. *Tetrahedron, 1988, 44*, 613.

Ph⌒COCl (O) → 1. HN⌒S / N-CN, NEt₃, CH₂Cl₂ / 2. BuNH₂, CH₂Cl₂, RT, 3h → Ph⌒CO-NHBu (O) **80%**

Iwata, C.; Watanabe, M.; Okamoto, S.; Fujimoto, M.; Sakae, M.; Katsurada, M.; Imanishi, T. *Heterocycles*, **1988**, *27*, 323.

$$N_3\text{-}(CH_2)_{14}COOH \xrightarrow[\text{PhH , reflux}]{SnCl_2 \text{ , PhSH , NEt}_3}$$

68%

+ 15% dimer

Barta, M.; Bou, V.; Garcia, J.; Urpi, F.; Vilarrasa, J. *J. Chem. Soc., Chem. Commun.*, **1988**, 270.

$$Ph\text{-}COOH \xrightarrow[140^\circ C]{PhCh_2NH_2 \text{ , MS 4Å , 2h}}$$

95%

Cossy, J.; Pale-Grosdemange, C. *Tetrahedron Lett.*, **1989**, *30*, 2771.

$$PhCOOH \xrightarrow[12h]{(Et_2N)_2Mg \text{ , CH}_2Cl_2 \text{ , } 65^\circ C}$$

66%

Sanchez, R.; Vest, G.; Despres, L. *Synth. Commun.*, **1989**, *19*, 2909.

SECTION 78: AMIDES FROM ALCOHOLS AND THIOLS

$$Ph\text{-}\overset{OH}{\underset{}{}}\text{-}Me \xrightarrow[2. \text{ NaNp , DME , RT}]{1. \text{ TsNHBOC , DEAD , THF} \atop Ph_3P \text{ , RT}}$$

Henry, J.R.; Marcin, L.R.; McIntosh, M.C.; Scola, P.M.; Harris Jr., G.D.; Weinreb, S.M. *Tetrahedron Lett.*, **1989**, *30*, 5709.

$$PhCH_2N=C=O \xrightarrow[45 \text{ min}]{BnOH \text{ , CuCl , DMF , RT}}$$

96%

Duggan, M.E.; Imagire, J.S. *Synthesis*, **1989**, 131.

92%

Nikifurov, A.; Jirovetz, L.; Buchbauer, G. *Liebigs Ann. Chem.*, *1989*, 489.

SECTION 79: AMIDES FROM ALDEHYDES

(3   :   1) 52%

Melnick, M.J.; Freyer, A.J.; Weinreb, S.M. *Tetrahedron Lett.*, *1988*, *29*, 3891.

96%

Shi, L.; Yang, J.; Wen, X.; Huang, Y.-Z. *Tetrahedron Lett.*, *1989*, *30*, 3949.

63%

Sisko, J.; Weinreb, S.M. *Tetrahedron Lett.*, *1989*, *30*, 3037.

SECTION 80: AMIDES FROM ALKYLS, METHYLENES AND ARYLS

SECTION 81: AMIDES FROM AMIDES

Conjugate reductions of unsaturated amides are listed in Section 74D (Alkyls from Alkenes).

Norman, M.H.; Heathcock, C.H. *J. Org. Chem.,* **1987,** *52,* 226.

Toshimitsu, A.; Terao, K.; Uemura, S. *J. Org. Chem.,* **1987,** *52,* 2018.

Stewart, J.D.; Fields, S.C.; Kochhar, K.S.; Pinnick, H.W. *J. Org. Chem.,* **1987,** *52,* 2110.

Hirama, M.; Iwakuma, T.; Itô, S. *J. Chem. Soc., Chem. Commun., 1987*, 1523.

Corley, E.G.; Karady, S.; Abramson *Tetrahedron Lett., 1988, 29*, 1497.

Becking, L.; Schäfer, H.J. *Tetrahedron Lett., 1988, 29*, 2797.

Dittami, J.P.; Ramanathan, H. *Tetrahedron Lett., 1988, 29*, 45.

(30 : 1)

58%

Jolly , R.S.; Livinghouse, T. *J. Am. Chem. Soc.*, *1988*, *110*, 7536.

Me_2NNH_2 , $AlMe_3$

$CHCl_3$-PhMe

16h , 40°C

82%

Benderly, A.; Stabchansky, S. *Tetrahedron Lett.*, *1988*, *29*, 739.

1. KOt-Bu , THF , -42°C
2. BuLi
3. MeI

85%

Comins, D.L.; Weglarz, M.A.; O'Connor, S. *Tetrahedron Lett.*, *1988*, *29*, 1751.

Cl , $PdCl_2$

MeCN , RT , 20h

73%

Iritani, K.; Matsubara, S.; Utimoto, K. *Tetrahedron Lett.*, *1988*, *29*, 1799.

$Mn(OAc)_3 \cdot H_2O$

AcOH , 70°C , 3h

Ketcha, D.M. *Tetrahedron Lett.*, *1988*, *29*, 2151.

Basha, A. *Tetrahedron Lett.*, *1988*, *29*, 2525.

Rishton, G.M.; Schwartz, M.A. *Tetrahedron Lett.*, *1988*, *29*, 2643.

Lenz, G.R.; Costanza, C. *J. Org. Chem.*, *1988*, *53*, 1176.

Katritzky, A.R.; Drewniaki, M.; Lue, P. *J. Org. Chem.*, *1988, 53*, 5854.

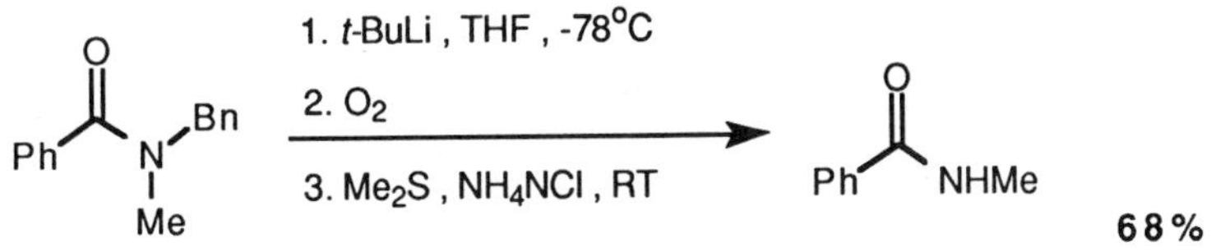

1. *t*-BuLi , THF , -78°C

2. O$_2$

3. Me$_2$S , NH$_4$NCl , RT

Ph—C(=O)—N(Bn)(Me) → Ph—C(=O)—NHMe   **68%**

Williams. R.M.; Kwast, E. *Tetrahedron Lett., 1989, 30*, 451.

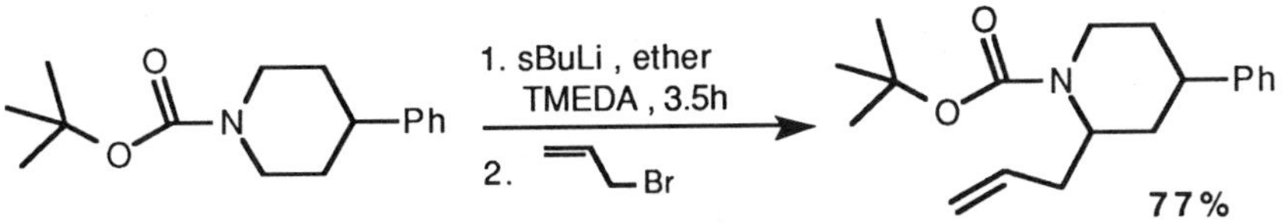

1. sBuLi , ether
TMEDA , 3.5h

2. allyl—Br   **77%**

Beak. P.; Lee, W.-K. *Tetrahedron Lett., 1989, 30*, 1197.

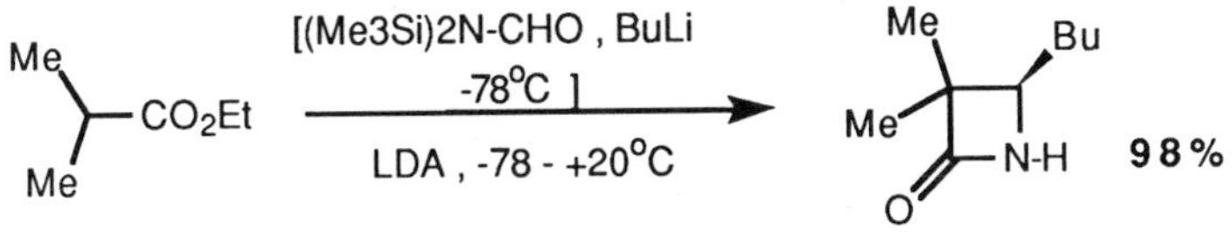

Me$_2$CH—CO$_2$Et

[(Me3Si)2N-CHO , BuLi
-78°C]
────────────
LDA , -78 - +20°C

98%

Uyehara. T.; Suzuki, I.; Yamamoto. Y. *Tetrahedron Lett., 1989, 30*, 4275.

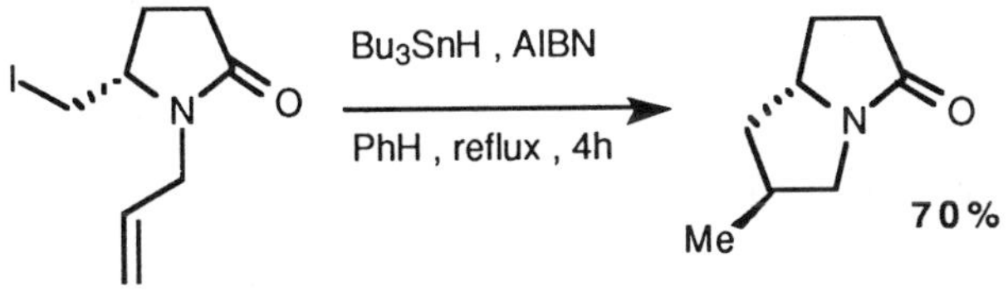

Bu$_3$SnH , AIBN
────────────
PhH , reflux , 4h

70%

Keusenkothen, P.F.; Smith. M.B. *Tetrahedron Lett., 1989, 30*, 3369.

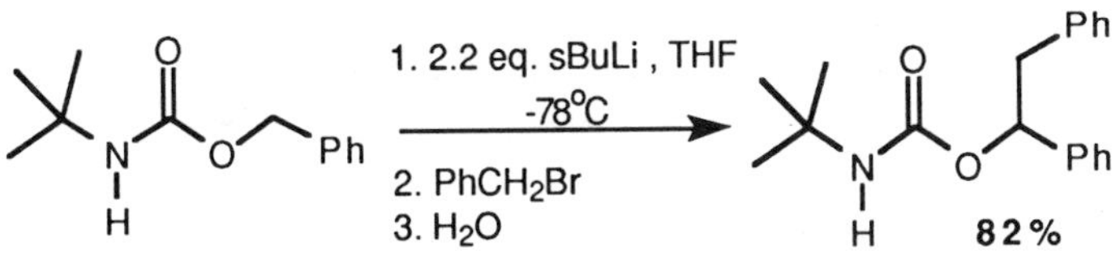

1. 2.2 eq. sBuLi , THF
-78°C
2. PhCH$_2$Br
3. H$_2$O

82%

Barner. B.A.; Mani, R.S. *Tetrahedron Lett., 1989, 30*, 5413.

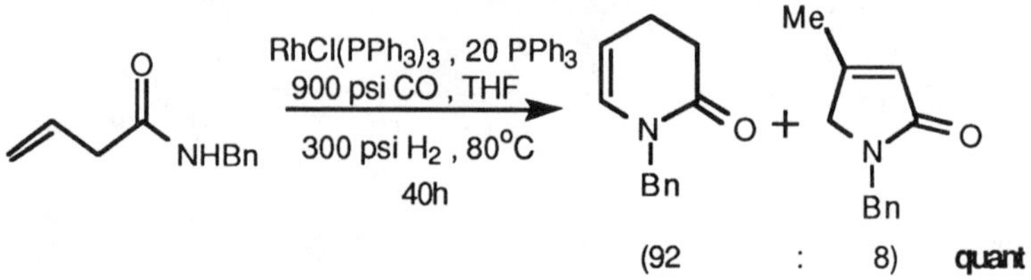

(92   :   8)   quant

Ojima. I.; Korda, A. *Tetrahedron Lett.*, *1989*, *30*, 6283.

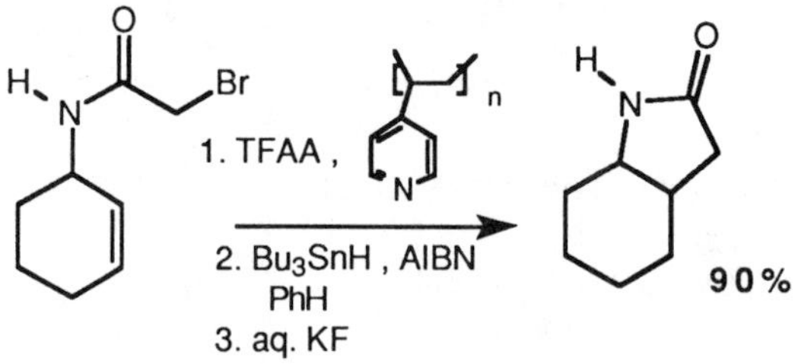

90%

Stork. G.; Mah, R. *Heterocycles*, *1989*, *28*, 723.

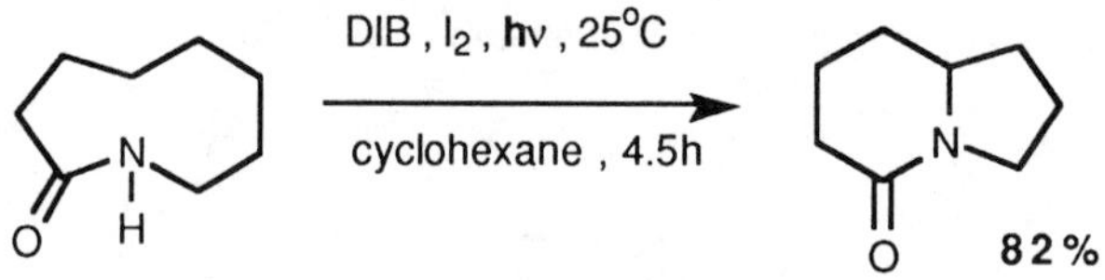

DIB , I₂ , hv , 25°C

cyclohexane , 4.5h

82%

DIB = (diacetoxyiodo) benzene

Dorta, R.L.; Francisco, C.G.; Suárez. E. *J. Chem. Soc., Chem. Commun.*, *1989*, 1168.

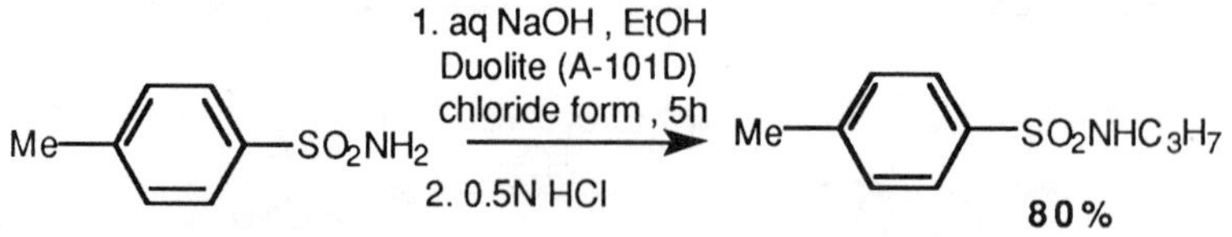

1. aq NaOH , EtOH
   Duolite (A-101D)
   chloride form , 5h

2. 0.5N HCl

80%

Sanghavi. N.M.; Parab, V.L.; Patravale, B.S.; Patel, M.N. *Synth. Commun.*, *1989*, *19*, 1499.

SECTION 82: AMIDES FROM AMINES

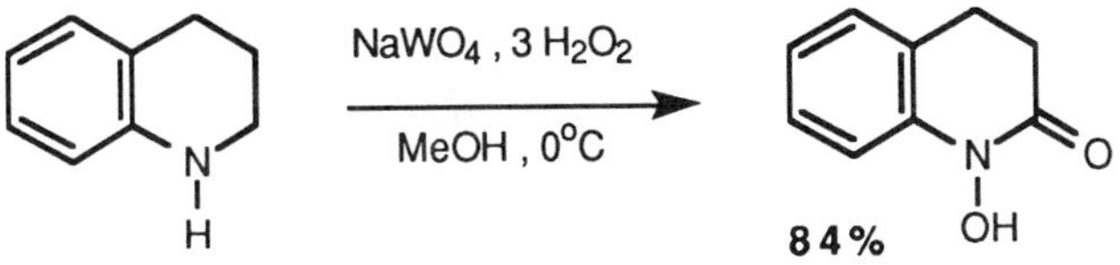

NaWO$_4$, 3 H$_2$O$_2$

MeOH , 0°C

84%

Murahashi, S.I.; Oda, T.; Sugahara, T.; Masui, Y. *J. Chem. Soc., Chem. Commun., 1987,* 1471.

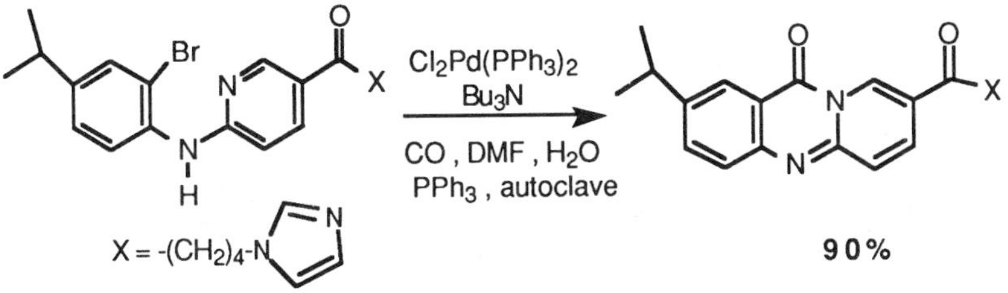

Cl$_2$Pd(PPh$_3$)$_2$
Bu$_3$N

CO , DMF , H$_2$O
PPh$_3$, autoclave

X = -(CH$_2$)$_4$-N

90%

Tilley, J.W.; Coffen, D.L.; Schaer, B.H.; Lind, J. *J. Org. Chem., 1987, 52,* 2469.

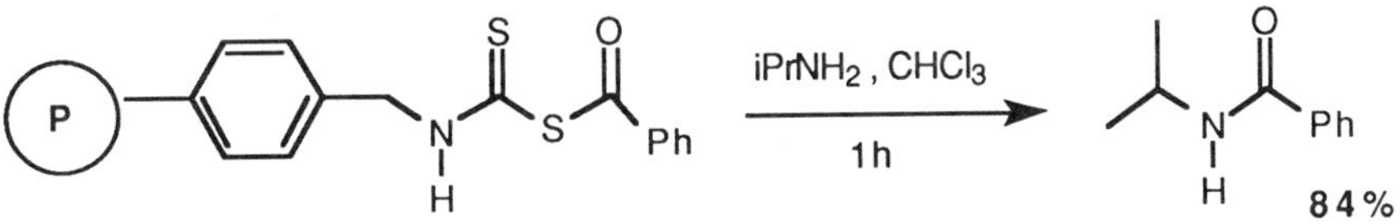

iPrNH$_2$, CHCl$_3$

1h

84%

Haridasan, V.K.; Ajayaghosh, A.; Pillai, V.N.R. *J. Org. Chem., 1987, 52,* 2662.

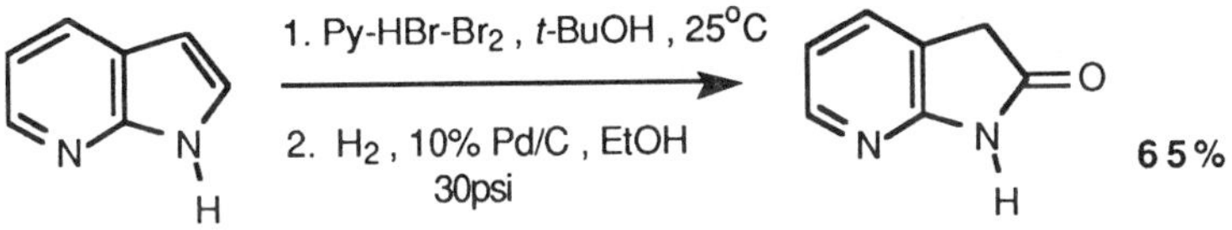

1. Py-HBr-Br$_2$, *t*-BuOH , 25°C

2. H$_2$, 10% Pd/C , EtOH
   30psi

65%

Marfat, A.; Carta, M.P. *Tetrahedron Lett., 1987, 28,* 4027.

C$_6$F$_5$OCHO , CHCl$_3$

PhCH$_2$NH$_2$ ⟶ PhCH$_2$NHCHO

RT , 30 min

90%

Kisfaludy, L.; Ötvös Jr., L. *Synthesis, 1987,* 510.

1. BuLi , -50 → -30°C

2. t-BuLi , -30 → +20°C

3. $(EtO)_2C=O$, -5 → +20°C

88%

Barluenga, J.; Fañanás, F.J.; Foubelo, F.; Yus, M. *Tetrahedron Lett., 1988, 29*, 4859.

$(PhIO)_n$, 2 eq. H_2O

36h

55%

Moriarty, R.M.; Vaid, R.K.; Duncan, M.P.; Ochiai, M.; Inenaga, M.; Nagao, Y. *Tetrahedron Lett., 1988, 29*, 6913.

$(PhIO)_n$, $CHCl_3$

RT , 2d

70%

Ochiai, M.; Inenaga, M.; Nagao, Y.; Moriarty, R.M.; Vaid, R.K.; Duncan, M.P. *Tetrahedron Lett., 1988, 29*, 6917.

1. I^- , THF , reflux

2. $Ni(CO)_4$, CO , THF reflux

37%

Chamchaang, W.; Pinhas, A.R. *J. Chem. Soc., Chem. Commun., 1988*, 710.

$Pd(OAc)_2$-dppp , CO

PhMe , 110°C , 20h autoclave

Murahashi, S.; Imada, Y.; Nishimura, K. *J. Chem. Soc., Chem. Commun., 1988*, 1578.

Wenkert, E.; Angell, E.C. *Synth. Commun.*, *1988*, *18*, 1331.

quant

Calet, S.; Urso, F., Alper, H. *J. Am. Chem. Soc.*, *1989*, *111*, 931.

49%

Krafft, M.E. *Tetrahedron Lett.*, *1989*, *30*, 539.

trans

82%

Sugano, Y.; Naruto, S. *Chem. Lett.*, *1989*, 1331.

REVIEW:

"Amine Dealkylations with Acyl Chlorides"

Cooley, J.H.; Evain, E.J. *Synthesis*, *1989*, 1.

SECTION 83: AMIDES FROM ESTERS

Arai, K.; Shaw, C.; Nozawa, K.; Kawai, K.; Nakajima. S. *Tetrahedron Lett.*, *1987*, *28*, 441.

Högberg. T.; Ström, P.; Ebner, M.; Rämsby, S. *J. Org. Chem.*, *1987*, *52*, 2033.

Wada, M.; Aiura, H.; Akiba. K. *Tetrahedron Lett.*, *1987*, *28*, 3377.

Rajeeswari, S.; Jones, R.J.; Cava. M.P. *Tetrahedron Lett.*, *1987*, *28*, 5099.

Gotor. V.; Brieva, R.; Rebelledo, F. *Tetrahedron Lett.*, *1988*, *29*, 6973.

Ph—S(=O)—CH$_2$CO$_2$Et $\xrightarrow[\text{8 KBar , 40°C}]{\begin{array}{c}\text{N-H , MeCN}\end{array}}$ Ph—S(=O)—CH$_2$—C(=O)—N(pyrrolidine)
4d
90%

Matsumoto, K.; Hashimoto, S.; Okamoto, T.; Tanikaga, R.; Uryu, T.; Munakata, M. *Chemistry Express*, *1988*, *3*, 615.

SECTION 84: AMIDES FROM ETHERS, EPOXIDES AND THIOETHERS

MeO—C$_6$H$_5$ $\xrightarrow[\text{reflux}]{\text{HgBr}_2 \text{ , PhMe , 12h}}$ MeO—C$_6$H$_4$—C(=O)—NEt$_2$

76%

Fujiwara, S.; Ogawa, A.; Kambe, N.; Ryu, I.; Sonoda, N. *Tetrahedron Lett.*, *1988*, *29*, 6121.

$\xrightarrow[\substack{\text{Co}_2(\text{CO})_8 \text{ , CO , 15h} \\ \text{RT}}]{\text{BnNHSiMe}_3 \text{ , PhH}}$ Bn—N(H)—C(=O)—(CH$_2$)$_4$—OSiMe$_3$

60%

Tsuji, Y.; Kobayashi, M.; Okuda, F.; Watanabe, Y. *J. Chem. Soc., Chem. Commun.*, *1989*, 1253.

SECTION 85: AMIDES FROM HALIDES AND SULFONATES

Ph$_2$Mg $\xrightarrow[\text{2. aq. H}_2\text{SO}_4]{\text{1. DMF , PhMe , 0} \rightarrow \text{5°C}}$ Ph—C(=O)—NMe$_2$ **80%**

Screttas, C.G.; Steele, B.R. *J. Org. Chem.*, *1988*, *53*, 5151.

SECTION 86: AMIDES FROM HYDRIDES

NO ADDITIONAL EXAMPLES

SECTION 87: AMIDES FROM KETONES

>95 ee

Georg, G.I.; Guan, X.; Kant, J. *Tetrahedron Lett., 1988, 29,* 403.

Hoffman, R.V.; Salvador, J.M. *Tetrahedron Lett., 1989, 30,* 4207.

1. 1 atm CO , K_2CO_3
   5% Pd(PPh$_3$)$_4$,
   100°C , 12h , NMP
2. 3 eq THF•Mg$_2$Cl$_2$O
   Tl-N=C=O, NMP , 24h
   120°C

NMP = N-methyl-2-pyrrolidinone

Uozumi, Y.; Kawasaki, N.; Mori, E.; Mori, M.; Shibasaki, M. *J. Am. Chem. Soc., 1989, 111,* 3725.

REVIEW:

"Syntheses of Macrocyclic Compounds by Ring Enlargement"

Stauch, H.; Hesse, M. *Tetrahedron, 1988, 44,* 1573.

SECTION 88: AMIDES FROM NITRILES

MnO_2 , SiO_2 , hexanes

H_2O , reflux , 16h

quant

Liu, K.-T.; Shih, M.-H.; Huang, H.-W.; Hu, C.-J. *Synthesis, 1988*, 715.

1. [Dibal-H , PhMe , BuLi]
2. Me_3SiCl
3. Me   Me

   LiO   OEt   , -78°C→ RT
4. 0.25M H_2SO_4

60%

Andreoli, P.; Cainelli, G.; Contento, M.; Giacomini, D.; Martelli, G.; Panunzio, M. *J. Chem. Soc., Perkin Trans. I, 1988*, 945.

Ph-C≡N

30% H_2O_2 , DMSO

K_2CO_3 , 5 min

75%

Katritzky, A.R.; Pilarski, B.; Urogdi, L. *Synthesis, 1989*, 949.

$PhCH_2C≡N$

1. AcOH , $Hg(OAc)_2$
   reflux , 70h
2. ice/H_2O

77%

Plummer, B.F.; Menendez, M.; Songster, M. *J. Org. Chem., 1989, 54*, 718.

SECTION 89: AMIDES FROM ALKENES

1. $ClSO_2NCO$, ether ,
   pressure bottle , RT
2. $NaHSO_3$, ether ,
   NaOH , 50°C

73%

Hauser, F.M.; Ellenberger, S.R. *Synthesis, 1987*, 324.

REVIEW:

"The Ester Iminium Condensation Route to β-Lactams"

Hart, D.J.; Ha, D.-C. *Chem. Rev.*, *1989*, *89*, 1447.

SECTION 90: AMIDES FROM MISCELLANEOUS COMPOUNDS

Baldwin, S.W.; Aubé, J. *Tetrahedron Lett.*, *1987,.28*, 179.

Gennari, C.; Venturini, I.; Gislon, G.; Schimperna, G. *Tetrahedron Lett.*, *1987*, *28*, 227.

Zhang, Y.; Jiang, J.; Chen, Y. *Tetrahedron Lett.*, *1987*, *28*, 3815.

Tokitoh, N.; Okazaki, R. *Bull. Chem. Soc., Jpn.*, *1987*, *60*, 3291.
Tokitoh, N.; Okazaki, R. *Bull. Chem. Soc., Jpn.*, *1988*, *61*, 735.

(4.7   :   1) 59%

Aubé, J.; Burgett, P.M.; Wang, Y. *Tetrahedron Lett.*, *1988*, *29*, 151.

1. HCl , CH$_2$Cl$_2$, -78°C

2. P(OMe)$_3$, RT , 12h

3. P(OMe)$_3$, MeOH, 25°C , 24h

86%

Krafft, M.E.; Wilson, L.J.; Onan, K.D. *Tetrahedron Lett.*, *1988*, *29*, 6421.

Rh(OAc)$_2$, CH$_2$Cl$_2$, RT

1 h

86%

Doyle, M.P.; Shanklin, M.S.; Pho, H.Q.; Mahapatro, S.N. *J. Org. Chem.*, *1988*, *53*, 1017.

1% Rh$_2$(OAc)$_4$

PhH , reflux

98%

Doyle, M.P.; Shanklin, M.S.; Oon, S.-M.; Pho, H.Q.; van der Heide, F.R.; Veal, W.R. *J. Org. Chem.*, *1988*, *53*, 3384.

REVIEWS:

"Synthesis and Reactions of β-Sultams: A Review"

Chanet-Ray, J.; Vessiere, R. *Org. Prep. Proceed. Int., 1986, 18*, 157.

"Amides from N-Acyl Imines and Related Heterodienes in [4+2]-Cycloaddition Reactions"

Weinreb, S.M.; Scola, P.M. *Chem. Rev., 1989, 89*, 1525.

CHAPTER 7
PREPARATION OF AMINES

SECTION 91: AMINES FROM ALKYNES

NO ADDITIONAL EXAMPLES

SECTION 92: AMINES FROM ACID DERIVATIVES

NO ADDITIONAL EXAMPLES

SECTION 93: AMINES FROM ALCOHOLS AND THIOLS

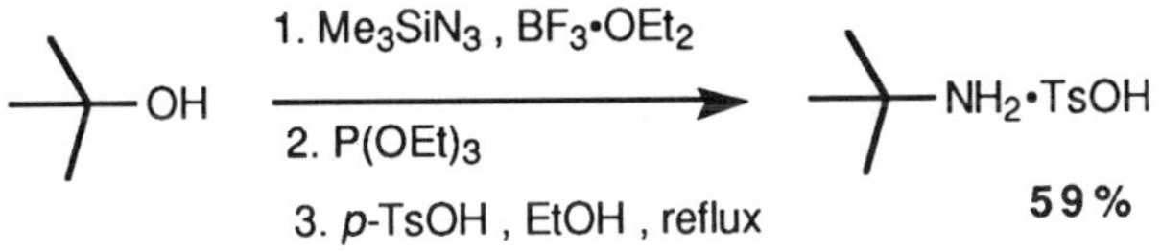

OH
1. Me3SiN3 , BF3•OEt2
2. P(OEt)3
3. p-TsOH , EtOH , reflux
NH2•TsOH
59%

Koziara, A.; Zwierzak, A. *Tetrahedron Lett.*, ***1987***, *28*, 6513.

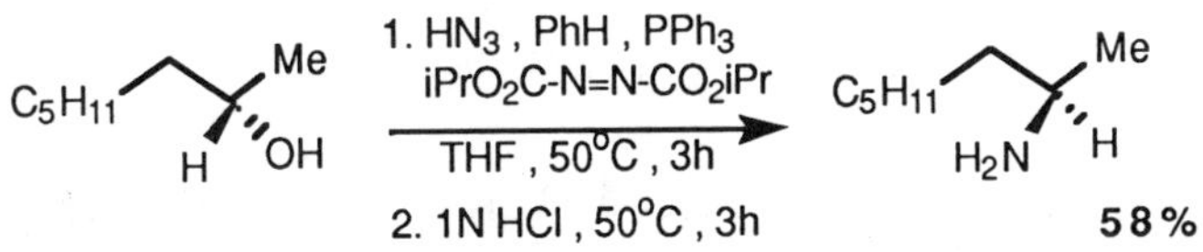

C5H11
Me
H
OH
1. HN3 , PhH , PPh3
iPrO2C-N=N-CO2iPr
THF , 50°C , 3h
2. 1N HCl , 50°C , 3h
C5H11
Me
H2N
H
58%

Fabiano, E.; Golding, B.T.; Sadeghi, M.M. *Synthesis*, ***1987***, 190.

SECTION 94: AMINES FROM ALDEHYDES

$$C_7H_{15}CHO \xrightarrow[\text{NEt}_3]{\text{BuNH}_2, \text{Al}_2\text{Te}_3, \text{aq. THF}} \text{BuNHC}_8H_{17}$$

96%

Kambe, N.; Inagaki, T.; Miyoshi, N.; Ogawa, A.; Sonoda, N. *Chem. Lett.*, *1987*, 1275.

77%

Siddiqui, M.A.; Snieckus, V. *Tetrahedron Lett.*, *1988*, *29*, 5463.

1. iPrNH$_2$
2. NBS
3. NaOMe, MeOH
4. NaBH$_4$, EtOH

67%

DeKimpe, N.; Stanoeva, E.; Verhé, R.; Schamp, N. *Synthesis*, *1988*, 587.

PhCHO

1. [Ti(OiPr)$_4$, LiNEt$_2$, ether, -78°C]
   20°C
2. PhCH$_2$MgCl, ether
   RT
3. H$_2$O

92%

Takahashi, H.; Tsubuki, T.; Higashiyama, K. *Synthesis*, *1988*, 238.

Related Methods: Section 102 (Amines from Ketones)

SECTION 95: AMINES FROM ALKYLS, METHYLENES AND ARYLS

iPrNH$_2$, mDCNB , hν

aq. MeCN , 12 h

NHiPr

88%

mDCNB = *m*-dicyanobenzene   74% conversion

Yasuda, M.; Yamashita, T.; Shima, K.; Pac, C. *J. Org. Chem.*, *1987, 52,* 753.

PhSO$_2$—

NMP , TMGA

80°C , 1h

PhSO$_2$

H

N$_3$

96%

NMP = N-methyl-2-pyrrolidinone
TMGA = N,N,N',N'-tetramethylguanidinium azide

Gaoni, Y. *Tetrahedron Lett.*, *1988, 29*, 1591.

(CH$_2$)$_3$ SiMe$_3$

1. Ph-Te-O-CF$_3$, RT

BF$_3$ •OEt$_2$, ClCH$_2$CH$_2$Cl

2. PhNHMe , 65°C

(CH$_2$)$_3$ NMePh

89%

Hu, N.X.; Aso, Y.; Otsubo, T.; Ogura, F. *Tetrahedron Lett.*, *1988, 29*, 4949.

SECTION 96: AMINES FROM AMIDES

Ph-C(S)-N(morpholine)

1. BuLi , ether , RT

2. LiAlH$_4$

3. aq. NaOH

Ph-CH(Bu)-N(morpholine)

94%

Tominaga, Y.; Kohra, S.; Hosomi, A. *Tetrahedron Lett.*, *1987, 28*, 1529.

Bartoli, G.; Bosco, M.; Dalpozzo, R.; Todesco, P.E. *J. Chem. Soc., Chem. Commun.,* **1988**, 807.

Belletire, J.L.; Fry, D.F. *Synth. Commun.,* **1988**, *18*, 29.

Ferringa, B.L.; Jansen, J.F.G.A. *Synthesis,* **1988**, 184.

Markov, V.I.; Dmitrikova, L.V.; Dubina, V.L.; Burmistrov, K.S. *J. Org. Chem., U.S.S.R.,* **1988**, *24*, 2222.

Zhou, X.-J.; Huang, Z.-Z. *Synth. Commun.,* **1989**, *19*, 1347.

Related Methods: Section 105A (Protection of Amines)

SECTION 97: AMINES FROM AMINES

Dai-Ho, G.; Mariano, P.S. *J. Org. Chem.*, *1987*, *52*, 704.

Grieco, P.A.; Parker, D.T.; Fobare, W.F.; Ruckle, R. *J. Am. Chem. Soc.*, *1987*, *109*, 5859.

Kotlsuki, H.; Ushio, Y.; Ochi, M. *Heterocycles*, *1987*, *26*, 1771.

Grieco, P.A.; Bahsas, A. *J. Org. Chem.*, *1987*, *52*, 1378.

Barton, D.H.R.; Finet, J.-P.; Khamsi, J. *Tetrahedron Lett.*, *1987*, *28*, 887.
Barton, D.H.R.; Finet, J.-P.; Khamsi, J. *Tetrahedron Lett.*, *1989*, *30*, 937.

(1   :   9) 62%

Tsuji, Y.; Huh, K.-T.; Watanabe, Y. *J. Org. Chem.*, *1987, 52*, 1673.

(96   :   4) 84%

Nakano, M.; Sato, Y. *J. Org. Chem.*, *1987, 52*, 1844.

Kang, J.; Kim, K.S. *J. Chem. Soc., Chem. Commun., 1987*, 897.

synthetic equivalent of non-stabilized
nitrile ylid

Tsuge, O.; Kanemasa, S.; Yamada, T.; Matsuda, K. *J. Org. Chem., 1987, 52*, 2523.

Misner, J.W. J. Org. Chem., 1987, 52, 3166.

Campbell, A.L.; Pilipauskas, D.R.; Khanna, I.K.; Rhodes, R.A. Tetrahedron Lett., 1987, 28, 2331.

Larock, R.C.; Babu, S. Tetrahedron Lett., 1987, 28, 5291.

modified Ullman Synthesis

Gauthier, S.; Fréchet, J.M.J. Synthesis, 1987, 383.

1. t-BuCH$_2$OH , Py , 0°C

2. BuLi , THF , -78°C

61%

Hanessian, S.; Kagotani, M. *Synthesis, 1987*, 409.

Ph-NHMe $\xrightarrow[\text{RT}]{\text{(HCHO)}_n \text{ , NaBH}_4 \text{ , TFA , THF}}$ Ph-NMe$_2$

83%

Gribble, G.W.; Nutaitis, C.F. *Synthesis, 1987*, 709.

$C_8H_{17}NH_2$ $\xrightarrow[\text{2. MeI \quad 3. BuNH}_2]{\text{1.}}$ $C_8H_{17}NHMe$

74%

Katritzky, A.R.; Drewniak, M.; Aurrecoechea, J.M. *J. Chem. Soc., Perkin Trans. I, 1987*, 2539.

1. $\left[\text{/\\/\\}\right]_2$CuMgBr •BF$_3$
THF , -70°C

2. SiO$_2$, TFA , CH$_2$Cl$_2$

3. PhSeCl , CH$_2$Cl$_2$

4. Ph$_3$SnH , PhMe reflux

28%

Wada, M.; Aiura, H.; Akiba, K. *Heterocycles, 1987, 25*, 929.

$\xrightarrow[\text{PbBr}_2 \text{ / Al}^0 \text{ / BF}_3 \text{ •OEt}_2]{\text{/\\=/\\-Br , ether , 10h}}$

84%

Tanaka, H.; Yamashita, S.; Ikemoto, Y.; Torii, S. *Chem. Lett., 1987*, 673.

1. LDA - TMEDA , THF , -78°C

2. $C_7H_{15}Br$

3. $NaBH_4$, NaOH , reflux

4. H_2 , 10% Pd/C

56%

Arseniyadis, S.; Huang, P.Q.; Husson, H-P. *Tetrahedron Lett.*, *1988*, *29*, 631.

1. *n*-BuLi , THF ,
-78°C - RT

2. H_2O

83%

Pearson, W.H.; Szura, D.P.; Harter, W.G. *Tetrahedron Lett.*, *1988*, *29*, 761.

Ph_2Bi ⟍ Ph

Cu(OAc)$_2$, CH_2Cl_2 , RT

51%

Barton, D.H.R.; Ozbalik, N.; Ramesh, M. *Tetrahedron Lett.*, *1988*, *29*, 857.

1. NEt_3 , Me_3SiCl

2. KOH , 60°C , 10^{-2} torr

VGSR

73%

VBSR = vacuum gas-solid reactions

Guillemin, J.-C.; Ammi, L.; Denis, J.-M. *Tetrahedron Lett.*, *1988*, *29*, 1287.

1. MeLi , ether , -78°C , 6.5h

2. *t*-BuBr

70%

Danks, T.N.; Thomas, S.E. *Tetrahedron Lett.*, *1988*, *29*, 1425.

Bu$_3$SnH , PhH , hv (10 mM)

85%

$\alpha:\beta$ = 4:1

Lathbury, D.C.; Parsons, P.J.; Pinto, I. *J. Chem. Soc., Chem. Commun.,* **1988**, 81.

1. 2.5 eq *t*-BuLi , THF

2.

3. reflux

51%

Reed, J.N.; Rotchford, J.; Strickland, D. *Tetrahedron Lett.,* **1988**, *29*, 5725.

1. LDA , PhCH=NPh , THF ,
   -55°C

2. *t*-BuOK

3. EtMgBr , THF

78%

Satoh, T.; Oohara, T.; Yamakawa, K. *Tetrahedron Lett.,* **1988**, *29*, 4093.

PhLi , BF$_3$ •OEt$_2$

44%

(Z:E = 47:53)

Rodrigues, K.E.; Basha, A.; Summers, J.B.; Brooks, D.W. *Tetrahedron Lett.,* **1988**, *29*, 3455.

$\alpha{:}\beta = 10{:}1$

Martin, S.F.; Yang, C.-P.; Laswell, W.L.; Rüeger, H. *Tetrahedron Lett.*, *1988*, *29*, 6685.

PhNMeEt $\xrightarrow{\text{1. CH}_2\text{Cl}_2\,,\,-78^\circ\text{C,}}$ PhNHEt
           2. H$_2$O 3. aq. HCl 4. aq. NaOH   65%

Hunter, D.H.; Racok, J.S.; Rey, A.W.; Ponce, Y.Z. *J. Org. Chem.*, *1988*, *53*, 1278.

67% ee , S   **quant**

Kang, G.-J.; Cullen, W.R.; Fryzuk, M.D.; James, B.R.; Kutney, J.P. *J. Chem. Soc., Chem. Commun.*, *1988*, 1466.

PhCH$_2$NHPh $\xrightarrow{\text{PhIO , RuCl}_2(\text{PPh}_3)\,,\,1\text{h}}$ Ph-CH=N-Ph
                                                      77%

Müller, P.; Gilabert, D.M. *Tetrahedron, 1988, 44*, 7171.

PhNH$_2$ $\xrightarrow[\substack{\text{MeOH , dioxane , 180}^\circ\text{C} \\ 15\text{h}}]{\text{RuCl}_3 \cdot n\,\text{H}_2\text{O , 3 eq. P(OBu)}_3}$ PhNMe$_2$
                                                      80%

Huh, K.-T.; Tsuji, Y.; Kobayashi, M.; Okuda, F.; Watanabe, Y. *Chem. Lett., 1988*, 449.

Hoffman, R.V.; Buntain, G.A. *J. Org. Chem.*, *1988*, *53*, 3316.

(87   :   13)   64%

(4   :   96)   50%

Tokuda, M.; Yamada, Y.; Sugionme, H. *Chem. Lett.*, *1988*, 1289.

65%

Barton, D.H.R.; Donnelly, D.M.X.; Finet, J.-P.; Guiry, P.J. *Tetrahedron Lett.*, *1989*, *30*, 1377.
Barton, D.H.R.; Yadav-Bhatnagar, N.; Finet, J.-P.; Khamsi, J. *Tetrahedron Lett.*, *1987*, *28*, 3111.

93%

Santamaria, J.; Ouchabane, R.; Rigaudy, J. *Tetrahedron Lett.*, *1989*, *30*, 2927.

Newcomb, M.; Marquardt, D.J. *Heterocycles*, *1989*, *28*, 129.

Hornback, J.M.; Murugaverl, B. *Tetrahedron Lett.*, *1989*, *30*, 5853.

Yankep, E.; Kapnang, H.; Charles, G. *Tetrahedron Lett.*, *1989, 30*, 7383.

Periasamy, M.; Devasagayaraj, A.; Satyanarayana, N.; Narayana, C. *Synth. Commun.*, *1989*, *19*, 565.

Murray, R.W.; Singh, M. *Synth. Commun.*, *1989*, *19*, 3509.

58%

Roskamp, E.J.; Dragovich, P.S.; Hartung Jr., J.B.; Pedersen, S.F. *J. Org. Chem.*, *1989*, *54*, 4736.

REVIEW:

"Transition Metals in the Synthesis and Functionalization of Indoles"

Hegedus, L.S. *Angew. Chem. Int. Ed., Engl*, *1988*, *27*, 1113.

SECTION 98: AMINES FROM ESTERS

91%

Murahashi, S.; Imada, Y.; Tankguchi, Y. *Tetrahedron Lett.*, *1988*, *29*, 2973.

80%

77% ee

Murahashi, S.; Taniguchi, Y.; Imada, Y.; Tanigawa, Y. *J. Org. Chem.*, *1989*, *54*, 3292.

SECTION 99: AMINES FROM ETHERS, EPOXIDES AND THIOETHERS

NO ADDITIONAL EXAMPLES

SECTION 100: AMINES FROM HALIDES AND SULFONATES

1. Me$_3$SiN$_3$, PhH

SnCl$_4$, 20°C

2. P(OEt)$_3$, RT
2 HCl

37%

Koziara, A.; Osowska-Pacewicka, K.; Zawadzki, S.; Zwierzak, A. *Synthesis, 1987*, 487.

MeOCH$_2$N(SiMe$_3$)$_2$

1. BnMgBr , ether

0°C→ reflux

2. aq. HCl

PhCH$_2$NH$_3^+$ Cl$^-$

82%

Bestmann, H.J.; Wölfel, G.; Mederer, K. *Synthesis, 1987*, 848.

O_2N——⟨ ⟩——Cl

n-PrNH$_2$, THF

7.2 KBar , 20h

50°C

O_2N——⟨ ⟩——NH*n*-Pr

93%

Ibata, T.; Isogami, Y.; Toyoda, J. *Chem. Lett., 1987*, 1187.

F_3C——⟨ ⟩——Cl

N-H , THF

8 KBar , 100°C

4d

F_3C——⟨ ⟩——N⟨ ⟩

quant

Matsumoto, K.; Uchida, T.; Okamoto, T.; Halshimoto, S. *Chemistry Express, 1987, 2,* 551.

1. PhMgBr

2. aq. NH$_3$ 3. HCl/ether

N=PPh$_3$

PhCH$_2$NH$_3^+$ Cl$^-$

82%

Katritzky, A.R.; Jiang, J.; Urogdi, L. *Tetrahedron Lett., 1989, 30*, 3303.

REVIEW:

"Electrophilic Amination of Carbanions"

Erdik, E.; Ay, M. *Chem. Rev., 1989, 89*, 1947.

SECTION 101: AMINES FROM HYDRIDES

$$
\text{MeO} - \text{C}_6\text{H}_4 - \text{CH(CH}_3)\text{CH}_2\text{CH}_3
\xrightarrow[\substack{\text{2. H}_2\text{ , Pd , BaSO}_4 \\ \text{EtOH} \\ \text{3. HCl}_{(g)}\text{ , EtOH}}]{\substack{\text{1. Me}_3\text{SiCl , CHCl}_3 \text{ ,} \\ \text{DDQ , 120h}}}
\text{MeO} - \text{C}_6\text{H}_4 - \text{C(CH}_3)_2 - \text{NH}_2 \cdot \text{HCl}
$$

58%

Guy, A.; Lemor, A.; Doussot, J.; Lemaire, M. *Synthesis, 1988*, 900.

SECTION 102: AMINES FROM KETONES

(3   :   1)   **35%**

Baccolini, G.; Sandali, C. *J. Chem. Soc., Chem. Commun., 1987*, 788.

Pienemann, T.; Schäfer, H.-J. *Synthesis, 1987*, 1005.

64%

92% ee , S

Sakito, Y.; Yoneyoshi, Y.; Suzukamo, G. *Tetrahedron Lett., 1988, 29*, 223.

Related Methods: Section 94 (Amines from Aldehydes)

SECTION 103: AMINES FROM NITRILES

BuLi , PhH , reflux

24h

43%

Gallulo, V.; Dimas, L.; Zezza, C.A.; Smith, M.B. Org. Prep. Proceed. Int., *1989*, *21*, 297.

SECTION 104: AMINES FROM ALKENES

PPh$_3$, aq. THF

(10   :   90)   **78%**

Knouzi, N.; Vaultier, M.; Toupet, L.; Carrie, R. Tetrahedron Lett., *1987*, *28*, 1757.

NEt$_3$, NO , NaBH$_4$,
RT , 5h

Co(DH)$_2$(Py)Cl

DH = dimethyl glyoxime

45%

Okamoto, T.; Kobayashi, K.; Oka, S.; Tanimoto, S. J. Org. Chem., *1987*, *52*, 5089.

1. BH$_3$ •THF , 0°C
2. Me$_3$Si-N$_3$
3. MeOH , reflux ,
   40h

64%

Kabalka, G.W.; Goudgaon, N.M.; Liang, Y. Synth. Commun., *1988*, *18*, 1363.

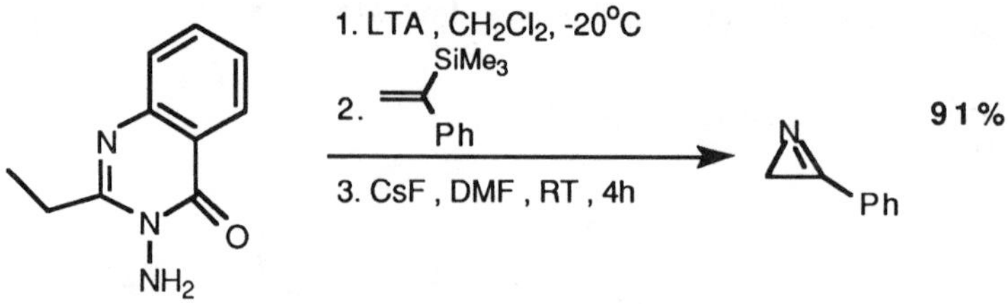

1. LTA , CH$_2$Cl$_2$, -20°C

2. (SiMe$_3$ / Ph alkene)

3. CsF , DMF , RT , 4h

91%

Atkinson, R.S.; Kelly, B.J. *Tetrahedron Lett.*, *1989*, *30*, 2703.

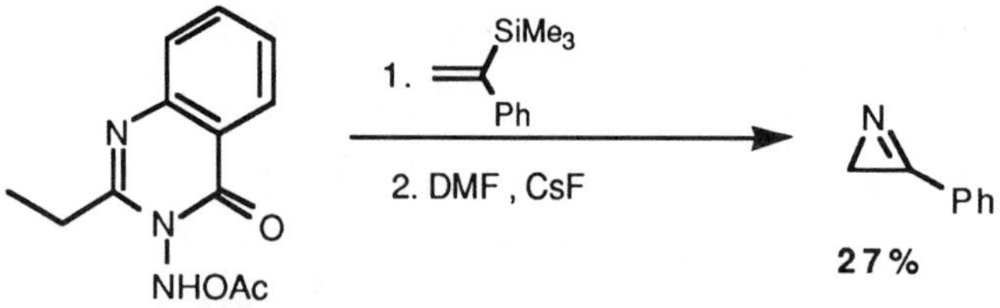

1. (SiMe$_3$ / Ph alkene)

2. DMF , CsF

27%

Atkinson, R.S.; Kelly, B.J. *J. Chem. Soc., Chem. Commun.*, *1989*, 836.

SECTION 105: AMINES FROM MISCELLANEOUS COMPOUNDS

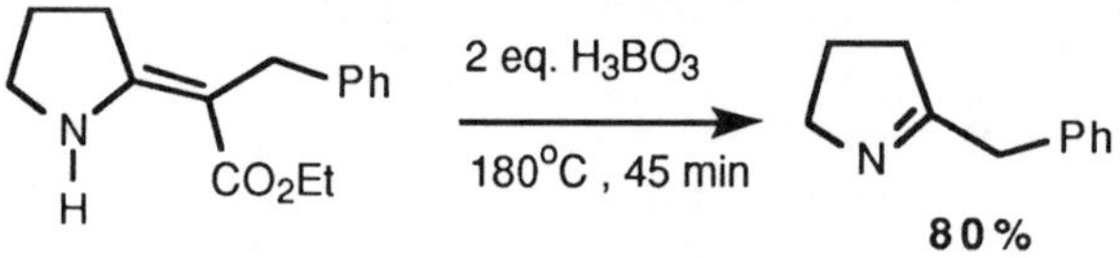

2 eq. H$_3$BO$_3$

180°C , 45 min

80%

Bacos, D.; Celerier, J.-P.; Lhommet, G. *Tetrahedron Lett.*, *1987*, *28*, 2353.

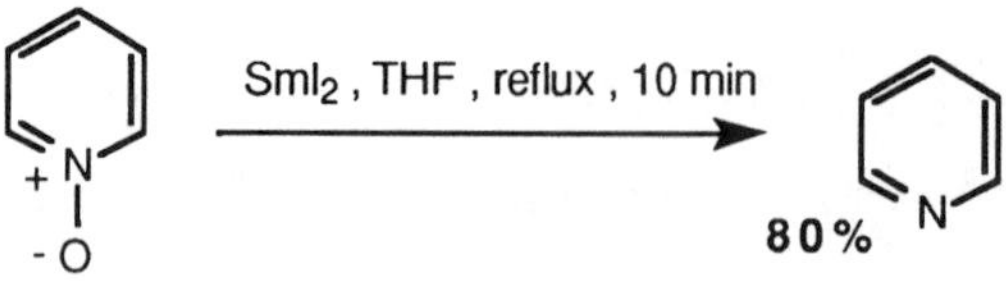

SmI$_2$, THF , reflux , 10 min

80%

Zhang, Y.; Lin, R. *Synth. Commun.*, *1987*, *17*, 329.

90%

Malinowski, M.; Kaczmarek, L. *Synthesis*, *1987*, 1013.

1. TiCl$_4$, THF , 3h
reflux
2. aq. NH$_4$OH

80%

Malinowski, M. *Synthesis*, *1987*, 732.

Bu$_3$N-O $\xrightarrow[\text{1h}]{\text{Bu}_6\text{Sn}_2 \text{ , THF , 50}^\circ\text{C}}$ Bu$_3$N

84%

Jousseaume, B.; Chanson, E. *Synthesis*, *1987*, 55.

Ph-N=N-Ph $\xrightarrow[\text{FeCl}_3\text{ , reflux}]{\text{Me}_3\text{SiCl/Li}^\circ\text{/THF}}$ Ph-N(SiMe$_3$)$_2$

77%

Kira, M.; Naga, S.; Nishimura, M.; Sakurai, H. *Chem. Lett.*, *1987*, 153.

PhN$_3$ $\xrightarrow{\text{Mg}^\circ \text{ , MeOH , 15 min}}$ PhNH$_2$

98%

Maiti, S.N.; Spevak, P.; Reddy, A.V.N. *Synth. Commun.*, *1988*, *18*, 1201.

BCl$_3$, PhH , 25°C
10 min

93%

Spagnolo, P.; Zanirato, P. *J. Chem. Soc., Perkin Trans. I*, *1988*, 2615.

(1   :   >50)   88%

Wuts, P.G.M.; Jung, Y.-W. *J. Org. Chem.*, *1988*, *53*, 1957.

73%

Saito, S.; Nakajima, H.; Inaba, M.; Moriwake, T. *Tetrahedron Lett.*, *1989*, *30*, 837.

1. BnNH$_2$, THF , reflux

2. LiAlH$_4$, THF , reflux

>96% ee , S   78%

Lohray, B.B.; Gao, Y. Sharpless, K.B. *Tetrahedron Lett.*, *1989*, *30*, 2623

PhCH$_2$N$_3$ $\xrightarrow[\text{Montmorillonite}]{\text{PPh}_2\text{PdCl}_2 \text{ , H}_2 \text{ , EtOH , 15 min}}$ PhCH$_2$NH$_2$
86%

Sharma, G.V.M.; Chandrasekhar, S. *Synth. Commun.*, *1989*, *19*, 3289.

NH$_4^+$ HCO$_2^-$

10% Pd/C , RT
MeOH , 10 min   94%

Balicki, R. *Synthesis*, *1989*, 645.

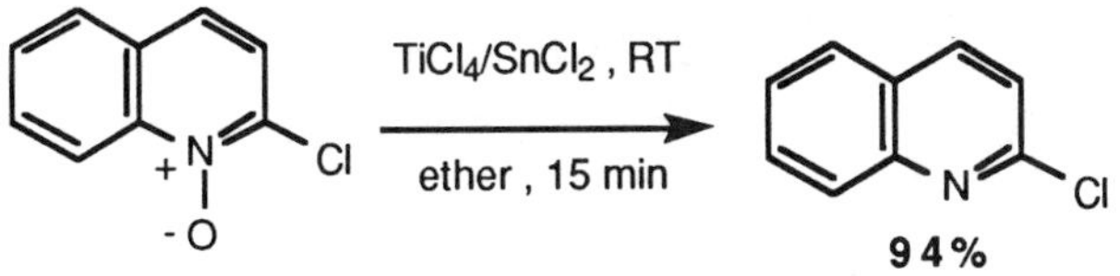

TiCl4/SnCl2 , RT
ether , 15 min
Cl
Cl
N
94%

Balicki, R.; Kaczmarek, L.; Malinowski, M. *Synth. Commun.*, ***1989***, *19*, 897.

AMINES FROM NITRO COMPOUNDS

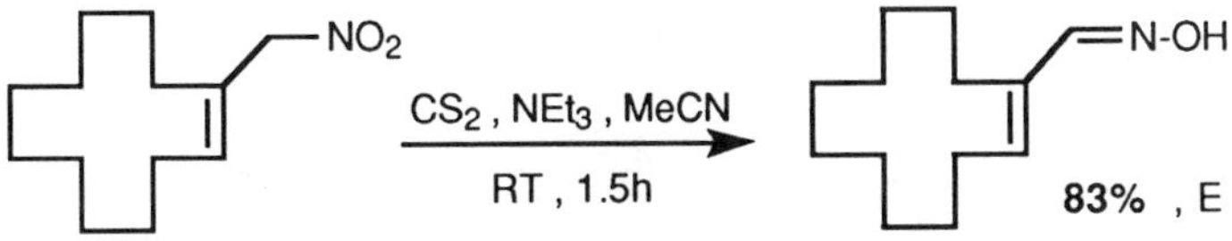

NO2
CS2 , NEt3 , MeCN
RT , 1.5h
N-OH
83% , E

Barton, D.H.R.; Fernandez, I.; Richard, C.S.; Zard, S.Z. *Tetrahedron*, ***1987***, *43*, 551.

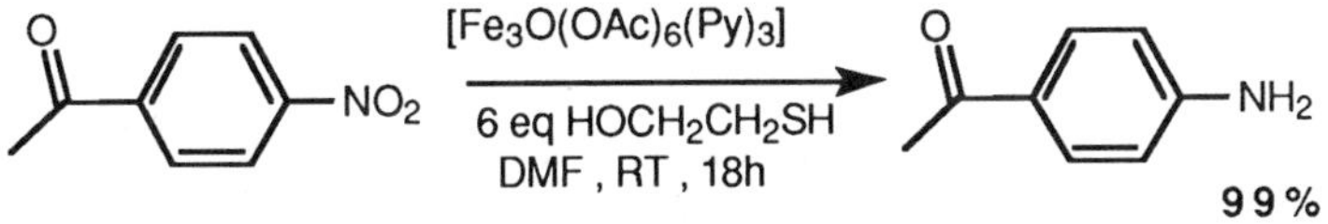

O
NO2
[Fe3O(OAc)6(Py)3]
6 eq HOCH2CH2SH
DMF , RT , 18h
O
NH2
99%

Murata, S.; Miura, M.; Nomura, M. *Chem. Lett.*, ***1988***, 361.

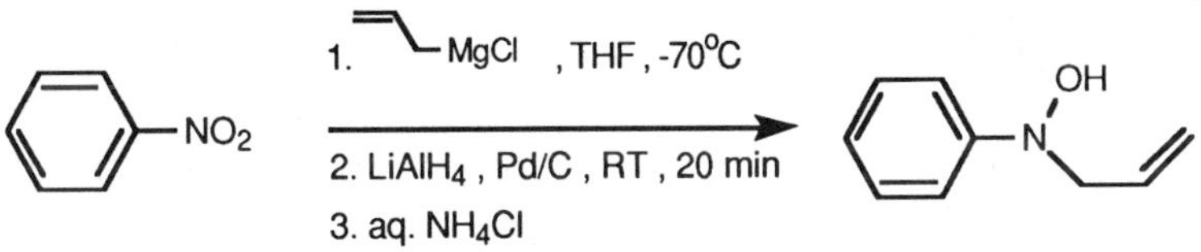

NO2
1. MgCl , THF , -70°C
2. LiAlH4 , Pd/C , RT , 20 min
3. aq. NH4Cl
OH
N

Bartoli, G.; Marcantoni, E. *Tetrahedron Lett.*, ***1988***, *29*, 2251.

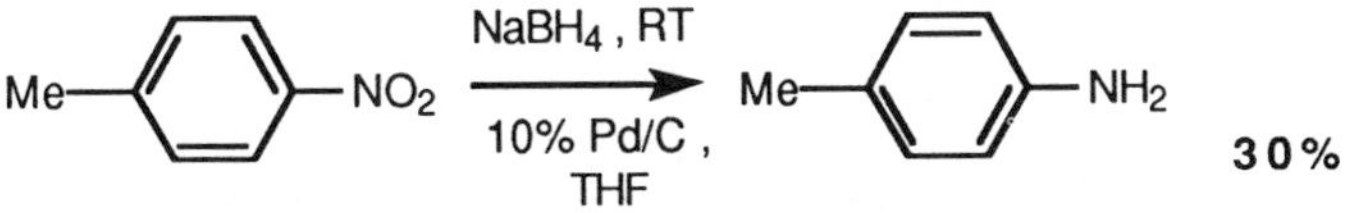

Me
NO2
NaBH4 , RT
10% Pd/C ,
THF
Me
NH2
30%

Petrini, M.; Ballini, R.; Rosini, G. *Synthesis*, ***1987***, 713.

$$\text{(structure) } \xrightarrow[\text{dry } Na_2S \text{ , EtOH}]{} \text{(structure)} \quad 96\%$$

Huber, D.; Andermann, G.; Leclerc, G._Tetrahedron Lett., **1988**, *29*, 635.

$$\text{Ph}-NO_2 \xrightarrow[\text{TFA , THF , 25}^{\circ}\text{C}]{10 \text{ eq [HFe(CO)}_4]^-} \text{Ph}-NH_2 \quad 98\%$$

Gaus, P.L.; Gerritz, S.W.; Jeffries, P.M. *Tetrahedron Lett.*, **1988**, *29*, 5083.

$$C_6H_{13}\text{—CH=CH—}NO_2 \xrightarrow[\begin{array}{c}\text{2. pH 7-8}\end{array}]{\begin{array}{c}1. \text{ SnCl}_2 \cdot 2 \text{ H}_2\text{O , RT}\\ \text{EtOAc , 4h}\end{array}} C_6H_{13}\text{—CH}_2\text{—CH=N-OH} \quad 91\%$$

Kabalka, G.W.; Gondgaon, N.M. *Synth. Commun.*, **1988**, *18*, 693.

$$\text{(structure, m-dinitrobenzene)} \xrightarrow[\underset{\text{M}}{\text{—Si-(CH}_2)_3\text{NH}_2 \cdot \text{PdCl}_2}]{3 \text{ H}_2 \text{ , EtOH , 10 min}} \text{(m-phenylenediamine)} \quad 99\%$$

(M) = montmorillonite

Mukkanti, K.; Subba Rao, Y.V.; Choudary, B.M. *Tetrahedron Lett.*, **1989**, *30*, 251.

$$\text{(o-nitrotoluene)} \xrightarrow[\begin{array}{c}2. \text{ aq. NH}_4\text{Cl}\end{array}]{\begin{array}{c}1. \text{ 3 eq. } CH_2\text{=CH—MgBr}\\ \text{THF , -40}^{\circ}\text{C}\end{array}} \text{(7-methylindole)} \quad 67\%$$

Bartoli, G.; Palmieri, G.; Bosco, M.; Dalpozzo, R. *Tetrahedron Lett.*, **1989**, *30*, 2129.

REVIEWS:

"The Hetero-Cope Rearrangement in Organic Synthesis"

Blechert, S._Synthesis, **1989**, 71.

"Azides: Their Preparation and Synthetic Uses"

Scriven , E.F.V.; Turnbull , K. *Chem. Rev.*, *1988*, *88*, 297.

SECTION 105A: PROTECTION OF AMINES

$$\text{\Large \left.\right>\!\!-NH_2} \quad \xrightarrow[{-20 - +35^{\circ}C}]{Me_3SiCl \text{ , ether}} \quad \text{\Large \left.\right>\!\!-NHSiMe_3} \qquad \textbf{85\%}$$

Courtois, G.; Miginiac, L. *Tetrahedron Lett.*, *1987*, *28*, 1659.

quant

Yamada, T.; Goto, K.; Mitsuda, Y.; Tsuji, J. *Tetrahedron Lett.*, *1987*, *28*, 4557.

NHBn $\xrightarrow[\text{MeOH , <10 min}]{HCO_2NH_4 \text{ , 10\% Pd/C}}$ NH$_2$

with CO$_2$Et   **97%**

Ram, S.; Spicer, L.D. *Tetrahedron Lett.*, *1987*, *28*, 515.
Ram, S.; Spicer, L.D. *Synth. Commun.*, *1987*, *17*, 415.

90%

Pandey, G.; Sudha Rani, K. *Tetrahedron Lett.*, *1988*, *29*, 4157.

$$\text{Me}_2t\text{-BuSiO}\!-\!\!\!\!\bigcirc\!\!\!\!-\text{NHBn} \xrightarrow[\text{1 atm H}_2]{\substack{20\% \text{ Pd(OH)}_2 \\ \text{MeOH}}} \text{Me}_2t\text{-BuSiO}\!-\!\!\!\!\bigcirc\!\!\!\!-\text{NH}_2$$

quant

Yoshida, K.; Nakajima, S.; Wakamatsu, T.; Ban, Y.; Shibasaki, M. *Heterocycles, 1988, 27,* 1167.

$$\xrightarrow[\substack{\text{Hg cathode} \\ \text{Pt anode}}]{\text{e}^-\text{, Et}_4\text{NBr , MeCN}}$$

85%

Roemmele, R.C.; Rapoport, H. *J. Org. Chem., 1988, 53,* 2367.

$$\underset{\text{N}-\text{BOC}}{\bigcirc} \xrightarrow[\text{RT , 4h}]{\text{ZnBr}_2\text{ , CH}_2\text{Cl}_2} \underset{\text{N}-\text{H}}{\bigcirc}$$

89%

Nigam, S.C.; Mann, A.; Taddei, M.; Wermuth, C.-G. *Synth. Commun., 1989, 19,* 3139.

CHAPTER 8

PREPARATION OF ESTERS

SECTION 106: ESTERS FROM ALKYNES

Buchwald, S.L.; Fang, Q.; King, S.M. *Tetrahedron Lett.*, **1988**, *29*, 3445.

Tsuda, T.; Morikawa, S.; Sumiya, R.; Saegusa, T. *J. Org. Chem.*, **1988**, *53*, 3140.
Tsuda, T.; Sumiya, R.; Saegusa, T. *Synth. Commun.*, **1987**, *17*, 147.

SECTION 107: ESTERS FROM ACID DERIVATIVES

The following types of reactions are found in this section:

1. Esters from the reaction of alcohols with carboxylic acids, acid halides and anhydrides.
2. Lactones from hydroxy acids
3. Esters from carboxylic acids and halides, sulfoxides and miscellaneous compounds

1. Ph$_2$SnCl$_2$, EtOH , 2h
reflux

Ph$\diagup$CO$_2$H $\xrightarrow{\text{2. aq. NaHCO}_3}$ Ph$\diagup$CO$_2$Et

63%

Kumar, A.K.; Chattopadhyay, T.K. *Tetrahedron Lett.*, *1987*, *28*, 3713.

1. C$_3$H$_7$CO$_2$H , NEt$_3$, CH$_2$Cl$_2$

$\xrightarrow{\text{2. AgBF}_4}$ C$_3$H$_7$CO$_2$iPr

3. iPrOH

84%

Fukuota, S.; Takimoto, S.; Katsuki, T.; Yamaguchi, M. *Tetrahedron Lett.*, *1987*, *28*, 4711.

Ph$_2$I$^+$ Cl$^-$, *t*-BuOH , 70°C

$\xrightarrow{\text{5h}}$

Me$_2$N$\diagup$S$^-$ Na$^+$ → Me$_2$N$\diagup$SPh 86%

Chen, Z.-C.; Jin, Y.-Y.; Stang, P.J. *J. Org. Chem.*, *1987*, *52*, 4117.

1. 2 eq. LDA , THF

Ph$\diagdown$COOH

2. $\bigcirc$—CHO

3. PhSO$_2$Cl

4. MgBr

83%

Black, T.H.; DuBay III, W.J. *Tetrahedron Lett.*, *1987*, *28*, 4787.

PhCH$_2$COOH $\xrightarrow[\text{K}_2\text{CO}_3 \text{ , MeCN}]{\text{Cl}\diagup\text{C(O)Me}}$

90%

Wakharkar, R.D.; Deshpande, V.H.; Landge, A.B.; Upadhye, B.K. *Synth. Commun.*, *1987*, *17*, 1513.

PhCOOH

1. [structure: Ph-substituted thiadiazole dithiocarbonate reagent]

NEt$_3$, ETOAc , 0°C

2. DMAP , [furfuryl alcohol structure with OH]
RT , 2h

→ [structure: Ph-C(=O)-O-CH$_2$-furan]   85%

Takeda, K.; Tsuboyama, K.; Takayanagi, H.; Ogura, H. *Synthesis,* **1987,** 560.

Tol—[CH with two CO$_2^-$ Na$^+$ groups]

P$_2$S$_5$, BnEt$_3$NCl , 135°C
dichlorobenzene , 7h
18-crown-6

→ [thiophene with Tol substituent]   80%

Sastry, C.V.R.; Marwah, A.K.; Marwah, P.; Rao, G.S.; Shridhar, D.R. *Synthesis,* **1987,** 1024.

HO$_2$C-(CH$_2$)$_{10}$-COOH

HO(CH$_2$)$_{16}$OH , 48h
Pseudomonas sp. K-10

→ [macrocyclic dilactone structure with (CH$_2$)$_8$ and (CH$_2$)$_{16}$]   56%

also obtained 15% of dilactone

Guo, Z.-W.; Sih, C.J. *J. Am. Chem. Soc.,* **1988,** *110,* 1999.

HO—[C(=O)—(CH$_2$)$_{14}$—COOH]

Pseudomonas sp. AK
isooctane , 25°C , 144h

+ 26% of diolide

→ [lactone structure with (CH$_2$)$_{14}$—O]   66%

Guo, Z.; Ngooi, T.K.; Scilimati, A.; Fülling, G.; Sih, C.J. *Tetrahedron Lett.,* **1988,** *29,* 5583.

PhCH$_2$COOH

1. 2 eq. LDA
2. EtCHO
3. PhSO$_2$Cl
4. MgBr$_2$, ether

→ [lactone structure with Me, Ph substituents]   53%

Black, T.H.; Fields, J.D. *Synth. Commun.,* **1988,** *18,* 125.

$$C_5H_{11}-C(=O)-Cl \xrightarrow[\begin{array}{c}\text{2. } C_8H_{17}Br\text{ , PhH ,}\\ \text{aq. NaOH , }30^\circ C\\ Bu_4NBr\end{array}]{\begin{array}{c}\text{1. } Me-C(=S)-NH_2\text{ , PhH , }30^\circ C\\ \text{1h}\end{array}} C_5H_{11}-C(=O)-S-C_8H_{17}$$

89%

Takido, T.; Toriyama, M.; Itabashi, K. *Synthesis, 1988*, 404.

PhMe , reflux

≈10^{-4} M , 2h

60%

Boeckman Jr., R.K.; Pruitt, J.R. *J. Am. Chem. Soc., 1989, 111*, 8286.

$$PhCOOH \xrightarrow[\text{imidazolium salt, 2 OTf}^-]{BnOH\text{ , }MeNO_2} PhCO_2Bn$$

quant

Saha, A.K.; Schultz, P.; Rapoport, H. *J. Am. Chem. Soc., 1989, 111*, 4856.

$$PhCH_2COOH \xrightarrow[\begin{array}{c}\text{2. MeI , }CH_2Cl_2\text{ , 3h}\\ \text{3. BnOH}\end{array}]{\begin{array}{c}\text{1. (2-pyridyl-S-S-2-pyridyl)}\\ MS\text{ 4Å , }CH_2Cl_2\text{ , RT}\end{array}} PhCH_2CO_2Bn$$

97%

Ravi, D.; Mereyala, H.B. *Tetrahedron Lett., 1989, 30*, 6089.

1. LDA , THF , -78°C
2. Me_3SiCl
3. $(EtO)_2CHCH_2Br$
   $TiCl_4$, CH_2Cl_2 , -78°C
4. DBU , PhMe , 60°C

48%

Demnitz, F.W.J. *Tetrahedron Lett., 1989, 30*, 6109.

66%

Miura, M.; Itoh, K.; Nomura, M. *Chem. Lett.*, *1989*, 77.

86%

Matsuyama, H.; Nakamura, T.; Kamigata, N. *J. Org. Chem.*, *1989*, *54*, 5218.

REVIEW:

"Enzymes and Organic Solvents Were Generally Considered Incompatible"

Sih, C.J.; Chen, C.-S. *Angew. Chem. Int. Ed., Engl.*, *1989*, *28*, 695.

Further examples of the reaction RCO₂H + R'OH → RCO₂R' are included in Section 108 (Esters from Alcohols and Phenols) and in Section 30A (Protection of Carboxylic Acids).

SECTION 108: ESTERS FROM ALCOHOLS AND THIOLS

80%

Beebe, T.R.; Adkins, R.; Baldridge, R.; Hensley, V.; McMillen, D.; Morris, D.; Noe, R.; Ng, F.W.; Powell, G.; Spielberger, C.; Stolte, M. *J. Org. Chem.*, *1987*, *52*, 5472.
McDonald, C.E.; Beebe, T.R.; Beard, M.; McMillen, D.; Selski, D. *Tetrahedron Lett.*, *1989*, *30*, 4791.

67%

Rautenstrauch, V. *Helv. Chim. Acta*, *1987*, *70*, 593.

HO~~~OH $\xrightarrow[\text{mesitylene , 24h}]{\text{RuH}_2(\text{PPh}_3)_4 \text{ , 180}^\circ\text{C}}$ [lactone]

99%

Murahashi, S.; Naota, T.; Ito, K.; Maeda, Y.; Taki, H. *J. Org. Chem., 1987, 52,* 4319.

PhCH$_2$OH $\xrightarrow[\text{CH}_2\text{Cl}_2 \text{ , 25}^\circ\text{C}]{2 \text{ eq. HC(OEt)}_3 \text{ , MgCl}_2}$ Ph~O~CH(OEt)$_2$

94%

Perron, F.; Gahman, T.C.; Albizati, K.F. *Tetrahedron Lett., 1988, 29,* 2023.

[acetonide alcohol] $\xrightarrow{\begin{array}{l}\text{1. (COCl)}_2 \text{ , DMSO}\\ \quad \text{CH}_2\text{Cl}_2 \text{ , -78}^\circ\text{C}\\ \text{2. NEt}_3\\ \text{3. MeOH}\\ \text{4. Br}_2 \text{ , H}_2\text{O , NaHCO}_3\end{array}}$ [acetonide ester CO$_2$Me]

84%

Lichtenthaler, F.W.; Jarglis, P.; Lorenz, K. *Synthesis, 1988,* 790.

[homoallylic alcohol with Me, C$_6$H$_{13}$] $\xrightarrow[\begin{array}{c}\text{benzoquinone , 8h}\\ \text{40}^\circ\text{C , DMF (2% H}_2\text{O)}\end{array}]{\text{PdCl}_2 \text{ , (CuCl} \cdot \text{O}_2)}$ [tetrahydrofuran product, C$_6$H$_{13}$, Me, HO]

64%

Nokami, J.; Ogawa, H.; Miyamoto, S.; Mandai, T.; Wakabayashi, S.; Tsuji, J. *Tetrahedron Lett., 1988, 29,* 5181.

~~~OH $\xrightarrow[\text{2. PhCH}_2\text{Br}]{\text{1. CO , S}^\circ \text{ , DBU}}$ ~~~O–C(=O)–S–CH$_2$Ph

86%

Mizuno, T.; Nishiguchi, I.; Hirashima, T. *Tetrahedron Lett., 1988, 29,* 4767.

HO~~~OH $\xrightarrow[\begin{array}{c}\text{reflux , 24h}\\ \text{[pyridinium] N}^+(\text{CH}_2)_{15}\text{Me (PW}_{12}\text{O}_{40})^{3-}\end{array}]{3 \text{ eq. H}_2\text{O}_2 \text{ , }t\text{-BuOH}}$ [lactone]

>98%

Ishii, Y.; Yoshida, T.; Yamawaki, K.; Ogawa, M. *J. Org. Chem., 1988, 53,* 5549.

87%

Björkling, F.; Godtfredsen, S.E.; Kirki, O. *J. Chem. Soc., Chem. Commun., 1989*, 934.

60%

Kajigaeshi, S.; Kawamukai, H.; Fujisaki, S. *Bull. Chem. Soc., Jpn., 1989, 62*, 2585.

Further examples of the reaction ROH → RCO$_2$R' are included in Section 107 (Esters from Acid Derivatives) and in Section 45A (Protection of Alcohols and Phenols).

SECTION 109: ESTERS FROM ALDEHYDES

52%

Barluenga, J.; Fernández, J.R.; Yus, M. *J. Chem. Soc., Chem. Commun., 1987*, 1534.

85%

O'Connor, B.; Just, G. *Tetrahedron Lett., 1987, 28*, 3235.

Black, T.H.; DuBay III, W.J. *Tetrahedron Lett.*, *1988*, *29*, 1747.

Williams, D.R.; Klingler, F.D.; Allen, E.E.; Lichtenthaler, F.W. *Tetrahedron Lett.*, *1988*, *29*, 5087.

Okimoto, M.; Chiba, T. *J. Org. Chem.*, *1988*, *53*, 218.

Linderman, R.J.; McKenzie, J.R. *Tetrahedron Lett.*, *1989*, *30*, 3911.

$$C_7H_{15}CH_2CHO \xrightarrow[\text{MeOH , reflux , 4h}]{\text{2.5 eq. NIS , K}_2\text{CO}_3} C_7H_{15}CH_2CO_2Me$$

80%

McDonald, C.; Holcomb, H.; Kennedy, K.; Kirkpatrick, E.; Leathers, T.; Vanemon, P. *J. Org. Chem.*, *1989*, *54*, 1213.

Related Methods: Section 117 (Esters from Ketones)

SECTION 110: ESTERS FROM ALKYLS, METHYLENES AND ARYLS

No examples of the reaction R-R $\rightarrow$ RCO$_2$R' or R'CO$_2$R (R,R' = alkyl, aryl, etc.) occur in the literature. For the reaction R-H $\rightarrow$ RCO$_2$R' or R'CO$_2$R, see Section 116 (Esters from Hydrides).

NO ADDITIONAL EXAMPLES

SECTION 111: ESTERS FROM AMIDES

1. LDA
2. PhMn(CO)$_3^+$ PtF$_6^-$
   -78$\rightarrow$0°C
3. LiOBn , THF , 0°C
4. 2 eq. DDQ , MeCN
   reflux , 6h

42%

Miles, W.H.; Smiley, P.M.; Brinkman, H.R. *J. Chem. Soc., Chem. Commun., 1989*, 1897.

1. N$_2$O$_4$, NaOAc ,
   CH$_2$Cl$_2$, -10°C
2. CCl$_4$, RT , 6h

82%

Torra, N.; Urpí, F.; Vilarrasa, J. *Tetrahedron, 1989, 45*, 863.

Phth(CH$_2$)$_4$N-(CH$_2$)$_3$NHCHO
1. Ac$_2$O , NaNO$_2$
   AcOH , RT , 43h
2. 95°C
3. N$_2$H$_4$ •H$_2$O , DMF ,
   85°C , 27h
$\longrightarrow$ Phth(CH$_2$)$_4$N-(CH$_2$)$_3$OH

74%

Iwata, M.; Kuzuhara, H. *Chem. Lett., 1989*, 1195.

1. AcOH , Ac$_2$O
   NaNO$_2$ 0°C→RT
2. ice , H$_2$O

90%

Nikolaides, N.; Ganem. B. J. Org. Chem., **1989**, *54*, 5996.

SECTION 112: ESTERS FROM AMINES

NO ADDITIONAL EXAMPLES

SECTION 113: ESTERS FROM ESTERS

Conjugate reductions and conjugate alkylations of unsaturated esters are found in Section 74 (Alkyls from Alkenes).

1. LDA , THF , -78°C
2. AcCl , -78°C
3. L-Selectride , -20°C

82%

threo:erythro = 99:1

Schlessinger. R.; Tata, J.R.; Springer, J.P. J. Org. Chem., **1987**, *52*, 708.

PdCl$_2$(PPh$_3$)$_2$, CO

Ac$_2$O , NEt$_3$, PhH
160°C

76%

Koyasu, Y.; Matsuzaka, H.; Hiroe, Y.; Uchida, Y.; Hidai. M. J. Chem. Soc., Chem. Commun., **1987**, 575.

HO-(CH$_2$)$_{15}$CO2Me $\xrightarrow[\text{40°C , 72h}]{\textit{Psuedomonas sp. , PhH}}$

80%

Makita, A.; Nihira, T.; Yamada, Y. *Tetrahedron Lett., 1987, 28,* 805.

$\xrightarrow[\text{reflux , 1d}]{\text{PPL , EtOH , MS 4Å , THF}}$ CO$_2$Et + CH$_3$CHO

83%

PPL = pig pancreatic lipase

Degueil-Castaing, M.; DeJeso, B. *Tetrahedron Lett., 1987, 28,* 953.

IZn CO$_2$Et $\xrightarrow[\text{Me}_2\text{CC(=O)Me}]{\text{cat. CuCN , THF , 1h}}$ CO$_2$Et

OTs , 60°C

89%

Ochiai, H.; Tamaru, Y.; Tsubaki, K.; Yoshida, Z. *J. Org. Chem., 1987, 52,* 4418.

MeO (CH$_2$)$_4$ OMe $\xrightarrow{\begin{array}{l}\text{1. 2 eq. LDA , THF , -78°C}\\\text{2. CuCl}_2\text{ , -78°C , 3h}\\\text{3. }p\text{-TsOH , MeOH , RT , 20h}\end{array}}$ CO$_2$Me CO$_2$Me

49%

Babler, J.H.; Sarussi, S.J. *J. Org. Chem., 1987, 52,* 3462.

Me CO$_2$Et OAc , Pd(PPh$_3$)$_4$ $\xrightarrow[\text{NEt}_3\text{ , THF , RT , 1h}]{}$ Me CO$_2$Et

NC   NC   81%

Ito, Y.; Sawamura, M.; Matsuoka, M.; Matsumoto, Y.; Hayashi, T. *Tetrahedron Lett., 1987, 28,* 4849.

$$(68 \quad : \quad 32) \quad 59\%$$

Kunz, T.; Reissig, H.-U. *Angew. Chem. Int. Ed., Engl*, **1988**, *27*, 268.

77%

Miura, K.; Fugami, K.; Oshima, K.; Otimoto, K. *Tetrahedron Lett.*, **1988**, *29*, 5135.

quant

Barry, J.; Bram, G.; Petit, A. *Tetrahedron Lett.*, **1988**, *29*, 4567.

48%   36%

Lee, G.M.; Parvez, M.; Weinreb, S.M. *Tetrahedron*, **1988**, *44*, 4671.

94%

Bagler, V.; Doyle, M.P.; Taunton, J.; Claxton, E.E. *J. Org. Chem.*, **1988**, *53*, 6158.

1. LDA , THF , HMPA , -78°C
2. iPr-I , 4h , -78° - -40°C

(92   :   8)   95%

Fuji, K.; Node, M.; Tanaka, F.; Hosoi, S. *Tetrahedron Lett.*, *1989*, *30*, 2825.

Bu₃SnH , AIBN , PhH
reflux , 10h

50%

Belletire, J.L.; Mahmoodi, N.O. *Tetrahedron Lett.*, *1989*, *30*, 4363.

1% Rh(acam)₄
CH₂Cl₂ ,
RT

(>99   :   <1)   96%

Doyle, M.P.; Bagheri, V.; Pearson, M.M.; Edwards, J.D. *Tetrahedron Lett.*, *1989*, *30*, 7001.

Bu₃SnH , AIBN
PhMe , 80°C

72%

Yamamoto, M.; Uruma, T.; Iwasa, S.; Kohmoto, S.; Yamada, K. *J. Chem. Soc., Chem. Commun.*, *1989*, 1265.

SECTION 114: ESTERS FROM ETHERS, EPOXIDES AND THIOETHERS

56%   **94%**

Ghosh, S.; Raychaudmuri, S.R.; Salomon, R.G. *J. Org. Chem.*, *1987*, *52*, 83.

95%

Jefford, C.W.; Sledeski, A.W.; Boukouvalas, J. *Tetrahedron Lett.*, *1987*, *28*, 949.

1. BuLi , ether , 0°C

2. [BCl₃/B(OMe)₃]
   ether , 0°C
3. mCPBA , ether
   Na₂CO₃ , -78°C , 5 min

90%

4. H₂O , 0°C

Pelter, A.; Rowlands, M. *Tetrahedron Lett.*, *1987*, *28*, 1203.

1. BuLi

2. $\triangle$ OPh

3. H₃O⁺

4. p-TsOH

5. NEt₃ , CH₂Cl₂

74%

Carretero, J.C.; Ghosez, L. *Tetrahedron Lett.*, *1988*, *29*, 2059.

NBS , MeOH , 0°C→ RT

Brownbridge, P.; Jowett, I.C. *Synthesis, 1988*, 252.

Me-furan-SnBu₃ →[Pb(OAc)₄ , Ch₂Cl₂ / reflux] → product **95%**

Yamamoto, M.; Izukawa, H.; Saiki, M.; Yamada, K. *J. Chem. Soc., Chem. Commun.*, **1988**, 560.

67%

Fujiwara, M.; Imada, M.; Baba, A.; Matsuda, H. *J. Org. Chem.*, **1988**, *53*, 5974.

Co₂(CO)₈ , DME / 60 atm CO , 160°C , 2d   **80%**

Wang, M.-D.; Calet, S.; Alper, H. *J. Org. Chem.*, **1989**, *54*, 20.

SECTION 115: ESTERS FROM HALIDES AND SULFONATES

R-X ⟶ **R-CO₂R'**

HMPA , Et₂O , CuBr•SMe₂ , 0° - RT

59% [96% ee]

Nakamura, E.; Sekiya, K.; Kuwajima, I. *Tetrahedron Lett.*, **1987**, *28*, 337.

[Dibal-H , PdCl₂(PPh₃)₂ / BrZnCH₂CO₂t-Bu]

HMPA , RT , overnight / then aq. HCl   **48%**

Orsini, F.; Pelizzoni, F. *Synth. Commun.*, **1987**, *17*, 1389.

EtMgBr $\xrightarrow[\text{2.}]{\text{1. CS}_2\text{ , THF ,0}^\circ\text{C}}$ Et—C(=O)—SPh   **60%**

2. Cl—C(=O)—SPh

Beslin, P.; Dlubala, A.; Levesque, G. *Synthesis, 1987*, 835.

n-C$_6$H$_{13}$I $\xrightarrow[\text{hv (UV) , THF , K}_2\text{CO}_3]{\text{CO , PtCl}_2\text{ (AsPh}_3)_2\text{ , MeOH}}$ n-C$_6$H$_{13}$CO$_2$Me

55%

Kondo, T.; Tsuji, Y.; Watanabe, Y. *Tetrahedron Lett., 1988, 29*, 3833.

Ph-Cl $\xrightarrow{\begin{array}{c}\text{70 psi CO , 150}^\circ\text{C , MeOH}\\ \text{DMF , NaOAc , 20h}\end{array}}$ PhCO$_2$Me

78%

Ben-David, Y.; Portnoy, M.; Milstein, D. *J. Am. Chem. Soc., 1989, 111*, 8742.

n-C$_8$F$_{17}$—CH(I)—Bu $\xrightarrow[\text{cat. NEt}_3\text{ , 80}^\circ\text{C , 12h}]{\text{30 atm CO , 5% (PPh}_3)_2\text{PdCl}_2}$ n-C$_8$F$_{17}$—CH(CO$_2$Et)—Bu   **73%**

Urata, H.; Kosukegawa, O.; Ishii, Y.; Yugari, H.; Fuchikami, T. *Tetrahedron Lett., 1989, 30*, 4403.
Urata, H.; Ishii, Y.; Fuchikami, T. *Tetrahedron Lett., 1989, 30*, 4407.

CH$_2$=C(Me)—CH$_2$Br $\xrightarrow[\text{Na}_2\text{PdCl}_4\text{ , dppe , 2h}]{\text{CO , NaOEt , EtOH , 50}^\circ\text{C}}$ Me$_2$C=CH—CO$_2$Et   **87%**

Kiji, J.; Okano, T.; Konishi, H.; Nishiumi, W. *Chem. Lett., 1989*, 1873.

PhCH$_2$Cl $\xrightarrow[\begin{array}{c}\text{NaOAc , BnEt}_3\text{NCl}\\ \text{70}^\circ\text{C , 20h , }t\text{-BuOH}\end{array}]{\text{CO , Pd(PPh}_3)_2\text{Cl}_2\text{ , Ph}_3}$ Ph—CH$_2$—C(=O)—Ot-Bu   **60%**

Adapa, S.R.; Prasad, C.S.N. *J. Chem. Soc., Perkin Trans. I, 1989*, 1706.

PhCH$_2$Cl $\xrightarrow[\substack{\text{EtOH , Bu}_4\text{NI , NaHCO}_3 \\ \text{16 psi CO , 70}^\circ\text{C , 6h}}]{\text{(P)}-\text{PCl}_2\text{ / PPh}_3\text{ / PdCl}_2}$ PhCH$_2$CO$_2$Et

63%

Reddy, N.P.; Kantam, M.L.; Choudary, B.M. *Indian J. Chem.*, **1989**, *28B*, 105.

REVIEW:

"Nickel-Aluminum Alloy as a Reducing Agent"

Keefer, L.K.; Lunn, G.*Chem. Rev.*, **1989**, *89*, 459.

Related Methods: Section 25 (Acid Derivatives from Halides).

SECTION 116: ESTERS FROM HYDRIDES

This section contains examples of the reaction R-H → RCO$_2$R' or R'CO$_2$R
(R = alkyl, aryl, etc.).

MeO—⟨ ⟩—OTf $\xrightarrow[\substack{\text{dppp , DMSO-MeOH} \\ \text{45 min , 70}^\circ\text{C}}]{\text{CO , Pd(OAc)}_2\text{ , NEt}_3}$ MeO—⟨ ⟩—CO$_2$Me

95%

Dolle, R.E.; Schmidt, S.J.; Kruse, L.I. *J. Chem. Soc., Chem. Commun.*, **1987**, 904.

$\xrightarrow[\text{80}^\circ\text{C}]{\text{Pd(O}_2\text{CCF}_3)_2}$

O$_2$CCF$_3$

>50%

Gretz, E.; Oliver, T.F.; Sen, A. *J. Am. Chem. Soc.*, **1987**, *109*, 8109.

63%

Demir, A.S.; Jeganathan, A.; Watt. D.S. *J. Org. Chem.*, *1989*, *54*, 4020.

Also via: Section 26 (Acid Derivatives) and Section 41 (Alcohols).

SECTION 117: ESTERS FROM KETONES

64%

Milenkov, B.; Guggisberg, A.; Hesse. M. *Tetrahedron Lett.*, *1987*, *28*, 315.

70%

Fukuzawa. S.; Nakanishi, A.; Fujinami, T.; Sakai, S. *J. Chem. Soc., Perkin Trans. I*, *1988*, 1669.

(99   :   1) **83%**

Frenette. R.; Kakushima, M.; Zamboni, R.; Young, R.N. *J. Org. Chem.*, *1987*, *52*, 304.

Sml$_2$, THF-HMPA

92%

trans:cis = 90:10

Fukuzawa, S.; Iida, M.; Nakanishi, A.; Fujinami, T.; Sakai, S. *J. Chem. Soc., Chem. Commun., **1987**, 920.*

NCl , MeOH

PhH , reflux

CO$_2$Me

81%

Beebe, T.R.; Baldridge, R.; Beard, M.; Cooke, D.; DeFays, I.; Hensley, V.; Hua, D.; Lao, J.-C.; McMillen, D.; Morris, D.; Noe, R.; O'Bryan, E.; Spielberger, C.; Stotle, M.; Waller Jr., J. *J. Org. Chem., **1987**, 52,* 3165.

Actinobacter NCIB 9871

monooxygenase

80%
(>98% ee)

Taschner, M.J.; Black, D.J. *J. Am. Chem. Soc., **1988**, 110,* 6892.

7 eq. BF$_3$·OEt$_2$
5 eq. ethylene glycol

3h , RT

65%

Tanaka, M.; Suemune, H.; Sakai, K. *Tetrahedron Lett., **1988**, 29,* 1733.

BF$_3$ ·OEt$_2$, CH$_2$Cl$_2$, 1h

HOCH$_2$CH$_2$OH , RT

CO$_2$CH$_2$CH$_2$OH

87%

Nagumo, S.; Suemune, H.; Sakai, K. *Tetrahedron Lett., **1988**, 29,* 6927.

Ph—C(=O)—CH2—Ph →(mCBPA (solid state), RT)→ Ph—C(=O)—O—CH2—Ph   97%

(mCPBA in CHCl₃ , 46% yield)

Toda, F.; Yagi, M.; Kiyoshige, K. *J. Chem. Soc., Chem. Commun.*, **1988**, 958.

1. 2.4 eq. *t*-BuO₂CCH₂Li
   THF , -65°C
2. HBF₄ , -65°C → RT
3. TFAA , RT , 6h

52%

Dieter, R.K.; Fishpaugh, J.R. *J. Org. Chem.*, **1988**, *53*, 2031.

Ph—C(=O)—Ph →(LDA, THF; Cl—CH(OMe)—CH(OMe)₂)→   71%

Olah, G.A.; Wu, A.; Farooq, O.; Surya Prakash, G.K. *Synthesis*, **1988**, 537.

Ph—C(=O)—Me →(e⁻ , NaBr , 30°C , MeOH; Pt anode , Cu-Zn cathode)→ Ph-CO₂Me   74%

Nikishin, E.I.; Elinson, M.N.; Makhova, I.V. *Angew. Chem. Int. Ed., Engl*, **1988**, *27*, 1716.

Ph—C(=O)—CH₂CH₃ →(e⁻ , I₂ , CH(OMe)₃ , LiClO₄•3 H₂O; RT , Pt electrodes)→ Ph—CH(Me)—CO₂Me   85%

Shono, T.; Matsumura, Y.; Katoh, S.; Fujita, T.; Kamada, T. *Tetrahedron Lett.*, **1989**, *30*, 371.

Ph—CH₂CH₂—C(=O)—SiMe₃ →(e⁻ , Et₄NOTs , MeOH , RT; C anode and cathode)→ Ph—CH₂CH₂—C(=O)—OMe   80%

Yoshida, J.; Matsunaga, S.; Isoe, S. *Tetrahedron Lett.*, **1989**, *30*, 5293.

modified Baeyer-Villiger

88%

Koch, S.S.C.; Chamberlin, A.R. *Synth. Commun., 1989, 19*, 829.

REVIEWS:

"Enzymatic Baeyer-Villiger Oxidations by Flavin Dependent Monooxygenases"

Walsh, C.T.; Chen, Y.-C.J. *Angew. Chem. Int. Ed., Engl, 1988, 27*, 333.

"Syntheses of Macrocyclic Compounds by Ring Enlargement"

Stauch, H.; Hesse, M. *Tetrahedron, 1988, 44*, 1573.

Also via: Section 27 (Acid Derivatives).

SECTION 118: ESTERS FROM NITRILES

49%

Khanapure, S.P.; Biehl, E.R. *J. Org. Chem., 1987, 52*, 1333.

63%

Tiecco, M.; Tingoli, M.; Testaferri, L.; Bartoli, D. *Synth. Commun., 1989, 19*, 2817.

SECTION 119: ESTERS FROM ALKENES

Salomon, R.G.; Roy, S.; Salomon, M.F. *Tetrahedron Lett., 1988, 29*, 769.

Bäuml, E.; Tscheschlok, K.; Pock, R.; Mayr, H. *Tetrahedron Lett., 1988, 29*, 6925.

Stork, G.; Mah, R. *Tetrahedron Lett., 1989, 30*, 3609.

REVIEW:

"Transition Metal-Catalyzed Epoxidations"

Jørgensen, K.A. *Chem. Rev., 1989, 89*, 431.

Also via: Section 44 (Alcohols).

SECTION 120: ESTERS FROM MISCELLANEOUS COMPOUNDS

75%

Ochiai, M.; Iwaki, S.; Ukita, T.; Nagao, Y. *Chem. Lett., 1987*, 133.

50%

Bairgrie, L.M.; Leung-Toung, R. Tidwell, T.T. *Tetrahedron Lett., 1988, 29*, 1673.

REVIEW:

"Reactions of Carbon Dioxide with Carbon-Carbon Bond Formation Catalyzed by Transition Metal Complexes"

Braunstein, P.; Matt, D.; Nobel, D. *Chem. Rev., 1988, 88*, 747.

CHAPTER 9

PREPARATION OF ETHERS, EPOXIDES AND THIOETHERS

SECTION 121: ETHERS, EPOXIDES AND THIOETHERS FROM ALKYNES

Bu—≡—CH(OH)—C$_6$H$_{13}$

1. CH$_2$Cl$_2$, -40°C, 2h, CH$_2$=CH—OEt

2. Bu$_3$SnCl, NaCNBH$_3$ AIBN, t-BuOH

3. p-TsOH . PhH , RT , 8h

→ furan with Bu and C$_6$H$_{13}$

35%

Srikrishna, A.; Sunderbabu, G. *Tetrahedron Lett.*, **1987**, *28*, 6393.

Ph—C(O)—CH$_2$—C≡C—C$_6$H$_4$—Cl

Pd(dba)$_2$, PPh$_3$
100°C ,
10h

→ Ph—furan—C$_6$H$_4$—Cl

33%

Sheng, H.; Lin, S.; Huang, Y. *Synthesis,* **1987**, 1022.

diyne with two Me groups

PhCHO , Ni(COD)$_2$
P(t-Bu)$_3$, 5h
20°C

→ bicyclic product with Me, Ph, O, Me

52%

Tsuda, T.; Kiyoi, T.; Miyane, T.; Saegusa, T. *J. Am. Chem. Soc.*, **1988**, *110*, 8570.

1. LDA , THF , Me$_3$SiCl , -78°C
2. mCPBA , 0°C
3. 5 Naq. HCl/ether
4. chromatography
5. aq. HBF$_4$/THF

58%

Obrecht, D. *Helv. Chim. Acta*, **1989**, *72*, 447.

SECTION 122: ETHERS, EPOXIDES AND THIOETHERS FROM ACID DERIVATIVES

(CF$_3$SO$_3$)$_2$Sn , MeNO$_2$

collidine , MS , 24h , RT

70%

Gnonlonfoun, N.; Zamarlik, H. *Tetrahedron Lett.*, **1987**, *28*, 4053.

SECTION 123: ETHERS, EPOXIDES AND THIOETHERS FROM ALCOHOLS AND THIOLS

1. AcOH , reflux , 15h
2. NH$_2$NH$_2$, EtOH , 60°C , 20 min

86%

Hu, N.X.; Aso, Y.; Otsubo, T.; Ogura, F. *Tetrahedron Lett.*, **1987**, *28*, 1281.

AlPO$_4$ - Al$_2$O$_3$, neat

distil

cis:trans = 1:1

84%

Costa, A.; Riego, J.M. *Synth. Commun.*, **1987**, *17*, 1373.

$$\underset{Ph}{\overset{OH}{\diagup}}\hspace{-0.5em}\underset{Me}{\diagdown} \xrightarrow[\text{RT , 4h}]{\text{ZnCl}_2 \text{ , ClCH}_2\text{CH}_2\text{Cl}} \underset{Me\ \ Me}{Ph\diagdown O\diagup Ph} \quad 91\%$$

Kim, S.; Chung, K.N.; Yang, S. *J. Org. Chem., 1987, 52,* 3917.

$$\text{PhSH} \xrightarrow[\text{NiBr}_2\text{(dppf) , Zn}^\circ \text{ , K}_2\text{CO}_3 \text{ , 10h}]{\text{N-methyl-2-pyrrolidinone , 25}^\circ\text{C}} \text{Ph-S-Ph} \quad 97\%$$

Takagi, K. *Chem. Lett., 1987,* 2221.

$$C_{15}H_{29}CH_2OH \xrightarrow[\text{CH}_2\text{Cl}_2 \text{ , cyclohexane}]{\underset{\text{Cl}_3\text{C}}{\overset{NH}{\|}}\text{C}-\text{O}t\text{-Bu}\ ,\ \text{BF}_3\cdot\text{OEt}_2} C_{15}H_{29}CH_2Ot\text{-Bu} \quad 69\%$$

Armstrong, A.; Brackenridge, I.; Jackson, R.F.W.; Kirk, J.M. *Tetrahedron Lett., 1988, 29,* 2483.

Brown, D.S.; Ley, S.V.; Vile, S. *Tetrahedron Lett., 1988, 29,* 4873.

Dorta, R.L.; Francisco, C.G.; Freire, R.; Suárez, E. *Tetrahedron Lett., 1988, 29,* 5429.

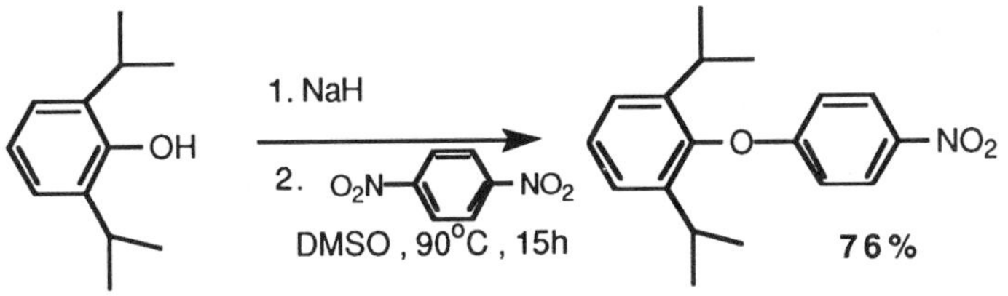

OH
1. NaH
2. O_2N–C6H4–NO_2
DMSO, 90°C, 15h
O
NO_2
76%

Sammes, P.G.; Thetford, D.; Voyle, M. *J. Chem. Soc., Perkin Trans. I*, ***1988***, 3229.

J. Chem. Soc., Chem. Commun., ***1987***, 1373

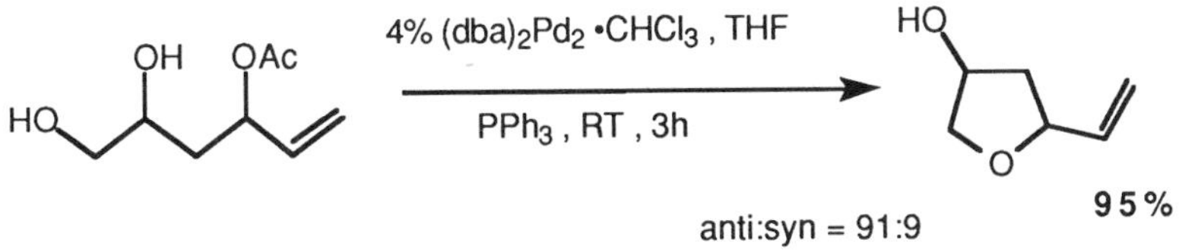

OH
OAc
HO
4% $(dba)_2Pd_2$ •$CHCl_3$, THF
PPh_3, RT, 3h
HO
O
anti:syn = 91:9
95%

Trost, B.M.; Tenaglia, A. *Tetrahedron Lett.*, ***1988***, *29*, 2927.

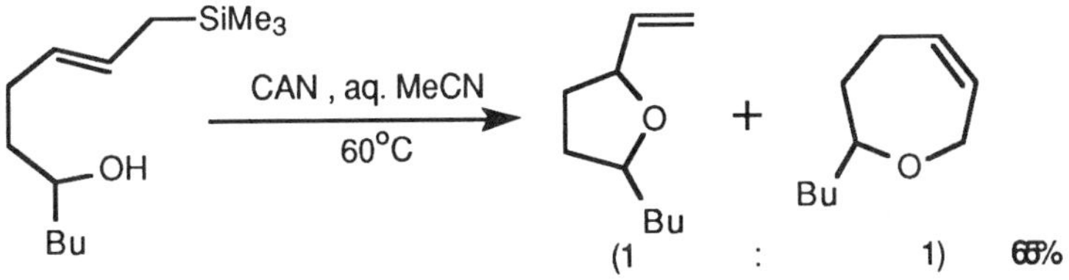

$SiMe_3$
CAN, aq. MeCN
60°C
OH
Bu
O
+
O
Bu
Bu
(1 : 1)

Wilson, S.R.; Augelli-Szafran, C.E. *Tetrahedron*, ***1988***, *44*, 3983.

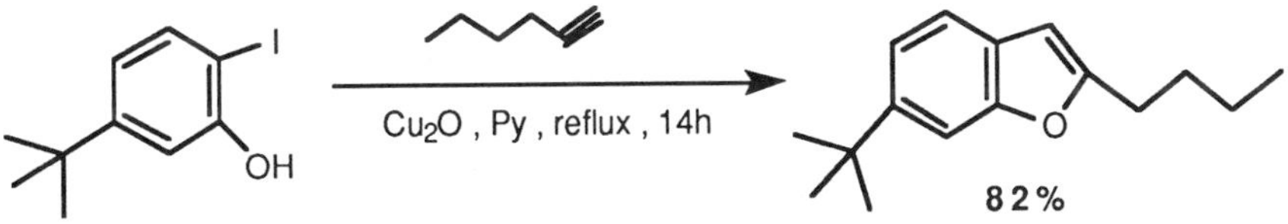

I
OH
Cu_2O, Py, reflux, 14h
O
82%

Doad, G.J.S.; Barltrop, J.A.; Petty, C.M.; Owen, T.C. *Tetrahedron Lett.*, ***1989***, *30*, 1597.

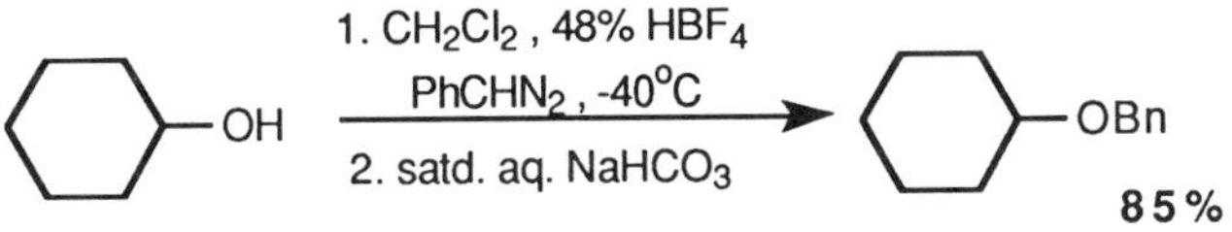

OH
1. CH_2Cl_2, 48% HBF_4
$PhCHN_2$, -40°C
2. satd. aq. $NaHCO_3$
OBn
85%

Liotta, L.J.; Ganem, B. *Tetrahedron Lett.*, ***1989***, *30*, 4759.

2.6 eq PhSSPh
1.8 eq. Bu₃P

THF , reflux ,
15h

HO ... → PhS ...

70%

Cleary, D.G. Synth. Commun., **1989**, 19, 737.

1. OAc , MeCN
LiPdCl₃ , 0°C

2. NaH , reflux .
5h

53%

Larock, R.C.; Song, H. Synth. Commun., **1989, 19**, 1463.

SECTION 124: ETHERS, EPOXIDES AND THIOETHERS FROM ALDEHYDES

ClCH₂CN ,<27°C

t-BuOK , t-BuOH

70%

cis   :   trans   = 69:31
(59%ee : 56% ee)

Baldoli, C.; Del Buttero, P.; Lincandro, E.; Maiorana, S.; Papagni, A. J. Chem. Soc., Chem. Commun., **1987**, 762.

K₂CO₃ , MeCN
H₂O

79% , E

Oikoua, K.; Borredon, M.E.; Delmas, M.; Gaset, A. Synth. Commun., **1987, 17**, 1593.

$$PhCHO \xrightarrow[\substack{Et_3SiH , Bu_4N^+ClO_4^- \\ \text{Pt electrodes}}]{\substack{e^- , CH_2Cl_2 , LiClO_4}} PhCH_2OBn$$

91%

Torii, S.; Takagishi, S.; Inokuchi, T.; Okumoto, H. *Bull. Chem. Soc., Jpn.,* **1987**, *60*, 775.

$$PhCHO \xrightarrow[\substack{KOH , THF (tr. H_2O) , RT \\ 17h}]{\substack{Ph_3As^+CH_2CH_2OH \ Br^-}}$$

E:Z = 89:11

83%

Shi, L.; Wang, W.; Huang, Y.-Z. *Tetrahedron Lett.,* **1988**, *29*, 5295.

$$Tol\text{-}CHO \xrightarrow[\substack{100^oC , 10 \ min}]{\substack{BrCH_2CO_2Me , PPh_3 , Zn^o}} Tol\text{-}CH=CHCO_2Me$$

86%

Shen, Y.; Xin, Y.; Zhao, J. *Tetrahedron Lett.,* **1988,** *29,* 6119.

98%

Sugimura, H.; Osumi, K. *Tetrahedron Lett.,* **1989,** *30,* 1571.

SECTION 125: ETHERS, EPOXIDES AND THIOETHERS FROM ALKYLS, METHYLENES AND ARYLS

NO ADDITIONAL EXAMPLES

SECTION 126: ETHERS, EPOXIDES AND THIOETHERS FROM AMIDES

Me—C(=S)—NH$_2$
$\xrightarrow[\text{2. iBuBr , Bu}_4\text{NBr ,}\\\text{PhH , aq. NaOH}]{\text{1. BuBr , CHCl}_3\text{ , reflux}\\\text{1h}}$
Bu—S—CH$_2$CH(Me)$_2$

88%

Takido, T.; Itabashi, K. *Synthesis, 1987*, 817.

SECTION 127: ETHERS, EPOXIDES AND THIOETHERS FROM AMINES

NO ADDITIONAL EXAMPLES

SECTION 128: ETHERS, EPOXIDES AND THIOETHERS FROM ESTERS

⬡—CO$_2$Et
$\xrightarrow[\text{BBr}_2\cdot\text{SMe}_2\text{ , CH}_2\text{Cl}_2\\\text{RT , 8h , ZnI}_2]{\text{Me}_2\text{CHCMe}_2\text{BHSPh}}$
⬡—CH$_2$—SPh

82%

Kim, S.; Kim, S.S. *Tetrahedron Lett., 1987, 28*, 1913.

Ph—C(=O)—SMe
$\xrightarrow[\text{PhCH}_2\text{Ph}]{\text{Pd(PCy}_3)_2\text{ , 140}^\circ\text{C , 9h}}$
Ph—S—Me

quant

Osakada, K.; Yamamoto, T.; Yamamoto, A. *Tetrahedron Lett., 1987, 28*, 6321.

$\xrightarrow[\text{2. 3 eq., Me}_2\text{Zn ,}\\\text{CH}_2\text{Cl}_2]{\text{1. 2 eq. BF}_3\cdot\text{OEt}_2}$

(8.1   :   1)   **82%**

Tomooka, K.; Matsuzawa, K.; Suzuki, K.; Tsuchihashi, G. *Tetrahedron Lett., 1987, 28*, 6339.

Ph—CH₂—O—C(=S)—SEt →[Pd(PPh₃)₄ , dppe , THF / 65°C , 7h]→ Ph—CH₂—SEt   **92%**

Lu, X.; Ni, Z. *Synthesis,* *1987*, 66.

(δ-valerolactone) →[⟋SiMe₃ , 0.05 TrSbCl₆ / OSiMe₂t-Bu / ⟋OEt , CH₂Cl₂ , / -78°C → RT]→ (tetrahydropyran with CO₂Et and allyl)   **87%**

Mukaiyama,T.; Homma, K.; Takenoshita, H. *Chem. Lett., 1988*, 1725.

Ph—C(=O)—O—CH₂CH₂CH₂—OSiMe₃ →[1. SbCl₅-Me₃SiCl , SnI₂ , -78°C / 2. Et₃SiH]→ (tetrahydrofuran)—Ph   **91%**

Homma, K.; Mukaiyama, T. *Chem. Lett., 1989*, 259.

MeO—⟨C₆H₄⟩—OAc →[2.2 eq. NaH , MeI / HMPA - THF / 25°C]→ MeO—⟨C₆H₄⟩—OMe   **98%**

Yamashita, A.; Toy, A. *Synth. Commun., 1989, 19*, 755.

(δ-valerolactone) →[1. LiN(TMS)₂ , THF / HMPA / 2. PhNTf₂ / 3. Bu₂CuLi]→ (dihydropyran with butyl)   **49%**

Tsuchima, K.; Araki, K.; Murai, A. *Chem. Lett., 1989*, 1313.

SECTION 129: ETHERS, EPOXIDES AND THIOETHERS FROM ETHERS, EPOXIDES AND THIOETHERS

Zn(BH$_4$)$_2$, Me$_3$SiCl

25°C , 2h , CH$_2$Cl$_2$
ether

87%

Katsuki, H.; Ushio, Y.; Yoshimura, N.; Ochi, M. *J. Org. Chem.*, *1987*, *52*, 2594.

solid KSCN (hydrated)
1 eq. H$_2$O

80%

Bouda, H.; Borredon, M.E.; Delmas, M.; Gaset, A. *Synth. Commun.*, *1987*, *17*, 943.

1. 2.2 BuLi , DMF , 0°C

2. MeI , LiCl , 0°C , 24h

3. TBAF , THF , 12h

Bures, E.J.; Keay, B.A. *Tetrahedron Lett.*, *1988*, *29*, 1247.

1. BuLi ,THF , °C

2. I⁀⁀⁀OBn

3. Me$_3$Sil , NaI , MeCN

4. H$_2$, 10% Pd/C

42%

Cox, P.; Mahon, M.F.; Molloy, K.C.; Lister, S.; Gallagher, T. *Tetrahedron Lett.*, *1988*, *29*, 1993.

PhI(PAc)$_2$, PhH

hv , I$_2$, 40°C

92%

Furuta, K.; Nagata, T.; Yamamoto, H. *Tetrahedron Lett.*, *1988*, *29*, 2215.

Miyoshi, N.; Hatayama, Y.; Ryu, I.; Kambe, N.; Murai, T.; Murai, S.; Sonoda, N. *Synthesis*, *1988*, 175.

Brown, D.S.; Ley, S.V. *Tetrahedron Lett.*, *1988*, *29*, 4869.

Simpkins, N.S. *Tetrahedron Lett.*, *1988*, *29*, 6787.

Gao, Y.; Sharpless, K.B. *J. Org. Chem.*, *1988*, *53*, 4114.

SECTION 130: ETHERS, EPOXIDES AND THIOETHERS FROM HALIDES AND SULFONATES

Vedejs, E.; Buchanan, R.A.; Conrad, P.C.; Meier, G.P.; Mullins, M.J.; Schaffhausen, J.G.; Schwartz, C.E. *J. Am. Chem. Soc.*, *1989*, *111*, 8421.

88%

α:β = 1:11

Suzuki, K.; Maeta, H.; Matsumoto, T. *Tetrahedron Lett., 1989, 30*, 4853.
        also with Cp$_2$ZrCl$_2$ •AgClO$_4$ (99% , 1.3:1 (α:β).
Matsumoto, T.; Maeta, H.; Suzuki, K.; Tsuchihashi, G. *Tetrahedron Lett., 1988, 29*,
3567, 3571.
Suzuki, K.; Maeta, H.; Suzuki, T.; Matsumoto, T. *Tetrahedron Lett., 1989, 30*, 6879.

Related Methods: Section 123 (Ethers from Alcohols).

SECTION 131: ETHERS, EPOXIDES AND THIOETHERS FROM HYDRIDES

59%

Shono, T.; Matsumura, Y.; Onomura, O.;Yamada, Y. *Synthesis, 1987*, 1099.

63%

Kim, S.; Park, J.H. *Chem. Lett., 1988*, 1323.

SECTION 132: ETHERS, EPOXIDES AND THIOETHERS FROM KETONES

Fukuzawa, S.; Fujinami, T.; Sakai, S. *J. Chem. Soc., Chem. Commun., 1987*, 919.

E:Z = 97:3

Morris, T.H.; Smith, E.H.; Walsh, R. *J. Chem. Soc., Chem. Commun., 1987*, 964.

Bouda, H.; Borredon, M.E.; Delmas, M.; Gaset, A. *Synth. Commun., 1987, 17,* 503.

Sassaman, M.B.; Kotian, K.D.; Prakash, G.K.S.; Olah, G.A. *J. Org. Chem., 1987, 52,* 4314.

DAMP = diethyl diazomethyl phosphonate

Buxton, S.R.; Holm, K.H.; Skattebøl, L. *Tetrahedron Lett., 1987, 28,* 2163.

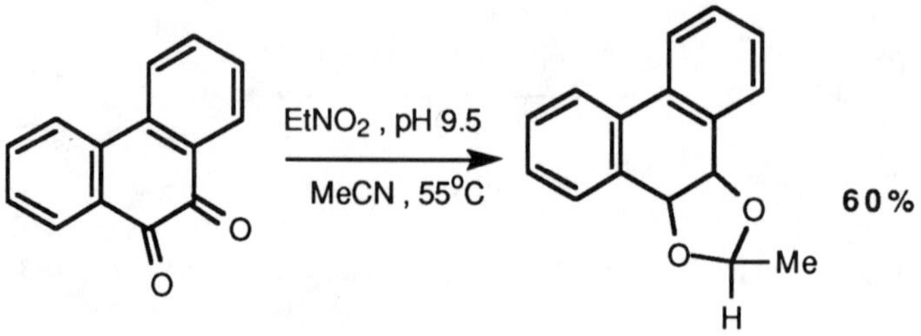

EtNO2 , pH 9.5
MeCN , 55°C
O
O
O
O
Me
H
60%

Itoh, S.; Nil, K.; Mure, M.; Ohshiro, Y. *Tetrahedron Lett.*, ***1987***, *28*, 3975.

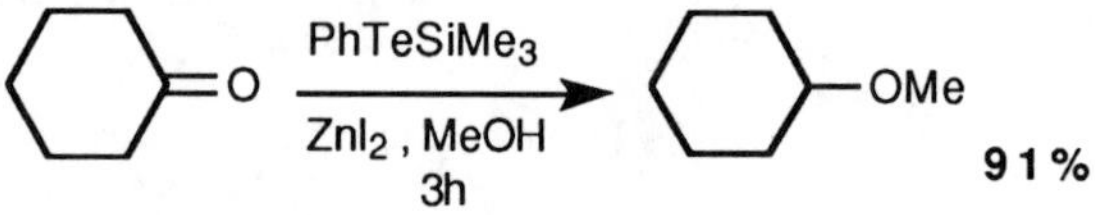

O
PhTeSiMe3
ZnI2 , MeOH
3h
OMe
91%

Nagakawa, K.; Osuka, M.; Sasaki, K.; Aso, Y.; Otsubo, T.; Ogura, F. *Chem. Lett.*, ***1987***, 1331.

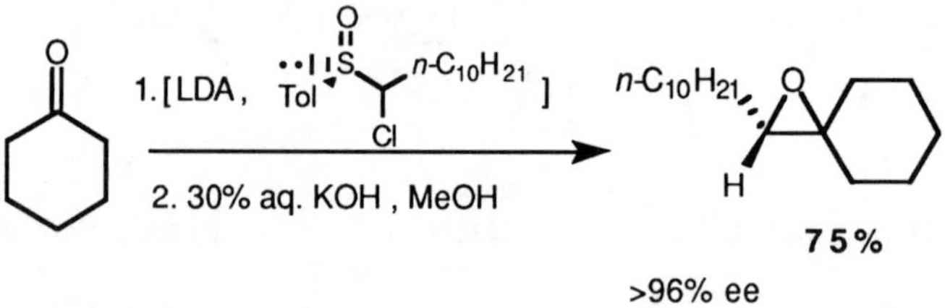

O
1. [LDA , Tol S n-C10H21 Cl]
2. 30% aq. KOH , MeOH
n-C10H21
O
H
75%
>96% ee

Satoh, T.; Oohara, T.; Yamakawa, K. *Tetrahedron Lett.*, ***1988***, *29*, 2851.

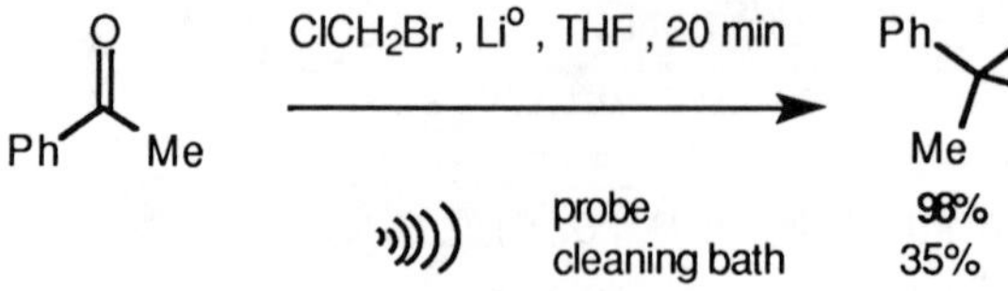

O
Ph
Me
ClCH2Br , Li° , THF , 20 min
probe
cleaning bath
Ph
O
Me
98%
35%

Einhorn, C.; Allavena, C.; Luche, J.-L. *J. Chem. Soc., Chem. Commun.*, ***1988***, 333.

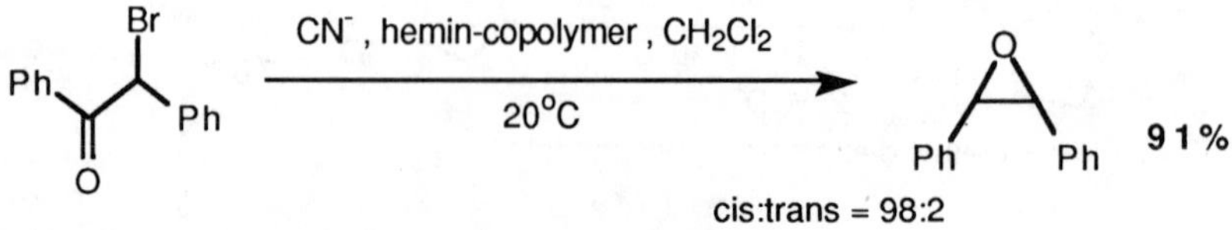

Br
Ph
Ph
O
CN⁻ , hemin-copolymer , CH2Cl2
20°C
O
Ph
Ph
91%
cis:trans = 98:2

Saito, K.; Harada, K. *Tetrahedron Lett.*, ***1989***, *30*, 1651.

Related Methods: Section 124 (Epoxides from Aldehydes).

SECTION 133: ETHERS, EPOXIDES AND THIOETHERS FROM NITRILES

NO ADDITIONAL EXAMPLES

SECTION 134: ETHERS, EPOXIDES AND THIOETHERS FROM ALKENES

PCMP = [π-C$_5$H$_5$N$^+$ (CH$_2$)$_{15}$CH$_3$]$_3$[PMo$_{12}$O$_{40}$]$^{3-}$ **79%**

Ishii, Y.; Yamawaki, K.; Yoshida, T.; Ura, T.; Ogawa, M. *J. Org. Chem.*, *1987*, *52*, 1868.

Brownbridge, P. *J. Chem. Soc., Chem. Commun.*, *1987*, 1280.

Saito, I.; Mano, T.; Nagata, R.; Matsuura, T. *Tetrahedron Lett.*, *1987*, *28*, 1901.

Reetz, M.T.; Seitz, T. *Angew. Chem. Int. Ed., Engl, 1987*, *26*, 1028.

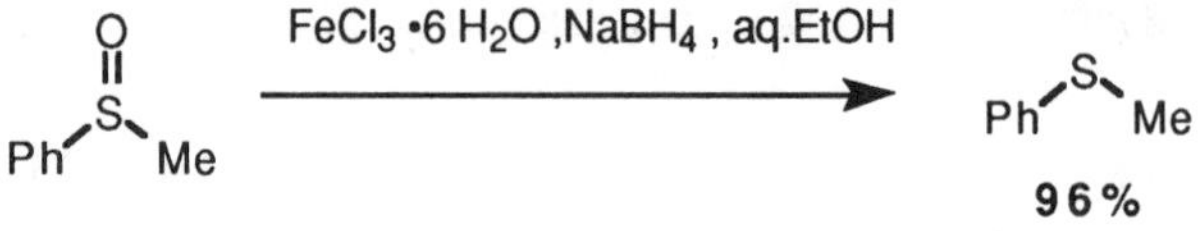

FeCl$_3$ •6 H$_2$O ,NaBH$_4$, aq.EtOH

96%

Lin, R.; Zhang, Y. *Synth. Commun.*, **1987**, *17*, 1403.

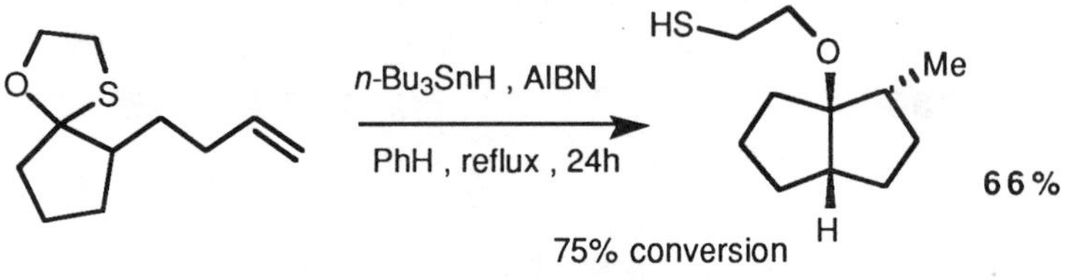

n-Bu$_3$SnH , AIBN

PhH , reflux , 24h

75% conversion

66%

Yadov, V.K.; Fallis, A.G. *Tetrahedron Lett.*, **1988**, *29*, 897.

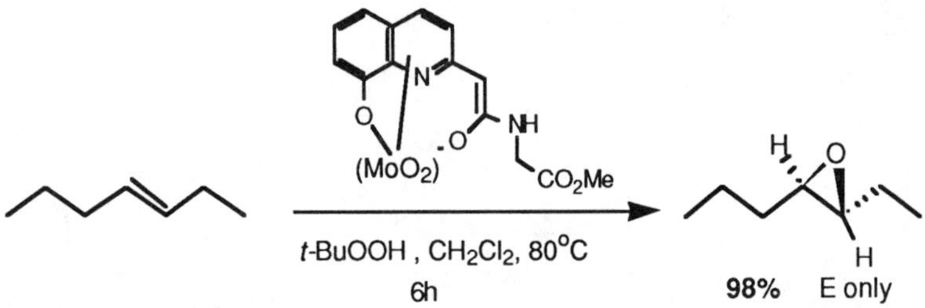

(MoO$_2$)

CO$_2$Me

t-BuOOH , CH$_2$Cl$_2$, 80°C
6h

98%   E only

Okamoto, Y.; Still, W.C. *Tetrahedron Lett.*, **1988**, *29*, 971.

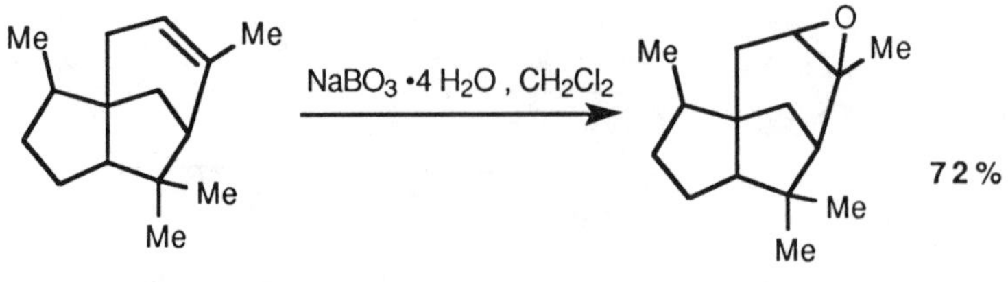

NaBO$_3$ •4 H$_2$O , CH$_2$Cl$_2$

72%

Xie, G.; Xu, L.; Hu, J.; Ma, S.; Hou, W.; Tao, F. *Tetrahedron Lett.*, **1988**, *29*, 2967.

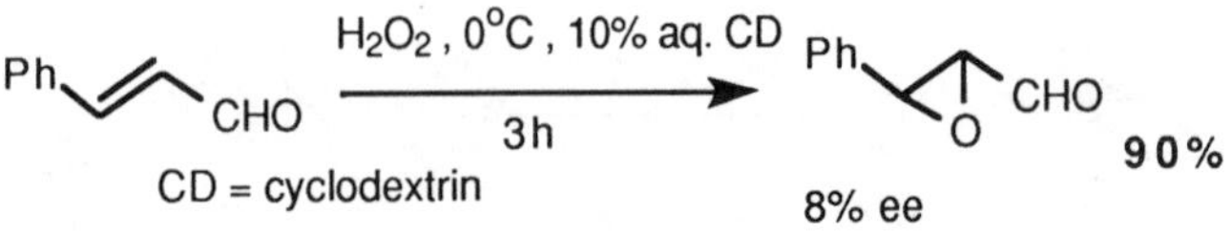

H$_2$O$_2$, 0°C , 10% aq. CD

3h

CD = cyclodextrin

8% ee

90%

Hu, Y.; Harada, A.; Takahashi, S. *Synth. Commun.*, **1988**, *18*, 1607.

71%

Danheiser, R.L.; Stoner, E.J.; Koyama, H.; Yamashita, D.S.; Klade, C.A. *J. Am. Chem. Soc.*, *1989*, *111*, 4407.

90%

Fringuelli, F.; Germani, R.; Pizzo, F.; Savelli, G. *Tetrahedron Lett.*, *1989*, *30*, 1427.

88% ee   56%

Takahashi, O.; Umezawa, J.; Furuhashi, K.; Takagi, M. *Tetrahedron Lett.*, *1989*, *30*, 1583.

99%

Adam, W.; Hadjiarapoglou, L.; Wang, X. *Tetrahedron Lett.*, *1989*, *30*, 6497.

88%

Reed, K.L.; Gupton, J.T.; Solarz, T.L. *Synth. Commun.*, *1989*, *19*, 3579.

(>99   :   1) 87%

Yamamoto, K.; Yamamoto, N. *Chem. Lett.,* **1989**, 1149.

REVIEW:

"Original Syntheses of Epoxides Involving Organoselenium Intermediates"

Krief, A.; Dumont, W.; Van Ende, D.; Halazy, S.; Labar, D.; Laboureur, J.-L.; Lê, T *Heterocycles,* **1989**, *28*, 1203.

SECTION 135: ETHERS, EPOXIDES AND THIOETHERS FROM MISCELLANEOUS COMPOUNDS

62%

anti:syn = 75:25

47%

>99% ee

Tsuboi, S.; Furutani, H.; Utaka, M.; Takeda, A. *Tetrahedron Lett.,* **1987**, *28*, 2709.

b86%

Singh, S.P.; Saxena, R.K. *Synth. Commun.,* **1987**, *17*, 575.

94%

Kim, K.S.; Jung, I.B.; Kim, Y.H.; Oae, S. *Tetrahedron Lett.,* **1989**, *30*, 1087.

Bn, Bn   AlCl₃ , NaI , RT , 6h
 \S/   ———————————————→   Bn\S/Bn
 ‖   **85%**
 O

Bhatt, M.; Babu, J.R. *Indian J. Chem.*, *1988*, *27B*, 259.

1. 3 eq. SnCl₄
2. Bu₃SnH

(5   :   1)   71%

Castañeda, A.; Kucera, D.J.; Overman, L.E. *J. Org. Chem.*, *1989*, *54*, 5691.

REVIEWS:

"Synthetic Routes to Tetrahydrofuran, Tetrahydropyran and Spiroketal Units of
Polyether Antibiotics and a Survey of Spiroketals of Other Natural Products"

Boivin, T.L. *Tetrahedron*, *1987*, *43*, 3309.

"α-Thioalkylation via Aldehydes and Thiols"

Massy, D.J.R. *Synthesis*, *1987*, 589.

"Reduction of Sulfoxides to Thioethers"

Madesclaire, M. *Tetrahedron*, *1988*, *44*, 6537.

SECTION 135A: PROTECTION OF ETHERS, EPOXIDES AND THIOETHERS

NO ADDITIONAL EXAMPLES

CHAPTER 10
PREPARATION OF HALIDES AND SULFONATES

SECTION 136: HALIDES AND SULFONATES FROM ALKYNES

NO ADDITIONAL EXAMPLES

SECTION 137: HALIDES AND SULFONATES FROM ACID DERIVATIVES

$$n\text{-}C_{17}H_{35}\overset{O}{\underset{}{-}}Cl \quad \xrightarrow[\text{2. } h\nu \text{ , } CCl_4 \text{ , RT (pyrex filter)}]{\text{1. } Ph_2C=N\text{-}OH} \quad n\text{-}C_{17}H_{35}\text{-}Cl$$

82%

Hasebe, M.; Tsuchiya, T. *Tetrahedron Lett.*, *1988*, *29*, 6287.

$$\text{COOH} \quad \xrightarrow[CCl_4]{PhI(OAc)_2 \text{ , 4h}} \quad \text{I}$$

85%

Singh, R.; Just, G. *Synth. Commun.*, *1988*, *18*, 1327.

$$\xrightarrow[150^\circ C]{\text{cat. } H_2SO_4}$$

60%

Basavaiah, D.; Dharma Rao, P.; Gowriswari, V.V.L. *Synth. Commun.*, *1988*, *18*, 1411.

COOH 1. Cl—C(=O)—C(=O)—Cl , DMF , RT , CH₂Cl₂

2. (pyridine-S-OH) , CF₃CH₂I
hν , cat. Py•NMe₂ , PhH

→ I   92%

Tsanaktsidis, J.; Eaton, P.E. *Tetrahedron Lett.*, **1989**, *30*, 6967.

SECTION 138: HALIDES AND SULFONATES FROM ALCOHOLS AND THIOLS

OH → 4% Aliquat 336®
48% HBr , 100°C
14h   → Br   99%

Dakka, G.; Sasson, Y. *Tetrahedron Lett.*, **1987**, *28*, 1223.

OTHP → Br₂ , diphos , CH₂Cl₂
23°C , 4h   → Br   96%

Schmidt, S.P.; Brooks, D.W. *Tetrahedron Lett.*, **1987**, *28*, 767.

n-C₈H₁₇OH
1. (iPr)—N=C=N—(Cy) , CuCl , 25°C
2. CF₃SO₃H , Bu₄NBr , CHCl₃
65°C , 7.5h
→ n-C₈H₁₇Br   83%

Collingwood, S.P.; Davies, A.P.; Golding, B.T. *Tetrahedron Lett.*, **1987**, *28*, 4445.

Me
(cyclopropyl)—CH(Ph)—OH
→ (iPr₂NH - KHF₂ - HF)ₙ•Py , CH₂Cl₂
→ Me / F / Ph / H   71%

trans:cis = 95:5

Kanemoto, S.; Shimizu, M.; Yoshioka, H. *Tetrahedron Lett.*, **1987**, *28*, 6313.

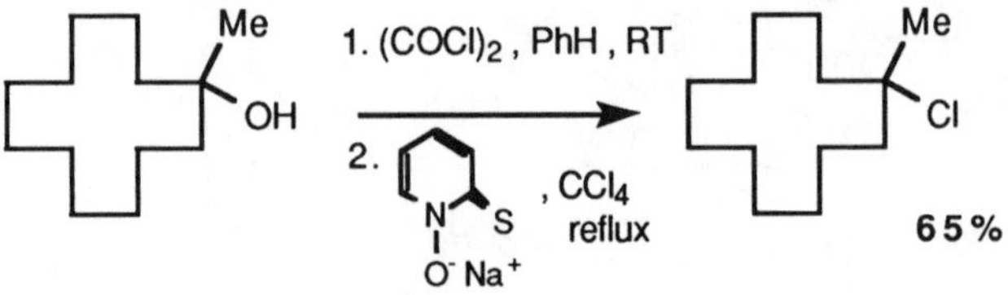

Crich, D.; Fortt, S.M. *Synthesis,* *1987*, 35.

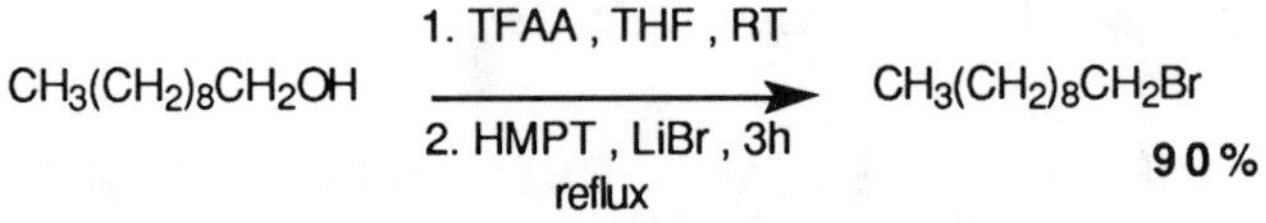

1. TFAA , THF , RT
2. HMPT , LiBr , 3h
reflux

$CH_3(CH_2)_8CH_2OH \longrightarrow CH_3(CH_2)_8CH_2Br$ 90%

Camps, F.; Gasol, V.; Guerrero, A. *Synthesis,* *1987*, 511.

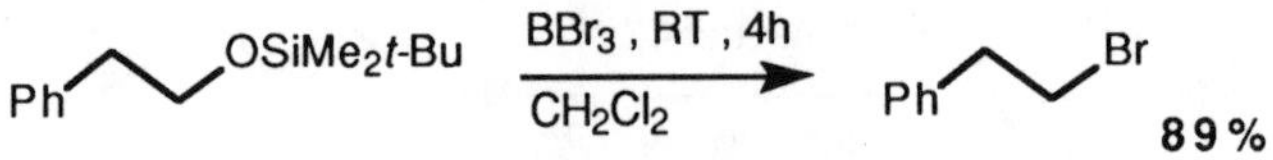

BBr$_3$, RT , 4h
CH_2Cl_2

89%

Kim, S.; Park, J.H. *J. Org. Chem.,* *1988*, *53*, 3111.

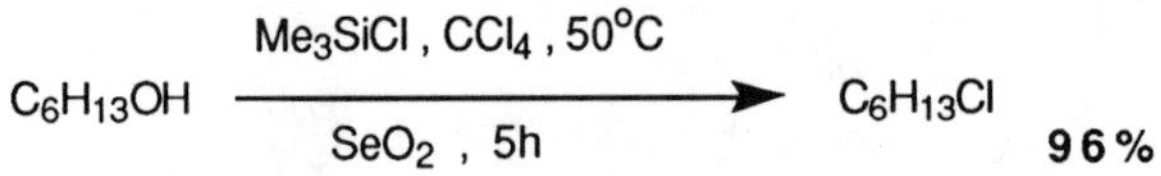

Me$_3$SiCl , CCl$_4$, 50°C
SeO$_2$, 5h

$C_6H_{13}OH \longrightarrow C_6H_{13}Cl$ 96%

Lee, J.G.; Kang, K.K. *J. Org. Chem.,* *1988*, *53*, 3634.

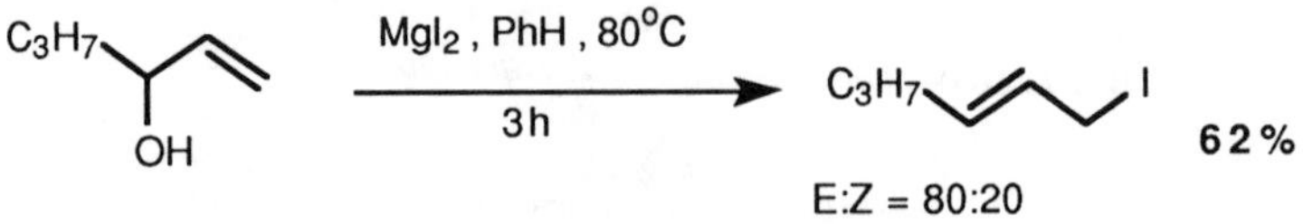

MgI$_2$, PhH , 80°C
3h

E:Z = 80:20   62%

Martinez, A.G.; Villalobos, A.C.; Ruiz, M.O. *Synthesis,* *1988*, 58.

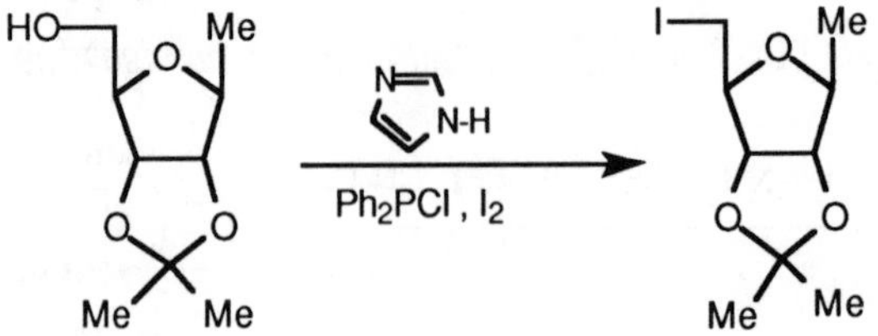

N=
N-H
Ph$_2$PCl , I$_2$

Classon, B.; Liu, Z. *J. Org. Chem.,* *1988*, *53*, 2126.

McAdam, D.P.; Stick, R.V. *Aust. J. Chem.*, *1988*, *41*, 1988.

Wagner, A.; Heitz-M.-P.; Mioskowski, C. *Tetrahedron Lett.*, *1989*, *30*, 557.

Munyemana, F.; Frisque-Hesbain, A.-M.; DeVos, A.; Ghosez, L. *Tetrahedron Lett.*, *1989*, *30*, 3077.

Ernst, B.; Winkler, T. *Tetrahedron Lett.*, *1989*, *30*, 3081.

82% (inversion)

Hanessian, S.; Kagotani, M.; Komaglou, K. *Heterocycles*, **1989**, *28*, 1115.

82%

Kishan Reddy, Ch.; Perisamy, M. *Tetrahedron Lett.*, **1989**, *30*, 5663.

89%

Amrollah-Madjdabadi, A.; Pham, T.N.; Ashby, E.C. *Synthesis*, **1989**, 614.

SECTION 139: HALIDES AND SULFONATES FROM ALDEHYDES

1. PhSH , H$^+$
2. HgF$_2$
3. Na^o , EtOH

38%

Purrington, S.T.; Pittman, J.H. *Tetrahedron Lett.*, **1988**, *29*, 6851.

PhCHO $\xrightarrow{\text{SOCl}_2 , \text{HMPA} , \text{RT}}$ PhCHCl$_2$ **92%**

Khurana, J.M.; Mehta, S. *Indian J. Chem.*, **1988**, *27B*, 1128.

PhCHO $\xrightarrow[\text{15 min , 0} \to 50^\circ\text{C}]{\text{Me}_3 \cdot \text{BH}_3 \text{, 2 eq Br}_2 \text{, CHCl}_3}$ PhCh$_2$Br

85%

also for ketones

LeCorre, M.; Gheerbrant, E.; LeDeit, H. *J. Chem. Soc., Chem. Commun., 1989*, 313.

SECTION 140: HALIDES AND SULFONATES FROM ALKYLS, METHYLENES AND ARYLS

For the conversion R-H → R-Halogen, see Section 146 (Halides from Hydrides).

69%   **26%**

Brower, K.R. *J. Org. Chem., 1987, 52*, 798.

70%

Rozen, S.; Gal, C. *J. Org. Chem., 1987, 52*, 2769.

87%

Srebnik, M.; Mechoulam, R.; Yona, I. *J. Chem. Soc., Perkin Trans. I, 1987*, 1423.

$$\text{PhOMe} \xrightarrow[\text{100}^\circ\text{C , 3h}]{\text{CuCl}_2 \text{ , Al}_2\text{O}_3 \text{ , PhCl}}$$

93%   +   3%

Kodomari. M.; Takahashi, S.; Yoshiltomi, S. *Chem. Lett., 1987*, 1901.

$$\xrightarrow[\text{CCl}_4 \text{ , reflux}]{\text{Br}_2 \text{ , HgO , H}_2\text{SO}_4 \text{ (conc)}}$$

Khan, S.A.; Munawar, M.A.; Siddiq. M. *J. Org. Chem., 1988, 53*, 1799.

$$\text{PhCH}_3 \xrightarrow[\substack{\text{EtOH , -78}^\circ\text{C ,} \\ \text{10 min}}]{[\text{Br}_2/\text{CFCl}_3/\text{F}_2]}$$

47%   +   47%

Rozen. S.; Brand, M.; Lidor, R. *J. Org. Chem., 1988, 53*, 5545.

$$\xrightarrow[\text{Pt electrodes}]{\text{CH(OMe)}_3 \text{ , I}_2 \text{ , LiClO}_4 \text{ , e}^-}$$

(30   :   70)   85%

Shono. T.; Matsumura, Y.; Katoh, S.; Ikeda, K.; Kamada, T. *Tetrahedron Lett., 1989, 30*, 1649.

Konishi, H.; Aritomi, K.; Okano, T.; Kiji, J. *Bull. Chem. Soc., Jpn., 1989, 62,* 591.

SECTION 141: HALIDES AND SULFONATES FROM AMIDES

NO ADDITIONAL EXAMPLES

SECTION 142: HALIDES AND SULFONATES FROM AMINES

Guziec Jr., F.S.; San Filippo, L.J. *Synthesis, 1988,* 547.

SECTION 143: HALIDES AND SULFONATES FROM ESTERS

Shimizu, I.; Ishii, H. *Chem. Lett., 1989,* 577.

SECTION 144: HALIDES AND SULFONATES FROM ETHERS, EPOXIDES AND THIOETHERS

NBS , BaCO3 , 3h

CCl₄

81%

Hashimoto, H.; Kawa, M.; Saito, Y.; Datel, T.; Horito, S.; Yoshimura, J. *Tetrahedron Lett.*, **1987**, **28**, 3505.

$$PhCH_2OBn \xrightarrow[\text{CDCl}_3 \text{ (NMR tube)}]{[PhSiH_2I / HI]} 2 \text{ eq } PhCH_2I$$

quant

Keinan, E.; Perez, D. *J. Org. Chem.*, **1987**, *52*, 4846.

$$C_{17}H_{35}CH_2OCH_3 \xrightarrow[\substack{\text{NaI , MeCN , 8h} \\ \text{reflux}}]{Cl_2P(=O)OPh} C_{17}H_{35}CH_2I$$

97%

Liu, H.-J.; Wisniewski, V. *Synth. Commun.*, **1988**, *18*, 119.

SECTION 145: HALIDES AND SULFONATES FROM HALIDES AND SULFONATES

1. Mg

2.

3.

67%

Yankep. E.; Charles, G. *Tetrahedron Lett.*, **1987**, *28*, 427.

1. KI , DMF , 150°C
Ni⁰ (powder) , 24h

2. 3% aq. HCl

86%

Yang, S.H.; Li, C.S.; Cheng, C.H. *J. Org. Chem.*, **1987**, *52*, 691.

$$C_{12}H_{25}OTs \xrightarrow[\text{THF , HMPT}]{Bu_4N^+ \ HF_2^- \ , \ 14h} C_{12}H_{25}F \quad 96\%$$

Bosch, P.; Camps, F.; Chamorro, E.; Gasol, V.; Guerrero, A. *Tetrahedron Lett., 1987, 28*, 4733.

$$\xrightarrow[\text{1 h}]{AgBF_4 \ , \ CH_2Cl_2 \ , \ RT} \quad 60\%$$

Bloodworth, A.J.; Bowyer, K.J.; Mitchell, J.C. *Tetrahedron Lett., 1987, 28*, 5347.

$$C_7H_{15}CH_2Cl \xrightarrow[\text{98}^\circ C \ , \ 6h]{LiBr \ , \ 5\% \ Aliquat-336} C_7H_{15}CH_2Br$$

88%

Loupy, A.; Pando, C. *Synth. Commun., 1988, 18*, 1275.

$$Me_3SiBr \xrightarrow[\text{-10}^\circ C \ , \ 3h]{CH_2N_2 \ , \ ZnBr \ , \ ether} Me_3SiCH_2Br \quad 60\%$$

Lee, J.G.; Ha, D.S. *Synthesis, 1988*, 318.

$$Ph\diagup\!\!\!\diagdown_{Br} \xrightarrow[\text{120}^\circ C]{KI \ , \ CuI \ , \ HMPT \ , \ 10h} Ph\diagup\!\!\!\diagdown_{I} \quad 85\%$$

Suzuki, H.; Aihara, M.; Yamamoto, H.; Takemoto, Y.; Ogawa, T. *Synthesis, 1988*, 236.

$$PhCCl_3 \xrightarrow[\substack{Zn \ anode \ , \ stainless \\ steel \ cathode}]{\substack{BnBr \ , \ RT \ , \ TMU \ , \ THF \\ TBAF \ , \ Bu_4NI \ , \ e^-}}$$

70%   10%

Nédélec, J.-Y.; Mouloud, H.A.H.; Folest,J.-C.; Périchon, J. *J. Org. Chem., 1988, 53*, 4720.
Nédélec, J.Y.; Ait-Haddou-Mauloud, H.; Folest, J.C.; Périchon, J. *Tetrahedron Lett., 1988, 29*, 1699.

$$Cl-\langle\text{ring}\rangle-NO_2 \xrightarrow[\text{DMSO , reflux , 5h}]{\text{*5\% freeze-dried KF}} F-\langle\text{ring}\rangle-NO_2$$

98%

* a freeze-dried 5% aq. solution of KF

Kimura, Y.; Suzuki, H. *Tetrahedron Lett.*, *1989*, *30*, 1271.

$$\langle\text{cyclohexyl}\rangle-Cl \xrightarrow[\text{CH}_2\text{Cl}_2 \text{ , 25°C}]{\text{HBr , FeBr}_3} \langle\text{cyclohexyl}\rangle-Br$$

99%

Yoon, K.B.; Kochi, J.K. *J. Org. Chem.*, *1989*, *54*, 3028.
Yoon, K.B.; Kochi, J.K. *J. Chem. Soc., Chem. Commun.*, *1987*, 1013.

SECTION 146: HALIDES AND SULFONATES FROM HYDRIDES

α-Halogenations of aldehydes, ketones and acids are found in Sections 338 (Halide-Aldehyde), 369 (Halide-Ketone), 359 (Halide-Esters) and 319 (Halide-Acids).

$$Ph\text{-}CH_3 \xrightarrow[\text{AIBN , 70°C , 4h}]{\text{BnMe}_3\text{N}^+ \text{ ICl}_4^- \text{ , CCl}_4} Ph\text{-}CH_2Cl + Ph\text{-}CHCl_2$$

77%   11%

Kajigaeshi, S.; Kakinami, T.; Moriwaki, M.; Tanaka, T.; Fujisaki, S. *Tetrahedron Lett.*, *1988*, *29*, 5783.

$$\langle\text{naphthalene}\rangle \xrightarrow[\text{130°C , 2h}]{\text{PhCl , CuCl}_2\text{-Al}_2\text{O}_3} \langle\text{1-chloronaphthalene}\rangle + \langle\text{1,4-dichloronaphthalene}\rangle$$

(94   :   6)   94%

Kodomari, M.; Satoh, H.; Yoshitami, S. *J. Org. Chem.*, *1988*, *53*, 2093.

SECTION 147: HALIDES AND SULFONATES FROM KETONES

Ph—CH₂—C(=O)—CH(CO₂Et)—CH₃

$$\xrightarrow[\text{20°C , 3h}]{\text{NCS , EtOH , NaOEt}}$$

Ph—CH₂—CHCl—CO₂Et   86%

Mignani, G.; Morel, D.; Grass, F. *Tetrahedron Lett.*, **1987**, *28*, 5505.

SECTION 148: HALIDES AND SULFONATES FROM NITRILES

NO ADDITIONAL EXAMPLES

SECTION 149: HALIDES AND SULFONATES FROM ALKENES

For halocyclopropanations, see Section 74E (Alkyls from Alkenes).

Ph—CH=CH—Ph

$$\xrightarrow[\text{HF , 20°C}]{\text{CsSO}_4\text{F , CH}_2\text{Cl}_2}$$

Ph—CHF—CHF—Ph

syn:anti = 65:35

Stavber, S.; Zupan, M. *J. Org. Chem.*, **1987**, *52*, 919.

$$\xrightarrow[\text{36°C , 2h}]{\text{I}_2/\text{Al}_2\text{O}_3 \text{ , pet ether}}$$

83%

Stewart, L.J.; Gray, D.; Pagni, R.M.; Kabalka, G.W. *Tetrahedron Lett.*, **1987**, *28*, 4497.

$Cl(CF_2)_2$-I

$$\xrightarrow[\text{cat. SmI}_2]{\text{, RT}}$$

$Cl(CF_2)_2$—CH₂—CHI—CH₂CH₂CH₃   55%

Lu, X.; Ma, S.; Zhu, J. *Tetrahedron Lett.*, **1988**, *29*, 5129.

$$\xrightarrow[\text{0.5 H}_2\text{O , RT , 1h}]{\text{Me}_3\text{SiCl , NaI , MeCN}}$$

96%

Irfune, S.; Kibayashi, T.; Ishii, Y.; Ogawa, M. *Synthesis*, **1988**, 366.

92%

Barluenga, J.; González, J.M.; Campos, P.J.; Asensio, G. *Angew. Chem. Int. Ed., Engl,* **1988**, *27*, 1546.

SECTION 150:   HALIDES AND SULFONATES FROM MISCELLANEOUS COMPOUNDS

99%

Moerlein, S.M. *J. Org. Chem.*, **1987**, *52*, 664.

b73%

Bay, E.; Timony, P.E.; Leone-Bay, A. *J. Org. Chem.*, **1988**, *53*, 2858.

REVIEW:

"Advances in the Synthesis of Iodoaromatic Compounds"

Merkushev, E.B. *Synthesis,* **1988**, 923.

CHAPTER 11
PREPARATION OF HYDRIDES

This chapter lists hydrogenolysis and related reactions by which functional groups are replaced by hydrogen: e.g. $RCH_2X \rightarrow RCH_2\text{-H}$ or R-H.

SECTION 151: HYDRIDES FROM ALKYNES

NO ADDITIONAL EXAMPLES

SECTION 152: HYDRIDES FROM ACID DERIVATIVES

This section lists examples of decarboxylations ($RCO_2H \rightarrow$ R-H) and related reactions.

NO ADDITIONAL EXAMPLES

SECTION 153: HYDRIDES FROM ALCOHOLS AND THIOLS

This section lists examples of the hydrogenolysis of alcohols and phenols (ROH $\rightarrow$ R-H).

Sakai, T.; Miyata, K.; Utaka, M.; Takeda, A. *Tetrahedron Lett.*, *1987*, *28*, 3817.

Krafft, M.A.; Crooks III, W.J. *J. Org. Chem.*, *1988*, *53*, 432.

$$Ph_3C\text{-}OH \xrightarrow[\text{RT}]{Me_2SiCl_2\,,\,NaI\,,\,MeCN} Ph_3C\text{-}H$$

quant

Wiggins, J.M. *Synth. Commun.*, **1988**, *18*, 741.

(>95   :   <5)   **92%**

Orfanopoulos, M.; Smoniu, I. *Synth. Commun.*, **1988**, *18*, 833.

73%

Kasuda, K.; Inanaga, J.; Yamaguchi, M. *Tetrahedron Lett.*, **1989**, *30*, 2945.

80%

Ho, K.M.; Lam, C.H.; Luh, T.-Y. *J. Org. Chem.*, **1989**, *54*, 4474.

Also via: Section 160 (Halides and Sulfonates).

SECTION 154: HYDRIDES FROM ALDEHYDES

For the conversion RCHO → R-Me, etc., see Section 64 (Alkyls from Aldehydes).

NO ADDITIONAL EXAMPLES

SECTION 155:   HYDRIDES FROM ALKYLS, METHYLENES AND ARYLS

quant

Rabideau, P.W.; Marcinow, Z. *Tetrahedron Lett.*, *1988*, *29*, 3761.

SECTION 156: HYDRIDES FROM AMIDES

NO ADDITIONAL EXAMPLES

SECTION 157: HYDRIDES FROM AMINES

This section lists examples of the conversion RNH_2 (or R_2NH) $\rightarrow$ R-H.

NO ADDITIONAL EXAMPLES

SECTION 158: HYDRIDES FROM ESTERS

This section lists examples of the reactions RCO_2R' $\rightarrow$ R-H and RCO_2R' $\rightarrow$ R'H.

88%

Nozaki, K.; Oshima, K.; Utimoto, K. *Tetrahedron Lett.*, *1988*, *29*, 6125.

58%

Greenspoon,.N.; Keinan, E. *J. Org. Chem.*, *1988*, *53*, 3723.

Pennanen, S.I. *Synth. Commun.*, **1988**, *18*, 1097.

Bianco, A.; Passacantilli, P.; Righi, G. *Tetrahedron Lett.*, **1989**, *30*, 1405.

Mandai, T.; Imaj, M.; Takada, H.; Kawata, M.; Nokami, J.; Tsuji, J. *J. Org. Chem.*, **1989**, *54*, 5395.

SECTION 159: HYDRIDES FROM ETHERS, EPOXIDES AND THIOETHERS

This section lists examples of the reaction R-O-R' → R-H.

Holton, R.A.; Crouse, D.J.; Williams, A.D.; Kennedy, R.M. *J. Org. Chem.*, **1987**, *52*, 2317.

Schultz-van Itter, N.; Steckhan, E. *Tetrahedron*, **1987**, *43*, 2475.

Chan, M.-C.; Cheng, K.-M.; Ho, K.M.; Ng, C.T.; Yam, T.M.; Wang, B.S.L.; Luh, T.-Y. *J. Org. Chem.*, **1988**, *53*, 4466.

Comasseto, J.V.; Ferraz, H.M.C.; Brandt, C.A.; Gaeta, K.K. *Tetrahedron Lett.*, **1989**, *30*, 1209.

SECTION 160: HYDRIDES FROM HALIDES AND SULFONATES

This section lists the reductions of halides and sulfonates, R-X → R-H.

cat. = polyethylene bound crown ether

Bergbreiter, D.E.; Blanton, J.R. *J. Org. Chem.*, **1987**, *52*, 472.

5% Mo(CO)$_6$, PhH
2 eq. PhSiH$_3$, NaHCO$_3$
23% PPh$_3$, reflux , 1h

77%

Perez, D.; Greenspoon, N.; Keinan, E. *J. Org. Chem.*, *1987*, *52*, 5570.

Zn^o , TMEDA , 5 eq. AcOH
EtOH , RT

84%

Danheiser, R.L.; Savariar, S. *Tetrahedron Lett.*, *1987*, *28*, 3299.

hv (350 nm) , NaBH$_4$
MeCN , H$_2$O

96%

Kropp, M.; Schuster, G.B. *Tetrahedron Lett.*, *1987*, *28*, 5295.

HOCH$_2$SO$_2$Na•H$_2$O , EtOH
reflux , 30 min

83%

Harris, A.R. *Synth. Commun.*, *1987*, *17*, 1587.

PPh$_2$H/t-BuOK
DMSO , 4h

84%

Meijs, G.F. *J. Org. Chem.*, *1987*, *52*, 3923.

NaHS , SnCl$_2$
aq. THF , 2h
reflux

90%

Ono, A.; Murayama, T.; Kamimura, J. *Synthesis*, *1987*, 1093.
Ono, A.; Kamimura, J.; Suzuki, N. *Synthesis*, *1987*, 406.

$$C_{10}H_{21}I \xrightarrow[\text{RT , 5 min}]{\text{SmI}_2 \text{ , iPrOH , THF-HMPA}} C_{10}H_{21}\text{-H} \quad >95\%$$

Inanaga, J.; Ishikawa, M.; Yamaguchi, M. *Chem. Lett., 1987*, 1485.

Harada, T.; Hara, D.; Hattori, K.; Oku, A. *Tetrahedron Lett., 1988, 29*, 3821.

Chatgilialoglu, C.; Griller, D.; Lesage, M. *J. Org. Chem., 1988, 53*, 3641.

Chung, C.K.; Ho, M.S.; Lun, K.S.; Wong, M.O.; Wong, H.N.C.; Tam, S.W. *Synth. Commun., 1988, 18*, 507.

Saiganesh, R.; Balasubramanian, K.K.; Venkatachalam, C.S. *Tetrahedron Lett., 1989, 30*, 1711.

$$C_{16}H_{33}Br \xrightarrow[\left(MeO-\bigcirc-\overset{O}{\overset{\|}{C}}O\right)_2 \text{ , monoglyme}]{NaBH_4 \text{ , } 10\% \text{ } (Me_3Si)_3SiH \text{ , } h\nu \text{ } (254 \text{ nm})} C_{16}H_{33}H$$

85%

Lesage, M.; Chatgilialoglu, C.; Giller, D. *Tetrahedron Lett.*, *1989*, *30*, 2733.

6 eq. Bu₃SnH
PhMe , AIBN

89%

Smith III, A.B.; Hale, K.J.; McCauley Jr., J.P. *Tetrahedron Lett.*, *1989*, *30*, 5579.

Et₃SiH , TBHN
t-dodecanethiol

96%

TBNH = di-*t*-butyl hyponitrite

Allen, R.P.; Roberts, B.P.; Willis, C.R. *J. Chem. Soc.*, *Chem. Commun.*, *1989*, 11387.

Mg° , MeOH , RT

$CH_3(CH_2)_8CH_3$

62%

Hutchins, R.O.; Suchismita; Zipkin, R.E.; Taffer, I.M. *Synth. Commun.*, *1989*, *19*, 1519.

CTAB (micelle) , Zn°
H₂O , RT

78%

Juršić, B.; Galoši, A. *Synth. Commun.*, *1989*, *19*, 1649.

Bu₃SnH - BEt₃ , PhMe
-78°C , 30m in

98%

Miura, K.; Ichinose, Y.; Nozaki, K.; Fugami, K.; Oshima, K.; Utimoto, K. *Bull. Chem. Soc., Jpn.*, *1989*, *62*, 143.

REVIEW:

"Tri-*n*-Butyltin Hydide as Reagent in Organic Synthesis"

Neumann, W.P. *Synthesis, 1987*, 665

SECTION 161: HYDRIDES FROM HYDRIDES

NO ADDITIONAL EXAMPLES

SECTION 162: HYDRIDES FROM KETONES

This section lists examples of the reaction R_2C-$(C=O)R \rightarrow R_2C$-H.

Ono, A.; Suzuki, N.; Kamimura, J. *Synthesis, 1987*, 736.

Ram, S.; Spicer, L.D. *Tetrahedron Lett., 1988, 29*, 3741.

Nishiyama, Y.; Hamanaka, S.; Ogawa, A.; Kambe, N.; Sonoda, N. *J. Org. Chem., 1988, 53*, 1326.

Cormier, R.A.; McCauley, M.D. *Synth. Commun., 1988, 18*, 675.

Haller-Bauer cleavage

82% retention

66%

Paquette, L.A.; Gilday, J.P. *J. Org. Chem.*, *1988*, *53*, 4972.

Lau, C.K.; Tardif, S.; Dufresne, C.; Scheigetz, J. *J. Org. Chem.*, *1989*, *54*, 491.

SECTION 163: HYDRIDES FROM NITRILES

This section lists examples of the reaction, R-C≡N → R-H (includes reactions of isonitriles (R-N≡C).

Ohsawa, T.; Mitsuda, N.; Nezu, J.; Oishi, T. *Tetrahedron Lett.*, *1989*, *30*, 845.

SECTION 164: HYDRIDES FROM ALKENES

NO ADDITIONAL EXAMPLES

SECTION 165:   HYDRIDES FROM MISCELLANEOUS COMPOUNDS

Ono, N.; Hashimoto, T.; Jun, T.X.; Kaji, A. *Tetrahedron Lett., 1987, 28,* 2277.

Kamimura, A.; Kurata, K.; Ono, N. *Tetrahedron Lett., 1989, 30,* 4819.

REVIEW:

"Synthetically Useful Extrusion Reactions of Organic Sulfur, Selenium and Tellurium Compounds"

Guziec Jr., F.S.; San Filippo, L.J. *Tetrahedron, 1988, 44,* 6241.

CHAPTER 12
PREPARATION OF KETONES

SECTION 166: KETONES FROM ALKYNES

Butler, I.R.; Cullen, W.R.; Lindsell, W.E.; Preston, P.N.; Rettig, S.J. *J. Chem. Soc., Chem. Commun.*, **1987**, 439.

Imi, K.; Imai, K.; Utimoto, K. *Tetrahedron Lett.*, **1987**, *28*, 3127.

Torii, S.; Inokuchi, T.; Hirata, Y. *Synthesis*, **1987**, 377.

73%

(trans:cis = 9:1)

Herndon, J.W.; Turner, S.U.; Schnatter, W.F.K. *J. Am. Chem. Soc., 1988, 110*, 3334.

salen = N,N'-ethylene *bis*-salicylidene iminato

Rihter, B.; Masnovi, J. *J. Chem. Soc., Chem. Commun., 1988*, 35.

Zweifel, G.; Shoup, T.M. *Synthesis, 1988*, 130.

Zibuck, R.; Seebach, D. *Helv. Chim. Acta, 1988, 71*, 237.

Smit, W.A.; Simonyan, S.O.; Tarasov, V.A.; Mikaelian, G.S.; Gybin, A.S.; Ibragimov, I.T.; Caple, R.; Froen, D.; Kreager, A. *Synthesis, 1989*, 472.

$$C_8H_{17}-C{\equiv}C-I-Ph \xrightarrow[\text{30 min}]{\text{NaOAc , THF , H}_2\text{O , RT}}$$

72%

Ochiai, M.; Kunishima, M.; Fuji, K.; Nagao, Y. *J. Org. Chem.*, *1989*, *54*, 4038.

SECTION 167: KETONES FROM ACID DERIVATIVES

$$\underset{Ph}{\overset{O}{\big\|}}Cl \xrightarrow[\text{THF , 65}^\circ\text{C}]{\text{Bu}_4\text{Pb , 1\% Pd(PPh}_3)4} \underset{Ph}{\overset{O}{\big\|}}Bu$$

99%

Yamada, J.; Yamamoto, Y. *J. Chem. Soc., Chem. Commun.*, *1987*, 1302.

$$\text{MeO-} \xrightarrow[\text{2. aq. HCl}]{\substack{\text{1. PhMgBr , Fe(acac)}_3 \\ \text{THF , RT}}} \text{MeO-}$$

80%

Cardellicchio, C.; Fiandanese, V.; Marchese, G.; Ronzini, L. *Tetrahedron Lett.*, *1987*, *28*, 2053.

NEt₃ , PhMe , reflux

43%

Snider, B.B.; Ron, E.; Burbaum, B.W. *J. Org. Chem.*, *1987*, *52*, 5413.
Kulkarni, Y.S.; Niwa, M.; Ron, E.; Snider, B.B. *J. Org. Chem.*, *1987*, *52*, 1568.

Ph , Pd(OAc)₂ , 130°C

PhCH₂NMe₂ , 4h

46%

Hori, K.; Ando, M.; Takaishi, N. *Tetrahedron Lett.*, *1987*, *28*, 5883.

BzCl , HMPA , 40°C

2% Pd[P(OTol)₃]Cl₂
1h

76%

Tamaru, Y.; Ochiai, H.; Nakamura, T.; Yoshida, Z. *Angew. Chem. Int. Ed., Engl*, **1987**, *26*, 1157.

SnMe₃   BzCl , 5% Pd(PPh₃)₄ , 100°C

1 atm CO , PhMe , 3h

75%

Renaldo, A.F. *Synth. Commun.*, **1987**, *17*, 1823.

N-Tos   1. PhLi , THF

2. H₂O

78%

Clerici, F.; Gelmi, M.L.; Rossi, L.M. *Synthesis*, **1987**, 1025.

1.
CHCl₃   , 0°C

2. NEt₃ , CHCl₃ , RT

COOH   62%

DeMesmaeker, A.; Veenstra, S.J.; Ernst, B. *Tetrahedron Lett.*, **1988**, *29*, 459.

1. AlCl₃ , CH₂Cl₂
-80°C , 15 min

2. CH₂Cl₂,
-80 - -20°C,
16h
3. H₂O

major   80%

Tubul, A.; Santelli, M. *J. Chem. Soc., Chem. Commun.*, **1988**, 191.

Fiandanesse, V.; Marchese, G.; Naso, F. *Tetrahedron Lett.*, *1988*, *29*, 3587.

Larock, R.C.; Lu, Y. *Tetrahedron Lett.*, *1988*, *29*, 6761.

Armesto, D.; Horspool, W.M.; Ortiz, M.J.; Perez-Ossorio, R. *Synthesis*, *1988*, 799.

Knochel, P.; Yeh, M.C.P.; Berk, S.C.; Talbert, J. *J. Org. Chem.*, *1988*, *53*, 2390.

Karaman, R.; Fry, J.L. *Tetrahedron Lett.*, *1989*, *30*, 6267.

Cahiez, G.; Laboue, B. *Tetrahedron Lett.*, *1989*, *30*, 7369.

$$C_8H_{17}-\overset{O}{\underset{}{C}}-Cl \quad \xrightarrow[\text{2. } H_3O^+]{\begin{array}{c}\text{1. 4 eq. } SmI_2\text{ , THF}\\ \text{18h}\end{array}} \quad C_8H_{17}-\overset{O}{\underset{}{C}}-C_8H_{17}$$

60%

Collin, J.; Dallemer, F.; Namy, J.L.; Kagan, H.B. *Tetrahedron Lett.*, *1989*, *30*, 7407.

AlCl₃ , CH₂Cl₂ , 3h
-15°C

61%

Bandodakar, B.S.; Nagendrappa, G. *Tetrahedron Lett.*, *1989*, *30*, 7461.

$$Ph_3C\text{-}COOH \quad \xrightarrow[\text{CH}_2\text{Cl}_2]{\begin{array}{c}Bu_4N\ HSO_4\text{ , 24h}\\ H_2O\text{ , NaOCl , pH 8-9}\end{array}} \quad Ph_3C\text{-}OH \quad + \quad \underset{Ph}{\overset{O}{\underset{}{C}}}\ Ph$$

6%   84%

Elmore, P.R.; Reed, R.T.; Terkle-Huslig, T.; Welch, J.S.; Young, S.M.; Landolt, R.G. *J. Org. Chem.*, *1989*, *54*, 970.

DCC , NEt₃
ether ,0°C

70%

Olah, G.A.; Wu, A.-h.; Farooq, O. *Synthesis*, *1989*, 568.

SECTION 168: KETONES FROM ALCOHOLS AND THIOLS

PCMP , 80°C , 24h

PhH

99%

$PCMP = [\pi\text{-}C_5H_5N^+ (CH_2)_{15}CH_3]_3[PMo_{12}O_{40}]^{3-}$

Ishii, Y.; Yamawaki, K.; Yoshida, T.; Ura, T.; Ogawa, M. *J. Org. Chem.*, *1987*, *52*, 1868.

88%

Morris Jr., P.E.; Kiely, D.E. *J. Org. Chem.*, *1987*, 52, 1149.

82%

Carlsen, P.H.J.; Bræden, J.E. *Acta Chem. Scand. B*, *1987*, B41, 313.

79%

Morimoto, T.; Hirano, M.; Ashiya, H.; Egashira, H. *Bull. Chem. Soc., Jpn.*, *1987*, 60, 4143.

86%

Taber, D.F.; Amedio Jr., J.C.; Jung, K.-C. *J. Org. Chem.*, *1987*, 52, 5621.

83%

Afonso, C.M.; Barros, M.T.; Maycock, C.D. *J. Chem. Soc., Perkin Trans. I*, *1987*, 1221.

Moiseenkov, A.M.; Cheskis, B.A.; Veselovskii, A.B.; Romanovich, A.,Ya.; Chizhov, O.S. *J. Org. Chem., U.S.S.R., 1987, 23,* 1646.

also converts 1° alcohols to aldehydes

Bortolini, O.; Campestrini, S.; DiFuria, F.; Modena, G.; Valle, G. *J. Org. Chem., 1987, 52,* 5467.

Lin, Y.; Ma, D.; Lu, X. *Tetrahedron Lett., 1987, 28,* 3115.

Linderman, R.J.; Graves, D.M. *Tetrahedron Lett., 1987, 28,* 4259.

Schobert, R. *Synthesis, 1987,* 741.

Kende, A.S.; Koch, K.; Smith, C.A. *J. Am. Chem. Soc.*, *1988*, *110*, 2210.

Liu, H.-J.; Nyangulu, J.M. *Tetrahedron Lett.*, *1988*, *29*, 3167.

Kanemoto, S.; Matsubara, S.; Takai, K.; Oshima, K.; Utimoto, K.; Nozaki, H. *Bull. Chem. Soc., Jpn.*, *1988*, *61*, 3607.

Morimoto, T.; Hirano, M.; Aikawa, Y.; Zang, X. *J. Chem. Soc., Perkin Trans. I*, *1988*, 2423.

Barak, G.; Dakka, J.; Sasson, Y. *J. Org. Chem.*, *1988*, *53*, 3553.

Ph—C(=O)—C(=O)—Ph →[SmI$_2$, quinidine , RT / PhH] Ph—CH(OH)—C(=O)—Ph

41%

40% ee , R

Takeuchi, S.; Ohgo, Y. *Chem. Lett., 1988*, 403.

(Me)(Me)C(OH)—C(OH)(Me)(Me) →[montmorillonite , Cu^{+2}] Me—C(=O)—C(Me)$_3$

100°C , 15h   30%

microwave (450W) , 15 min   **94%**

Gutierrez, E.; Loupy, A.; Bram, G.; Ruiz-Hitzky, E. *Tetrahedron Lett., 1989, 30*, 945.

Ph—CH(OH)—CH$_2$—CH$_2$—OH →[K$_2$MnO$_4$ •Al$_2$O$_3$• CuSO$_4$ •5 H$_2$O / BnEt$_3$N•HCl , 6h / 6% aq. NaOH , 25°C] Ph—C(=O)—CH$_2$—CH$_2$—OH

80%

Kim, K.S.; Chung, S.; Cho, I.H.; Hahn, C.S. *Tetrahedron Lett., 1989, 30*, 2559.

Me—CH(OMs)—C(OH)(C$_8$H$_{17}$)(cyclopropyl) →[Me$_3$Al , CH$_2$Cl$_2$, -70 → 0°C] cyclopropyl—C(=O)—CH(Me)—C$_8$H$_{17}$

81%

Shimazaki, M.; Hara, H.; Suzuki, K. *Tetrahedron Lett., 1989, 30*, 5443, 5447.

Ph—CH(OH)—C(=O)—Ph →[BaMnO$_4$, MeCN , reflux / 15 min] Ph—C(=O)—C(=O)—Ph

95%

Firouzabadi, H.; Mottaghinejad, E.; Seddighi, M. *Synthesis, 1989*, 378.

Singh, M.; Misra, R.A. *Synthesis*, **1989**, 403.

Kawada, K.; Gross, R.S.; Watt, D.S. *Synth. Commun.*, **1989**, *19*, 777.

Genet, J.P.; Pons, D.; Jugé, S. *Synth. Commun.*, **1989**, *19*, 1721.

Nishiguchi, T.; Asano, F. *J. Org. Chem.*, **1989**, *54*, 1531.
Nishiguchi, T.; Asano, F. *Tetrahedron Lett.*, **1988**, *29*, 6265.

Anelli, P.L.; Banfi, S.; Montanari, F.; Quici, S. *J. Org. Chem.*, **1989**, *54*, 2970.

REVIEW:

"Recent Applications of Oxochromium-Amine Complexes as Oxidants in Organic Synthesis. A Review"

Luzzio, F.A.; Guziec Jr., F.S. *Org. Prep. Proceed. Int.,* **1988,** *20,* 533.

Related Methods: Section 48 (Aldehydes from Alcohols and Phenols).

SECTION 169: KETONES FROM ALDEHYDES

Barrios, H.; Sandoval, C.; Ortíz, B.; Sánchez-Obregón, R.; Yuste, F. *Org. Prep. Proceed. Int.,* **1987,** *19,* 427.

Kondo, T.; Tsuji, Y.; Watanabe, Y. *Tetrahedron Lett.,* **1987,** *28,* 6229.

Tanino, K.; Katoh, T.; Kuwajima, I. *Tetrahedron Lett.,* **1988,** *29,* 1815, 1819.

Aoyama, T.; Shioiri, T. *Synthesis,* **1988,** 228.

$(MesS)_2B^--CHC_7H_{15}$

1. EtCHO , THF , -127°C

2. TFAA , THF , -78°C

C_8H_{17} —C(=O)— Et   72%

Pelter, A.; Smith, K.; Elgendy, S.; Rowlands, M. *Tetrahedron Lett.*, *1989*, *30*, 5643.

73% ee , S

78%

Taura, Y.; Tanaka, M.; Funakoshi, K.; Sakai, K. *Tetrahedron Lett.*, *1989*, *30*, 6349.

PhOH , PhMe , 50°C

Amberlyst-15

58%

Bunce, R.A.; Reeves, H.D. *Synth. Commun.*, *1989*, *19*, 1109.

SECTION 170: KETONES FROM ALKYLS, METHYLENES AND ARYLS

This section lists examples of the reaction, R-CH₂-R' → R(C=O)-R'.

NO ADDITIONAL EXAMPLES

SECTION 171: KETONES FROM AMIDES

Li , THF , -20°C

·))))

60%

Einhorn, J.; Einhorn, C.; Luche, J.-L. *Tetrahedron Lett.*, *1988*, *29*, 2183.

1. BuLi , ether , RT , 4h

2. dil aq. HCl

76%

Tominaga, Y.; Kohra, S.; Hosomi, A. *Tetrahedron Lett.*, **1987**, *28*, 1529.

BuLa(OTf)$_2$, ether , -30°C

30 min

94%

Collins, S.; Hong, Y. *Tetrahedron Lett.*, **1987**, *28*, 4391.

1. PhLi , -78°C

2. BuLi , -78°C - RT

60%

Hlasta, D.J.; Court, J.J. *Tetrahedron Lett.*, **1989**, *30*, 1773.

SECTION 172: KETONES FROM AMINES

1. PhCOCl , NaOH

2. PBr$_3$, Br$_2$

3. MeS-CH$_2$S(=O)Me , KH

4. H$_3$O$^+$

30%

>98% ee

Thurkauf, A.; Hillery, P.; Jacobson, A.E.; Rice, K.C. *J. Org. Chem.*, **1987**, *52*, 5466.

3 eq N$_2$O$_4$, THF , -40°C

CCl$_4$, 10 min

80%

Shim, S.B.; Kim, K.; Kim, Y.H. *Tetrahedron Lett.*, **1987**, *28*, 645.

NHSiMe$_3$

1. BuLi , -78°C

2. dry air , -40°C , 15 min

3. SiO$_2$, H$_2$O

84%

Chen, H.C.; Knochel, P. *Tetrahedron Lett.*, **1988**, *29*, 6701.

Ranu, B.C.; Sarkar, D.C. *J. Org. Chem.*, *1988*, *53*, 878.

Barluenga, J.; Aguilar, E.; Olano, B.; Fustero, S. *J. Org. Chem.*, *1988*, *53*, 1741.

TONSIL = Mexican bentonite clay

Cano, A.C.; Delgado, F.; Córdoba, A.A.; Márquez, C.; Alvarez, C. *Synth. Commun.*, *1988*, *18*, 2051.

Chidambaram, N.; Satyanarayana, K.; Chandrasekaran, S. *Synth. Commun.*, *1989*, *19*, 1727.

Firouzabadi, H.; Mottaghineiad, E.; Seddighi, M. *Synth. Commun.*, *1989*, *19*, 3469.

Balicki, R.; Kaczmarek, L.; Malinowski, M. *Liebigs Ann. Chem.*, *1989*, 1139.

SECTION 173: KETONES FROM ESTERS

Pogrebnoi, S.I.; Kalyan, Y.B.; Krimer, M.Z.; Smit, W.A. *Tetrahedron Lett., 1987, 28,* 4893.

Tsuji, J.; Yamada, T.; Minami, I.; Yuhara, M.; Nisar, M.; Shimizu, I. *J. Org. Chem., 1987, 52,* 2988.

Tajima, K. *Chem. Lett., 1987,* 1319.

Takai, K.; Kataoka, Y.; Okazoe, T.; Utimoto, K. *Tetrahedron Lett., 1988, 29,* 1065.

Sato, T.; Matsuoka, H.; Igarashi, T.; Minomura, M.; Murayama, E. *J. Org. Chem., 1988, 53,* 1207.

SECTION 174: KETONES FROM ETHERS, EPOXIDES AND THIOETHERS

NO ADDITIONAL EXAMPLES

SECTION 175: KETONES FROM HALIDES AND SULFONATES

Kosugi, M.; Sumiya, T.; Obara, Y.; Suzuki, M.; Sano, H.; Migita, T. *Bull. Chem. Soc., Jpn.,* **1987**, *60*, 767.

Ichikawa, J.; Sonoda, T.; Kobayashi, H. *Tetrahedron Lett.,* **1989**, *30*, 5437.

DMI = N,N-dimethyl-2-imidazolidinone

Hatanaka, Y.; Hiyama, T. *Chem. Lett.,* **1989**, 2049.

Yamashita, M.; Uchida, M.; Takashika, H.; Suemitsu, R. *Bull. Chem. Soc., Jpn.,* **1989**, *62*, 2728.

Related Methods: Section 177 (Ketones from Ketones).
                     Section 55 (Aldehydes from Halides).

SECTION 176: KETONES FROM HYDRIDES

This section lists examples of the replacement of hydrogen by ketonic groups, $R\text{-}H \rightarrow R(C=O)\text{-}R'$. For the oxidation of methylenes, $R_2CH_2 \rightarrow R_2C=O$, see section 170 (Ketones from Alkyls).

CrO_3, CH_2Cl_2

t-BuOOH, 7h

60%

Muzart, J. Tetrahedron Lett., **1987**, *28*, 2131.
Muzart, J. Tetrahedron Lett., **1987**, *28*, 4665.

1. $KMnO_4$, KOH
   CH_2Cl_2, RT
   aq. $Bu_4N\ HSO_4$

2. KOAc

80%

Gannon, S.M.; Krause, J.G. Synthesis, **1987**, 915.

PDC, t-BuOOH

PhH, 25°C,
6.5h

78%

82% conversion

Chidambaram, N.; Chandrasekaran, S. J. Org. Chem., **1987**, *52*, 5048.

PDC, t-BuOOH

25°C

74%

Schultz, A.G.; Taverns, A.G.; Harrington, R.E. Tetrahedron Lett., **1988**, *29*, 3907.

Kim, Y.H.; Kim, K.S.; Lee, H.Y. *Tetrahedron Lett.*, **1989**, *30*, 6357.

SECTION 177: KETONES FROM KETONES

This section contains alkylations of ketones and protected ketones, ketone transpositions and annulations, ring expansions and ring openings and dimerizations. Conjugate reductions and Michael alkylations of enone are listed in Section 74 (Alkyls from Alkenes).

For the preparation of enamines or imines from ketones, see Section 356 (Amine-Alkene).

E:Z = >99.5:0.5

Reich, H.J.; Holtan, R.C.; Borkowsky, S.L. *J. Org. Chem.*, **1987**, *52*, 312.

Satoh, T.; Itoh, M.; Yamakawa, K. *Chem. Lett.*, **1987**, 1949.

Tsuda, T.; Satomi, H.; Hayashi, T. Saegusa, T. *J. Org. Chem.*, **1987**, *52*, 439.

Gadwood, R.C.; Mallick, I.M.; DeWinter, A.J. *J. Org. Chem.*, *1987*, *52*, 774.

Krief, A.; Laboureur, J.L. *Tetrahedron Lett.*, *1987*, *28*, 1545.
Krief, A.; Laboureur, J.L.; Dumont, W. *Tetrahedron Lett.*, *1987*, *28*, 1549.

Tominaga, Y.; Kohra, S.; Hosomi, A. *Tetrahedron Lett.*, *1987*, *28*, 1529.
1545.

Uemura, S.; Ohe, K.; Sugita, N. *J. Chem. Soc., Chem. Commun.*, *1988*, 111.

Sato, T.; Watanabe, M.; Watanabe, T.; Onoda, Y.; Murayama, E. *J. Org. Chem.*, *1988*, *53*, 1894.

Bu₃SnH , AIBN , PhH

reflux , 3h

84%

trans:cis = 56:44

Boger, D.L.; Mathvink, R.J. *J. Org. Chem.*, *1988*, *53*, 3377.

1. TiCl₄ , Ti(OiPr)₄ , CH₂Cl₂
   -78°C
2. 1N HCl
3. K₂CO₃ , MeOH

83%

Engler, T.A.; Ali, M.H.; Velde, D.V. *Tetrahedron Lett.*, *1989*, *30*, 1761.

1. NaN(SiMe₃)₂
2. BnI , HMPA

89%

98% ee , S

Oppolzer, W.; Moretti, R.; Thomi, S. *Tetrahedron Lett.*, *1989*, *30*, 5603.

1. Me₃SnLi
2. Me₃SiOTf , -78°C , 3h

52%

trans:cis = 28:1

Sato, T.; Watanabe, T.; Hayata, T.; Tsukui, T. *J. Chem. Soc., Chem. Commun.*, *1989*, 153.

Ricci, A.; Degl'Innocenti, A.; Capperucci, A.; Reginato, G. *J. Org. Chem., 1989, 54*, 19.

Armesto, D.; Esteban, S.; Horspool, W.M.; Martin, J.-A.F.; Martínez-Alcazar, P.; Perez-Ossorio, R. *J. Chem. Soc., Perkin Trans. I, 1989*, 751.

Majerski, Z.; Vinković, V. *Synthesis, 1989*, 559.

REVIEW:

"One Carbon Ring Expansions of Bridged Bicyclic Ketones"

Krow, G.R. *Tetrahedron, 1987, 43*, 3.

Related Methods: Section 49 (Aldehydes from Aldehydes).

SECTION 178: KETONES FROM NITRILES

NO ADDITIONAL EXAMPLES

SECTION 179: KETONES FROM ALKENES

Tsuji, J.; Minato, M. *Tetrahedron Lett.*, *1987*, *28*, 3683.

(70   :   30) 48%

Welch, M.C.; Bryson, T.A. *Tetrahedron Lett.*, *1988*, *29*, 521.

Bäckvall, J.-E.; Hopkins, R.B. *Tetrahedron Lett.*, *1988*, *29*, 2885.

Nikishin, G.I.; Elinson, M.N.; Makhova, I.V. *Tetrahedron Lett.*, *1988*, *29*, 1603.

DCME = dichloromethylmethyl ether   β:α = 1:1.2

Akers, J.A.; Bryson, T.A. *Tetrahedron Lett.*, *1989*, *30*, 2187.

74%

Inoue, M.; Uragaki, T.; Kashiwagi, H.; Enomoto, S. *Chem. Lett., 1989*, 99.

Yusubov, M.S.; Filimonov, V.D. *J. Org. Chem., U.S.S.R., 1989, 25*, 199.

REVIEWS:

   "Intramolecular Cycloaddition Reactions of Ketones and Keteniminium Salts with Alkenes"

Snider, B.B. *Chem. Rev., 1988, 88*, 793.

   "Synthetic Applications of Intramolecular Enone-Olefin Photocycloadditions"

Crimins, M.T. *Chem. Rev., 1988, 88*, 1453.

See also: Section 134 (Ethers from Alkenes).
                   Section 174 (Ketones from Ethers).

SECTION 180: KETONES FROM MISCELLANEOUS COMPOUNDS

Conjugate reductions and reductive alkylations of enones are listed in Section 74 (Alkyls from Alkenes).

70%

Aizpurua, J.M.; Oiarbide, M.; Palomo, C. *Tetrahedron Lett., 1987, 28*, 5361.

1. MeLi
2. HCl
3. dichloromethyl ether
4. t-BuOLi
5. H₂O₂ , NaOH

71%

>99% ee , R

Brown, H.C.; Srebnik, M.; Bakshi, R.K.; Cole, T.E. *J. Am. Chem. Soc.*, *1987, 109,* 5420.

CH₂Cl₂ , NEt₃

CTAP

80%

Vankar, P.S.; Rathore, R.; Chandrasekaran, S. *Synth. Commun.*, *1987, 17,* 195.

EtCH₂NO₂

1. C_6H_{13} ... , RT
   t-BuOK , DMSO
2. Bu₃SnH , AIBN ,
   PhH , 80°C

58%

Ono, N.; Fujii, M.; Kaji, A. *Synthesis, 1987,* 532.

PhMe , TiCl₄ , RT

CH₂Cl₂ , H₂O , 2h
reflux

94%

Lee, K.; Oh, D.Y. *Tetrahedron Lett.*, *1988, 29,* 2977.

1. iBu₂Al⌇⌇⌇
   hexane , RT
2. 3N HCl
3. H₂ , Pd/C

Pecunioso, A.; Menicagli, R. *J. Org. Chem.*, *1988, 53,* 2614.

85%

Ballini, R.; Petrini, M. *Tetrahedron Lett.*, **1989**, *30*, 5329.

SECTION 180A: PROTECTION OF KETONES

99%

Pearson, W.H.; Cheng, M.-C. *J. Org. Chem.*, **1987**, *52*, 1353.

93%

Kim, Y.H.; Lee, H.K.; Chang, H.S. *Tetrahedron Lett.*, **1987**, *28*, 4285.

quant.

Fadel, A.; Yefsah, R.; Salaün, J. *Synthesis*, **1987**, 37.

80%

polystyrene backbone

Caputo, R.; Ferreri, C.; Palumbo, G. *Synthesis*, **1987**, 386.

Cossy, J. *Synthesis*, **1987**, 1113.

Liu, H.-J.; Wiszniewski, V. *Tetrahedron Lett.*, **1988**, *29*, 5471.

Thurkauf, A.; Jacobson, A.E.; Rice, K.C. *Synthesis*, **1988**, 233.

Ni, Z.-J.; Luh, T.-Y. *J. Chem. Soc., Chem. Commun.*, **1988**, 1011.

Shigemasa, Y.; Ogawa, M.; Sashiwa, H.; Saimoto, H. *Tetrahedron Lett.*, **1989**, *30*, 1277.

Machinaga, N.; Kibayashi, C. *Tetrahedron Lett.*, **1989**, *30*, 4165.

1. MeCN , AlI$_3$, PhH , RT
10 min
2. 10% aq.Na$_2$S$_2$O$_3$

90%

Sarmeh, P.; Barua, N.C. Tetrahedron Lett., **1989**, 30, 4703.

DMSO , H$_2$O , 180°C
10h

92%

Kametani, T.; Kondoh, H.; Honda, T.; Ishizone, H.; Suzuki, Y.; Mori, W. Chem. Lett., **1989**, 901.

SmCl$_3$, Me$_3$SiCl , THF
42h

99%

Ukaji, Y.; Koumoto, N.; Fujisawa, T. Chem. Lett., **1989**, 1623.

REVIEW:

"Synthetic Uses of the 1,3-Dithiane Grouping from 1977 to 1988"

Page, P.C.B.; van Niel, M.B.; Prodger, J.C. Tetrahedron, **1989**, 45, 7643.

See Section 362 (Ester-Alkene) for the formation of enol esters and Section 367 (Ether-Alkenes) for the formation of enol ethers. Many of the methods in Section 60A (Protection of Aldehydes) are also applicable to ketones.

CHAPTER 13
PREPARATION OF NITRILES

SECTION 181: NITRILES FROM ALKYNES

NO ADDITIONAL EXAMPLES

SECTION 182: NITRILES FROM ACID DERIVATIVES

NO ADDITIONAL EXAMPLES

SECTION 183: NITRILES FROM ALCOHOLS AND THIOLS

$$C_{13}H_{27}CH_2OH \xrightarrow[\substack{\text{2. NaCN , THF , HMPT}}]{\substack{\text{1. TFAA , CH}_2\text{Cl}_2\text{ , 2h}}} C_{13}H_{27}CH_2C{\equiv}N$$

one pot - two step   99%

Camps, F.; Gasol, V.; Guerrero, A. *Synth. Commun.*, *1988*, *18*, 445.

$$MeO\text{—}\langle\ \rangle\text{—}OH \xrightarrow[\substack{\text{2. KCN , Ni(PPh}_3)_4}]{\substack{\text{1. Tf}_2O\text{ , Py}}} MeO\text{—}\langle\ \rangle\text{—}C{\equiv}N$$

78%

Chambers, M.R.I.; Widdowson, D.A *J. Chem. Soc., Perkin Trans. I, 1989*, 1365.

SECTION 184: NITRILES FROM ALDEHYDES

Okimoto, M.; Chiba, T. *J. Org. Chem.*, *1988*, *53*, 218.

Reddy, P.S.N.; Reddy, P.P. *Synth. Commun.*, *1988*, *18*, 2179.

Said, S.B.; Skarzewski, J.; Młochowski, J. *Synthesis*, *1989*, 223.

Capdevielle, P.; Lavigne, A.; Maumy, M. *Synthesis*, *1989*, 451.

SECTION 185: NITRILES FROM ALKYLS, METHYLENES AND ARYLS

NO ADDITIONAL EXAMPLES

SECTION 186: NITRILES FROM AMIDES

MeO$_2$C, N$^-$, S, NEt$_3$$^+$ = Burgess' reagent

Claremon, D.A.; Phillips, B.T. *Tetrahedron Lett.*, **1988**, *29*, 2155.

$$Ph-C\equiv N \quad 91\%$$

P$_2$O$_5$, Me$_3$SiOTs, 3h
NEt$_3$, 50°C

Someswara Rao, C.; Rambabu, M.; Srinivasan, P.S. *Synth. Commun.*, **1989**, *19*, 1431.

SECTION 187: NITRILES FROM AMINES

1. TFAA, CH$_2$Cl$_2$, Ph$_3$PO/DCE
2. (with o-phenylenediamine NH$_2$/NH$_2$), PhCOOH, DCE, 30 min
3. aq. NaHCO$_3$

Tol-C≡N   94%

Hendrickson, J.B.; Hussoin, Md.S. *J. Org. Chem.*, **1987**, *52*, 4138.

Rh$_6$(CO)$_{16}$
CO, H$_2$O

Kaneda, K.; Doken, K.; Imanaka, T. *Chem. Lett.*, **1988**, 285.

350°C
Cs X-zeolite

85%

Rao, M.N.; Kumar, P.; Garyali, K. *Org. Prep. Proceed. Int.*, **1989**, *21*, 230

$$PhCH_2NH_2 \xrightarrow[60^{\circ}C \, , \, 24h]{CuCl \, , \, O_2 \, , \, Py \, , \, MS \, 4\text{Å}} Ph\text{-}C\equiv N \quad \textbf{96\%}$$

Capdevielle. P.; Lavigne, A.; Maumy, M. *Synthesis, 1989*, 453.

SECTION 188: NITRILES FROM ESTERS

$$PhCH_2CO_2Et \xrightarrow[\substack{2. \, 2 \, eq. \, PBr_3 \, , \, PhH \\ reflux \, , \, 5h}]{1. \, HONH_2 \cdot HCl} PhCH_2C\equiv N \quad \textbf{92\%}$$

Liguori, A.; Sindona, G.; Romeo, G.; Uccella, N. *Synthesis, 1987*, 168.

SECTION 189: NITRILES FROM ETHERS, EPOXIDES AND THIOETHERS

NO ADDITIONAL EXAMPLES

SECTION 190: NITRILES FROM HALIDES AND SULFONATES

$$PhZnCl \xrightarrow[\substack{(c\text{-}C_6H_{11})PPh_2 \, , \, THF \\ 20^{\circ}C \, , \, 1h}]{BrCH_2CN \, , \, 5\% \, Ni(acac)_2} PhCH_2C\equiv N \quad \textbf{68\%}$$

Frejd. T.; Klingstedt, T. *Synthesis, 1987*, 40.

Piers. E.; Fleming, F.F. *J. Chem. Soc., Chem. Commun., 1989*, 756.

$$PhOTs \xrightarrow[\substack{2\% \, [NiBr_2(PPh_3)_2] \, . \, PPh_3 \, , \\ Zn \, , \, 3h \, , \, 0°C}]{KCN \, , \, N\text{-methyl-2-pyrrolidinone}} Ph\text{-}C\equiv N$$

98%

Takagi, K.; Sakakibara, Y. *Chem. Lett., 1989*, 1957.

REVIEW:

"Cyanation of Aromatic Halides"

Bellis, G.P.; Romney-Alexander, T.M. *Chem. Rev., 1987, 87*, 779.

SECTION 191: NITRILES FROM HYDRIDES

NO ADDITIONAL EXAMPLES

SECTION 192: NITRILES FROM KETONES

also works with aldehydes

Yoneda, R.; Harusawa, S.; Kurihara, T. *Tetrahedron Lett., 1989, 30*, 3681.

Saito, K.; Kagabu, S. *Org. Prep. Proceed. Int., 1989, 21*, 354.

SECTION 193: NITRILES FROM NITRILES

Conjugate reductions and Michael alkylations of alkene nitriles are found in Section 74D (Alkyls from Alkenes).

NO ADDITIONAL EXAMPLES

SECTION 194: NITRILES FROM ALKENES

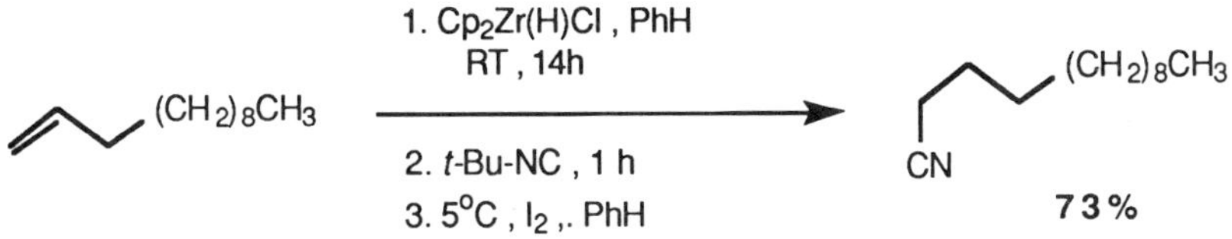

1. Cp$_2$Zr(H)Cl , PhH
   RT , 14h

2. *t*-Bu-NC , 1 h
3. 5°C , I$_2$,. PhH

(CH$_2$)$_8$CH$_3$ → (CH$_2$)$_8$CH$_3$, CN

73%

Buchwald, S.L.; LaMaire, S.J. *Tetrahedron Lett.*, *1987*, *28*, 295.

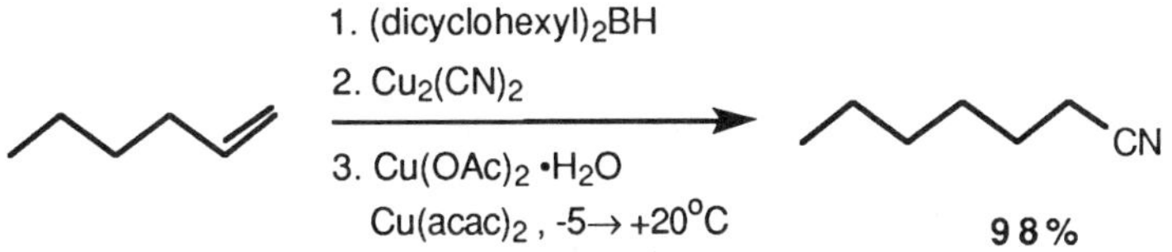

1. (dicyclohexyl)$_2$BH
2. Cu$_2$(CN)$_2$
3. Cu(OAc)$_2$ •H$_2$O
   Cu(acac)$_2$, -5→ +20°C

CN

98%

Masuda, Y.; Hoshi, M.; Arase, A. *J. Chem. Soc., Chem. Commun.*, *1989*, 266.

SECTION 195: NITRILES FROM MISCELLANEOUS COMPOUNDS

NO ADDITIONAL EXAMPLES

CHAPTER 14

PREPARATION OF ALKENES

SECTION 196: ALKENES FROM ALKYNES

Ph—≡—H $\xrightarrow[\text{[Et}_3\text{NCH}_2\text{Ph]}_3\text{ [IrCl}_6]]{\text{Et}_2\text{SiH , 1h , 120}^\circ\text{C}}$

$$\text{CH}_2=\text{C(Ph)SiEt}_2 \quad + \quad \text{PhCH=CHSiEt}_2$$

(20 : 80)

Iovel, I.G.; Goldberg, Y.Sh.; Shymanska, M.V.; Lukevics, E. *J. Chem. Soc., Chem. Commun.*, *1987*, 31.

$\xrightarrow[\text{2. SiO}_2 \text{ , CH}_2\text{Cl}_2]{\substack{\text{1. (Bu}_3\text{Sn)}_2 \text{ , PhH} \\ \text{AIBN}}}$

77%

Stork, G.; Mook Jr., R. *J. Am. Chem. Soc.*, *1987*, *109*, 2829.

$\xrightarrow[\text{PMHS}]{\substack{(\text{dba})_3\text{Pd}_2 \cdot \text{CHCl}_3 \\ \text{AcOH , PhH}}}$

96%

PMHS = polymetylhydrosiloxane

Trost, B.M.; Rise, F. *J. Am. Chem. Soc.*, *1987, 109*, 3161.

Clive, D.L.J.; Angoh, A.G.; Bennett, S.M. *J. Org. Chem.*, *1987*, *52*, 1339.

Negishi, E.; Swanson, D.R.; Cederbaum, F.E.; Takahashi, T. *Tetrahedron Lett.*, *1987*, *28*, 917.

Wang, K.K.; Yang, K.E. *Tetrahedron Lett.*, *1987*, *28*, 1003.
Wang, K.K.; Dhumrongvaraporn, S. *Tetrahedron Lett.*, *1987*, *28*, 1007.

Al-Hassan, M.I. *Synth. Commun.*, *1987*, *17*, 1413.

SiMe3
Me
285°C , neat
1.5h
SiMe3
Me

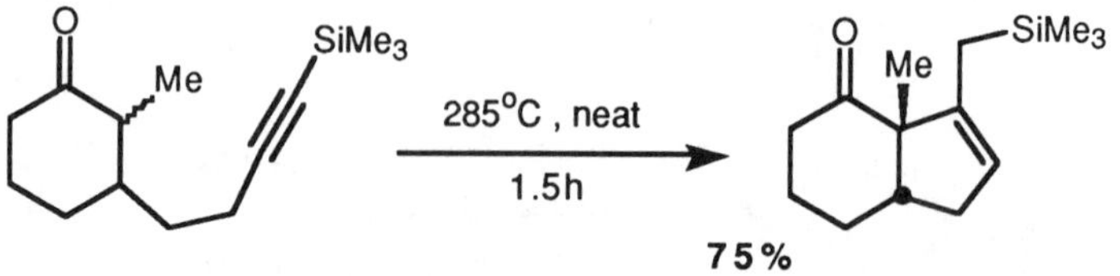

75%

Kende, A.S.; Hebeisen, P.; Newbold, R.C. *J. Am. Chem. Soc.*, ***1988***, *110*, 3315.

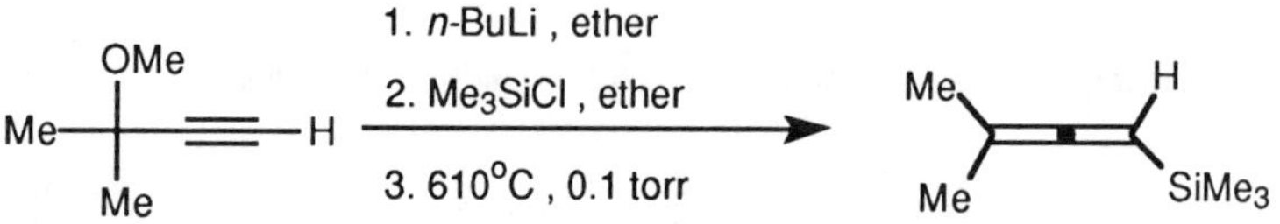

OMe
Me
Me
H
1. n-BuLi , ether
2. Me3SiCl , ether
3. 610°C , 0.1 torr
Me
Me
H
SiMe3

68%

Hopf, H.; Naujoks, E. *Tetrahedron Lett.*, ***1988***, *29*, 609.

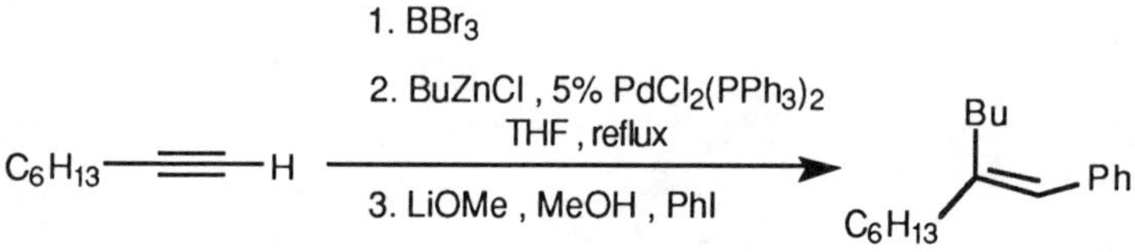

C6H13
H
1. BBr3
2. BuZnCl , 5% PdCl2(PPh3)2
THF , reflux
3. LiOMe , MeOH , PhI
Bu
Ph
C6H13

65%

97% E

Satoh, Y.; Serizawa, H.; Miyaura, N.; Hara, S.; Suzuki, A. *Tetrahedron Lett.*, ***1988***, *29*, 1811.

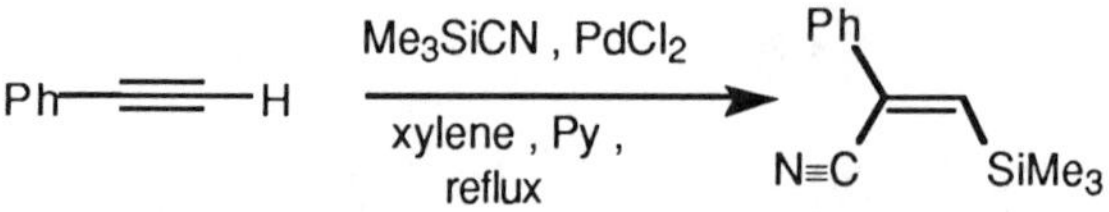

Ph
H
Me3SiCN , PdCl2
xylene , Py ,
reflux
Ph
N≡C
SiMe3

96%

Z:E = 95:5

Chatani, N.; Takeyasu, T.; Horluchi, N.; Hanafusa, T. *J. Org. Chem.*, ***1988***, *53*, 3539.

Bu
Bu
1.
BHBr •SMe 2
2.
Li
3. I2 , NaOH 4. NaOH
Bu
Bu

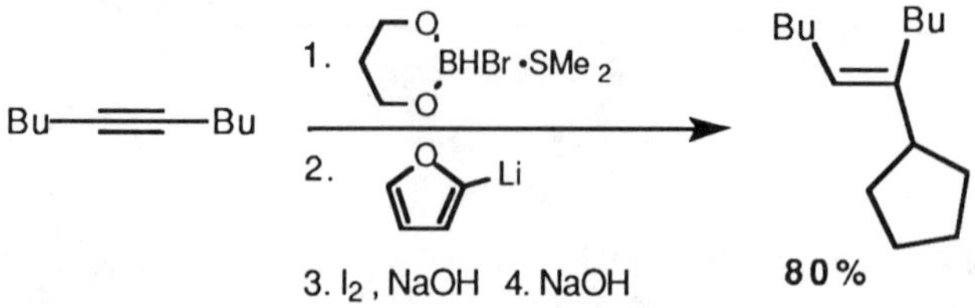

80%

Brown, H.C.; Bhat, N.G. *J. Org. Chem.*, ***1988***, *53*, 6009.

Elsevier. C.J.; Vermeer, P. *J. Org. Chem., 1989, 54,* 3726.

Barros, S.M.; Comasseto. J.V.; Berriet, J. *Tetrahedron Lett., 1989, 30,* 7353.

Chou. S.-S.P.; Kuo, H.-L.; Wang, C.-J.; Tsai, C.-Y.; Sun, C.-M. *J. Org. Chem., 1989, 54,* 868.

Sharma, G.V.M.; Choudary. B.M.; Sarma, M.R.; Rao, K.K. *J. Org. Chem., 1989, 54,* 2997.
Choudary. B.M.; Vasantha, G.; Sharma, M.; Bharathi, P. *Angew. Chem. Int. Ed., Engl., 1989, 28,* 465.

SECTION 197: ALKENES FROM ACID DERIVATIVES

Anderson, M.B.; Fuchs. P.L. *Synth. Commun., 1987, 17,* 621.

Meier, I.K.; Schwartz, J. *J. Am. Chem. Soc.*, *1989*, *111*, 3069.

Barluenga, J.; Fernández-Simon, F.; Concellón, J.M.; Yus, M. *J. Chem. Soc., Perkin Trans. I, 1989*, 77.

SECTION 198: ALKENES FROM ALCOHOLS AND THIOLS

Sarkar, T.K.; Ghosh, S.K. *Tetrahedron Lett.*, *1987*, *28*, 2061.

Tellier, F.; Sauvêtre, R.; Normant, J.-F. *Tetrahedron Lett.*, *1987, 28*, 3335.

King, J.L.; Posner, B.A.; Mak, K.T.; Yang, N.C. *Tetrahedron Lett.*, *1987, 28*, 3917.

82%

Garlaschelli, L.; Vidari, G. *Gazz. Chim. Ital., 1987, 117,* 251.

98%

Nishiguchi, T.; Machida, N.; Yamamoto, E. *Tetrahedron Lett., 1987, 28,* 4565.

quant.

Fitjer, L.; Majewski, M.; Kanschik, A. *Tetrahedron Lett., 1988, 29,* 1263.

1. $Cl_3CC(=O)CF_3$, PhH , 7h
   cat. hydroquinone
   cat. *p*-TsOH , reflux

2. PhH , aq. $NaHCO_3$ **75%**

Abdel-Baky, S.; Moussa, A. *Synth. Commun., 1988, 18,* 1795.

1. NEt_3 ,
   DMAP
   CH_2Cl_2 , 0°C

2. $MeSO_2Cl$

(2   :   1)   **85%**

Yadav, J.S.; Mysorekar, S.V. *Synth. Commun., 1989, 19,* 1057.

99%

Nishiguchi, T.; Kamio, C. *J. Chem. Soc., Perkin Trans. I,* **1989**, 707.

REVIEW:

"Carbonyl-Coupling Reactions Using Low-Valent Titanium"

McMurry, J.E. *Chem. Rev.,* **1989**, *89*, 1513.

SECTION 199: ALKENES FROM ALDEHYDES

Pelter, A.; Buss, D.; Colclough, E. *J. Chem. Soc., Chem. Commun.,* **1987**, 297.

Dormand, A.; El Boudadili, A.; Moise, C. *J. Org. Chem.,* **1987**, *52*, 688.

Okazoe, T.; Takai, K.; Utimoto, K. *J. Am. Chem. Soc.,* **1987**, *109*, 951.

Takai, K.; Kataoka, Y.; Okazoe, T.; Utimoto, K. *Tetrahedron Lett.*, *1987*, *28*, 1443.

(15 : 1) 77%

Iio, H.; Mizobuchi, T.; Tokoroyama, T. *Tetrahedron Lett.*, *1987*, *28*, 2379.

C-200 = activated barium hydroxide catalyst

Fuentes, A.; Marinas, J.M.; Sinisterra, J.V. *Tetrahedron Lett.*, *1987*, *28*, 2951.

Vedejs, E.; Marth, C. *Tetrahedron Lett.*, *1987*, *28*, 3445.

$$Ph_3As^+\text{-}CH_3\ I^- \quad \xrightarrow[\substack{\text{2. } C_6F_6 \text{, ether} \\ 0\rightarrow20^\circ C \\ \text{3. PhCHO}}]{\text{1. PhLi , ether , } 0^\circ C} \quad Ph\text{-}CH=CHC_6F_6$$

85%

Shen, Y.; Qiu, W. *Synthesis*, *1987*, 65.

$$C_8H_{19}CHO \xrightarrow[\substack{2.\ HCO_2NH_4\ ,\ DMF\ , \\ cat.\ Pd_2(dba)_3 \cdot CHCl_3 \\ PPh_3}]{\substack{1.\ HC{\equiv}CMgBr\ ,\ ClCO_2Me \\ THF\ ,\ 0°C}}$$

87%

Tsuji, J.; Sugiura, T.; Minami, I. *Synthesis, 1987*, 603.

$$Ph_2P \text{ } + \xrightarrow[\substack{2.\ PhCHO\ ,\ 20°C}]{\substack{1.\ 2\ n\text{-}BuLi\ ,\ THF\ ,\ 20°C}}$$

92%

Z:E = 12:88

McKenna, E.G.; Walker, B.J. *Tetrahedron Lett., 1988, 29*, 485.

$$Ph_3As^+ \xrightarrow[\substack{18h}]{\substack{PhCHO\ ,\ MeCN\ ,\ K_2CO_3}}$$

96%

Huang, Y.-Z.; Shi, L.-L.; Li, S.-W. *Synthesis, 1988*, 975.

$$Ph \xrightarrow[\substack{NiCl_2(PPh_3)_2\ ,\ reflux \\ overnight}]{\substack{Me_3SiCH_2MgCl\ ,\ PhH\text{-}ether}}$$

76%

Ni, Z.-J.; Luh, T.-Y. *J. Org. Chem., 1988, 53*, 2129.

$$Bu_2TeCH_2(CN)Cl \xrightarrow[\substack{2.\ PhCHO}]{\substack{1.\ t\text{-}BuOK}} PhCH{=}CHCN$$

78%

E:Z = 12:1

Huang, X.; Xie, L.; Wu, H. *J. Org. Chem., 1988, 53*, 4862.
Huang, X.; Xie, L.; Wu, H. *Tetrahedron Lett., 1987, 28*, 801.

$$\xrightarrow[\substack{THF\ ,\ RT\ ,\ overnight}]{\substack{n\text{-}PrCHO\ ,\ RuH_2(PPh_3)_4}}$$

83%

Naota, T.; Taki, H.; Mizuno, M.; Murahashi, S. *J. Am. Chem. Soc., 1989, 111*, 5954.

iBuO-CH₂CHO

1. (MeO)₂P(=O)—CH(CO₂Me)(NHCbz)
   CH₂Cl₂
   t-BuOK , -70 → +25°C

2. LDA , THF , -70°C

3. aq. NH₄Cl

iBuO—CH=CH—CH(CO₂Me)(NHCbz)

72%

Z:E = 6:1

Daumas, M.; Vo-Quang, L.; Vo-Quang, Y.; LeGoffic, F. *Tetrahedron Lett., 1989, 30,* 5121.

(cyclohexenyl)CHO

1. I—CH=CH—OEt—B
   pentane , -78°C
2. H₂O

(cyclohexenyl)—CH=CH—CO₂Et

80%

Satoh, Y.; Tayano, T.; Hara, S.; Suzuki, A. *Tetrahedron Lett., 1989, 30,* 5153.

Ph₂As⁺(Ph)(CH₂Ph) ⁻

n-C₅H₁₁CHO

THF-HMPA
-78°C

Ph—CH=CH—C₅H₁₁ + Ph—CH=CH—C₅H₁₁

(>99   :   1)   96%

Boub, A.B.; Mioskowski, C.; Bellamy, F. *Tetrahedron Lett., 1989, 30,* 5263.

Cl
 |
CH₃CH₂—CH—CH₂—CHO

1. Me₃SiCH₂MgCl , -78°C

2. Li° , -78 → +25°C

CH₃CH₂CH₂—CH=CH—CH₂—SiMe₃

65%

Barluenga, J.; Fernández-Simón, J.L.; Concellón, J.M.; Yus, M. *Tetrahedron Lett., 1989, 30,* 5927.

Me—C₆H₄—CHO

Ph₃As=CHSPh
THF-HMPA
MeCN , -78°C

Me—C₆H₄—CH=CH—SPh

67%

Boubia, B.; Mioskowski, C.; Manna, S.; Falck, J.R. *Tetrahedron Lett., 1989, 30,* 6023.
Chabert, P.; Mioskowski, C. *Tetrahedron Lett., 1989, 30,* 6031.

PhCHO $\xrightarrow{\begin{array}{c} 1. \text{(vinyl ether with OEt, OEt, Li)} \\ \hline 2. \text{HCl , acetone , reflux} \\ 3. \text{NaOH , RT} \end{array}}$ Ph‑(polyene chain)‑C(=O)‑Me   **70%**

Duhamel, L.; Ple, G.; Ramondenc, Y. *Tetrahedron Lett.*, *1989*, *30*, 7377.

$C_5H_{11}CHO$ $\xrightarrow[\text{SbBu}_3 , 120°C]{\text{ClCH}_2\text{CN , 3h}}$ C_5H_{11}‑CH=CH‑CN   **92%**

trans:cis = 62:38

Huang, Y.-Z.; Shen, Y.; Chen, C. *Synth. Commun.*, *1989*, *19*, 83.
Huang, Y.-Z.; Chen, C.; Shen, Y. *Synth. Commun.*, *1989*, *19*, 501.

(cyclohexyl)‑CHO $\xrightarrow[\begin{array}{c}\text{cat. Bu}_2\text{Te , 50°C} \\ \text{12h}\end{array}]{\begin{array}{c}\text{BrCH}_2\text{CO}_2\text{Me , THF} \\ \text{P(OPh)}_3 , \text{K}_2\text{CO}_3\end{array}}$ (cyclohexylidene)=CH‑CO$_2$Me   **74%**

Huang, Y.-Z.; Shi, L.-L.; Li, S.-W.; Wen, X.-Q. *J. Chem. Soc., Perkin Trans. I, 1989*, 2397.

Br‑C(CN)(CO$_2$Et) $\xrightarrow[\text{85°C , 5h}]{\text{AsBu}_3 , \text{PhCHO}}$ Ph‑CH=C(CN)(CO$_2$Et)   **87%**

Shen, Y.; Yang, B. *Synth. Commun.*, *1989*, *19*, 3069.

PhCHO $\xrightarrow[\begin{array}{c}\text{cat. Bu}_3\text{As , K}_2\text{CO}_3 , \text{THF} \\ \text{MeCN , RT , 30h}\end{array}]{\text{BrCH}_2\text{CO}_2\text{Me , P(OPh)}_3}$ Ph‑CH=CHCO$_2$Me   **86%**

E:Z = 99:1

catalytic Wittig Olefination

Shi, L.; Wang, W.; Wang, Y.; Huang, Y.-Z. *J. Org. Chem.*, *1989*, *54*, 2027.

REVIEWS:

"Reductions Promoted by Low Valent Transition Metal Complexes in Organic Synthesis"

Pons, J.-M.; Santelli, M. *Tetrahedron, 1988, 44,* 4295.

"The Applications of Low-Valent Titanium Reagents in Organic Synthesis"

Lenior, D. *Synthesis, 1989,* 883.

"The Wittig Olefination Reaction and Modifications Involving Phosphoryl-Stabilized Carbanions. Stereochemistry, Mechanism and Selected Synthetic Aspects"

Maryanoff, B.E.; Reitz, A.B. *Chem. Rev., 1989, 89,* 863.

Related Methods: Section 207 (Alkenes from Ketones).

SECTION 200: ALKENES FROM ALKYLS, METHYLENES AND ARYLS

This section contains dehydrogenations to form alkenes and unsaturated ketones, esters and amides. It also includes the conversion of aromatic rings to alkenes.
Reduction of aryls to dienes is found in Section 377 (Alkene-Alkene).
Hydrogenation of aryls to alkanes and dehydrogenations to form aryls are included in Section 74 (Alkyls from Alkenes).

Wilson, R.M.; Hengge, A. *J. Org. Chem., 1987, 52,* 2699.

Chaussard, J.; Combellas, G.; Thiebault, A. *Tetrahedron Lett., 1987, 28,* 1173.

MeO—⟨benzene⟩
$\xrightarrow[\text{CH}_2\text{Cl}_2 , -75^\circ\text{C} - \text{RT}]{\begin{array}{c}\text{(PhIO)}_n \text{ / BF}_3\text{·OEt}_2 \\ \text{—SiMe}_3\end{array}}$
MeO—⟨benzene⟩—CH₂CH=CH₂

95%

ortho:para = 1:4

Lee. K.; Kim, D.Y.; Oh. D.Y. *Tetrahedron Lett., 1988, 29*, 667.

REVIEW:

"The Metal-Ammonia Reduction of Aromatic Compounds"

Rabideau, P.W. *Tetrahedron, 1989, 45*, 1579.

SECTION 201: ALKENES FROM AMIDES

Related Methods: Section 65 (Alkyls from Alkyls).
                          Section 74 (Alkyls from Alkenes).

NO ADDITIONAL EXAMPLES

SECTION 202: ALKENES FROM AMINES

$\xrightarrow[\text{3. H}_3\text{O}^+]{\begin{array}{l}\text{1. HO}_3\text{SONO} \\ \text{2. LiAlH}_4\end{array}}$
Ph∼∼∼=∼∼∼=∼∼Ph

90%

Tufariello. J.J.; Milowsky, A.S.; Al-Nuri, M.; Goldstein, S. *Tetrahedron Lett., 1987, 28*, 263, 267.

⟨cyclopentane⟩=N-NH-TRIS
$\xrightarrow[\begin{array}{c}\text{2. TMEDA , ether} \\ -78^\circ\text{C} \rightarrow \text{RT} \\ \text{3. H}_2\text{O}\end{array}]{\begin{array}{l}\text{1. } J. \text{ Org Chem., 1978,} \\ 43 , 147.\end{array}}$
⟨bicyclic product with Me⟩

69%

(>10:1)

Chamberlin. A.R.; Bloom, S.H.; Vervini, L.A.; Fotsch, C.H. *J. Am. Chem. Soc., 1988, 110*, 4788.

92%

Vilsmaier, E.; Kristen, G.; Tetzlaff, C. *J. Org. Chem.*, *1988*, *53*, 1806.

95%

Murray, R.W.; Rajadhyaksha, S.N.; Mohan, L. *J. Org. Chem.*, *1989*, *54*, 5783.

SECTION 203: ALKENES FROM ESTERS

1. [dried CeCl$_3$·7 H$_2$O + Me$_3$SiCH$_2$MgCl]

−70°C → RT

2. HCl

>95%

Narayanan, B.A.; Bunnelle, W.H. *Tetrahedron Lett.*, *1987*, *28*, 6261.

Me$_3$SiCH$_2$MgCl

CeCl$_3$

74%

Lee, T.V.; Porter, J.R.; Roden, F.S. *Tetrahedron Lett.*, *1988*, *29*, 5009.

1. (AcO)$_3$Pb

CH$_2$Cl$_2$, RT, 3h

2. Ni(R) - W2, NEt$_3$, EtOH, RT, 30 min

47%

Hashimoto, S.; Miyazaki, Y.; Shinoda, T.; Ikegami, S. *Tetrahedron Lett.*, *1989*, *30*, 7195.

SECTION 204: ALKENES FROM ETHERS, EPOXIDES AND THIOETHERS

1. LDMAN , -45°C , THF

2. n-C_5H_{13}CHO , -45°C

3. 5 KOt-Bu , -45°C

4. reflux   **80%**

LDMAN = lithium 1-(dimethylamino)-naphthalenide

Cohen, T.; Jung, S.-H.; Romberger, M.L.; McCullough, D.W. *Tetrahedron Lett.,* **1988**, *29*, 25.

Bu_4CeLi , DME , TMEDA

-78 → -25°C

(90   :   10)   **75%**

Ukaji, Y.; Fujisawa, T. *Tetrahedron Lett.,* **1988**, *29*, 5165.

AlI_3 , MeCN/PhH

78%

Sarmah, P.; Barua, N.C. *Tetrahedron Lett.,* **1988**, *29*, 5815.

[Cp_2TiCl_2 , Mg° , THF , RT]

THF , -78°C→ RT

92%

Schobert, R. *Angew. Chem. Int. Ed., Engl,* **1988**, *27*, 855.

$\left(Br-\bigcirc- \right)_3 \overset{+}{N} \cdot SbCl_6$

0°C , CH_2Cl_2 , 1h

98%

Kamata, M.; Muruyama, K.; Miyashi, T. *Tetrahedron Lett.,* **1989**, *30*, 4129.

REVIEWS:

"Ether Cleavage with Organo-Alkali Metal Compounds and Alkali Metals"

Maercker, A. *Angew. Chem. Int. Ed., Engl, 1987, 26*, 972.

"Stereospecific Deoxygenation of Epoxides to Olefins"

Wong, H.N.C.; Fok, C.C.M.; Wong, T. *Heterocycles, 1987, 25*, 1345.

SECTION 205: ALKENES FROM HALIDES AND SULFONATES

McKean, D.R.; Parrinello, G.; Renaldo, A.F.; Stille, J.K. *J. Org. Chem., 1987, 52*, 422.

Trost, B.M.; Walchli, R. *J. Am. Chem. Soc., 1987, 109*, 3487.

Suárez, A.R.; Mazzieri, M.R. *J. Org. Chem., 1987, 52*, 1145.

Br
Br
1. 5
, THF
MgBr
reflux , 4h
2. H_3O^+
55%

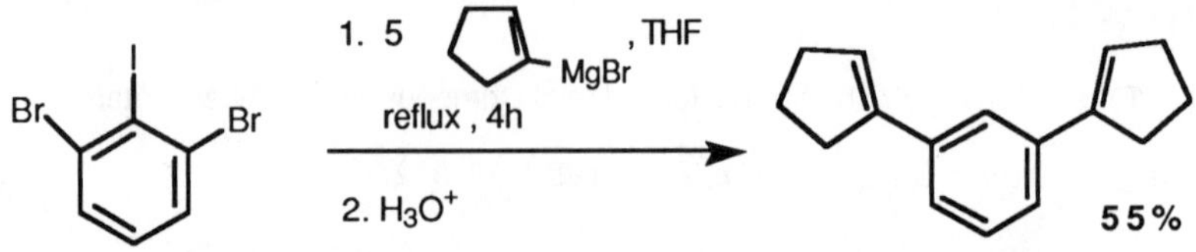

Hart, H.; Ghosh, T. *Tetrahedron Lett.*, ***1988***, *29*, 881.

I
Br
MgBr
1. THF ,
, RT
Me
2. 0°C ,
24h
Br
72%

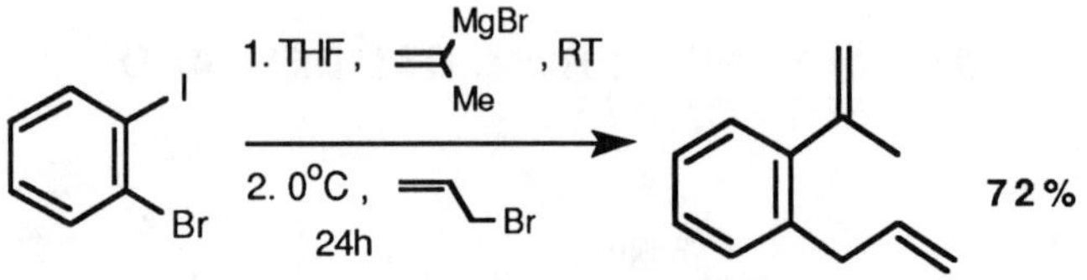

Hart, H.; Saednya, A. *Synth. Commun.*, ***1988***, *18*, 1749.

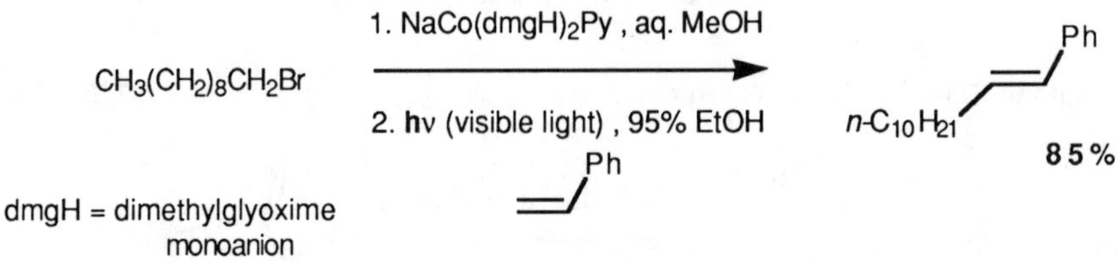

$CH_3(CH_2)_8CH_2Br$
1. $NaCo(dmgH)_2Py$, aq. MeOH
2. hν (visible light) , 95% EtOH
Ph
Ph
n-$C_{10}H_{21}$
85%
dmgH = dimethylglyoxime monoanion

Branchaud, B.P.; Meier, M.S.; Choi, Y. *Tetrahedron Lett.*, ***1988***, *29*, 167.

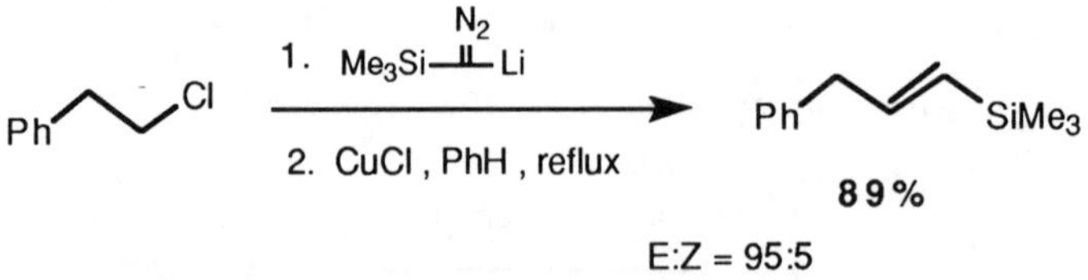

Ph
Cl
N_2
1. Me_3Si—Li
2. CuCl , PhH , reflux
Ph
$SiMe_3$
89%
E:Z = 95:5

Aoyama, T.; Shioiri, T. *Tetrahedron Lett.*, ***1988***, *29*, 6295.

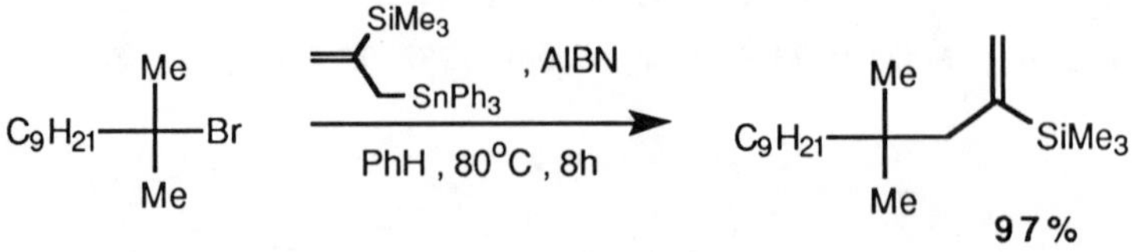

Me
C_9H_{21}
Br
Me
$SiMe_3$
, AIBN
$SnPh_3$
PhH , 80°C , 8h
Me
C_9H_{21}
$SiMe_3$
Me
97%

Lee, E.; Yu, S.-G.; Hur, C.-U.; Yang, S.-M. *Tetrahedron Lett.*, ***1988***, *29*, 6969.

TASF = tris(diethylamino)sulfonium
difluorotrimethyl silicate

Hatanaka, Y.; Hiyama, T. *J. Org. Chem.*, *1988, 53*, 918.

(13   :   1)   62%

Andersson, C.-M.; Hallberg, A. *J. Org. Chem.*, *1988, 53*, 2112.

Huang, X.; Hou, Y.Q. *Synth. Commun.*, *1988, 18*, 2201.

Corey, E.J.; Posner, G.H.; Atkinson, R.F.; Wingard, A.K.; Halloran, D.J.; Radzik, D.M.; Nash, J.J. *J. Org. Chem.*, *1989, 54*, 389.

$$C_{13}H_{27}CH_2CH_2CH_2Br \left[\begin{array}{l} 1.\ (PPh_3)_2NiCl_2\,, \\ THF\,,PPh_3\,, RT \\ \hline 2.\ BuLi \\ DBU \end{array} \right] \longrightarrow C_{13}H_{27}CH_2CH=CH_2\ +\ C_{13}H_{27}CH=CHCH_3$$

$$(90\quad :\quad 10)\ \textbf{78\%}$$

Henningsen, M.C.; Jeropoulos, S.; Smith, E.H. *J. Org. Chem.*, *1989, 54*, 3015.

SECTION 206: ALKENES FROM HYDRIDES

For conversions of methylenes to alkenes ($RCH_2R' \rightarrow RR'C=CH_2$), see Section 200 (Alkenes from Alkyls).

NO ADDITIONAL EXAMPLES

SECTION 207: ALKENES FROM KETONES

$$\underset{Ph}{\overset{O}{\parallel}} \quad \xrightarrow[\text{2. aq. HF , MeCN}]{\text{1. Me}_3\text{SiCH}_2\text{Li , CeCl}_3\text{ , THF , -78}^\circ\text{C}} \quad \underset{\textbf{78\%}}{Ph}$$

Johnson, C.R.; Tait, B.D. *J. Org. Chem.*, *1987, 52,* 281.

$$\underset{}{\overset{O}{\parallel}}SO_2Tol \xrightarrow[\substack{\text{2. }t\text{-BuO}_2CCl \\ \textbf{97\%} \\ Z:E = 10:1}]{\substack{\text{1. NaH , THF} \\ \overline{0^\circ C}}} \underset{}{\overset{OCOt\text{-Bu}}{}}SO_2Tol \xrightarrow[\substack{\text{THF , -78}^\circ C}]{2\ Bu_2Cu(CN)Li_2} \underset{\substack{\textbf{quant.} \\ Z:E = 58:1}}{\overset{Bu}{}}SO_2Tol$$

Giblin, G.M.P.; Simpkins, N.S. *J. Chem. Soc., Chem. Commun.*, *1987*, 207.

3 PhMgBr , 20h

5% NiCl₂(PPh₃)₂
ether/PhH , RT

68%

Ni, Z.-T.; Luh, T.-Y. *J. Chem. Soc., Chem. Commun.*, *1987*, 1515.

Ph₃CHCO₂Me , MeCN
reflux , 19h

78%

Garner, P.; Ramakanth, S. *J. Org. Chem.*, *1987, 52*, 2629.

C₃H₇CH=PPh₃

15 KBar , 48h
40°C

55%

for hindered ketones   +35% recovered ketone

Dauben, W.G.; Takasugi, J.J. *Tetrahedron Lett.*, *1987, 28*, 4377.

1. BuLi , THF
2. $$Ph \overset{O}{\underset{}{\overset{\|}{C}}} Me$$
3. KH , THF , reflux , 10 min
4. DMSO , 95°C

64%

Meijs, G.F.; Eichinger, P.C.H. *Tetrahedron Lett.*, *1987, 28*, 5559.

[Ph₃PCH₂C₃H₇/BuLi]

LiI , THF , -110°C

56%

+31% recovered acyl silane

Soderquist, J.A.; Anderson, C.L. *Tetrahedron Lett.*, *1988, 29*, 2425.

=PPh₃ , TDA-1 , THF

NaH , RT - 62°C

83%

TDA-1 = phase transfer catalyst
tris[2-(2-methoxyethoxy)ethyl]amine

Stafford, J.J.; McMurry, J.E. *Tetrahedron Lett.*, *1988, 29*, 2531.

Bicknell, A.J.; Burton, G.; Elder, J.S. Tetrahedron Lett., **1988**, 29, 3361.

1. sBuLi , THF , TMEDA

2. [structure] , -60 → -45°C

Me₃SiCl

3. Li⁺ C₁₀H₁₈⁻ , THF , overnight , -78 → 20°C

4. aq. NH₄Cl

Barluenga, J.; Fernández-Simón, J.L.; Concellón, J.M.; Yus, M. Synthesis, **1988**, 234.

Cp₂YCl , DME

80°C

93%

Qian, C.; Qiu, A. Tetrahedron Lett., **1988**, 29, 6931.

CH₂(AlEtCl)•2 THF

CH₂Cl₂ - hexane
25°C , 17h

70%

Piotrowski, A.M.; Malpass, D.B.; Boleslawski, M.P.; Eisch, J.J. J. Org. Chem., **1988**, 53, 2829.

Cp₂ZrCl₂ , CH₂Br₂

Zn°

Tour, J.M.; Bedworth, P.V.; Wu, R. Tetrahedron Lett., **1989**, 30, 3927.

W₂(OCH₂t-Bu)₆Py₂

PhMe , 0°C

57%

Chisholm, M.H.; Klang, J.A J. Am. Chem. Soc., **1989**, 111, 2324.

optimized procedure   **97%**

McMurry, J.E.; Lectka, T.; Rico, J.G. *J. Org. Chem.*, *1989*, *54*, 3748.

REVIEWS:

"Reductions Promoted by Low Valent Transition Metal Complexes in Organic Synthesis"

Pons, J.-M.; Santelli, M. *Tetrahedron*, *1988*, *44*, 4295.

"The Applications of Low-Valent Titanium Reagents in Organic Synthesis"

Lenior, D. *Synthesis*, *1989*, 883.

"The Wittig Olefination Reaction and Modifications Involving Phosphoryl-Stabilized Carbanions. Stereochemistry, Mechanism and Selected Synthetic Aspects"

Maryanoff, B.E.; Reitz, A.B. *Chem. Rev.*, *1989*, *89*, 863.

"Carbonyl-Coupling Reactions Using Low-Valent Titanium"

McMurry, J.E. *Chem. Rev.*, *1989*, *89*, 1513.

Related Methods: Section 199 (Alkenes from Aldehydes).

SECTION 208: ALKENES FROM NITRILES

NO ADDITIONAL EXAMPLES

SECTION 209: ALKENES FROM ALKENES

Ikenaga, K.; Kikukawa, K.; Matsuda, T. *J. Org. Chem.*, *1987*, *52*, 1276.

Takacs, J.M.; Anderson, L.G. *J. Am. Chem. Soc.*, *1987*, *109*, 2200.

catalyst = tris-(4-brmophenyl)-aminium
            hexachlorostibnate
Ar = 4-anisyl

Harirchian, B.; Bauld, N.L. *Tetrahedron Lett.*, *1987*, *28*, 927.

Snider, B.B.; Dombroski, M.A. *J. Org. Chem.*, *1987*, *52*, 5487.

Z:E = >20:1

Z:E = 0:100

91%

Ichinose, Y.; Nozaki, K.; Wakamatsu, K.; Oshima, K.; Utimoto, K. *Tetrahedron Lett.*, *1987*, *28*, 3709.

(1.5   :   1)   96%

Abelman, M.M.; Overman, L.E. *J. Am. Chem. Soc.*, *1988*, *110*, 2328.

85%

Mizumo, K.; Ikeda, M.; Toda, S.; Otsuji, Y. *J. Am. Chem. Soc.*, *1988*, *110*, 1288.

56%

other silyl derivatives were produced

Hori, Y.; Mitsudo, T.; Watanabe, Y. *Bull. Chem. Soc., Jpn.*, *1988*, *61*, 3011.

86%

Snider, B.B.; Buckman, B.O. *Tetrahedron*, *1989*, *45*, 6969.

Negishi, E.; Miller, S.R. *J. Org. Chem.*, *1989*, *54*, 6014.

Xiong, H.; Rieke, R.D. *J. Org. Chem.*, *1989*, *54*, 3247.

REVIEWS:

"Intramolecular, Stoichiometric (Li, Mg, Zn) and Catalytic (Ni, Pd, Pt) Metallo-Ene Reactions in Organic Synthesis"

Oppolzer, W. *Angew. Chem. Int. Ed., Engl.*, *1989*, *28*, 52.

"Cyclizations via Palladium Catalyzed Allylic Alkylation"

Trost, B.M. *Angew. Chem. Int. Ed., Engl.*, *1989*, *28*, 1173.

"Cation Radical Pericyclic Reactions"

Bauld, N.L.; Belleville, D.J.; Harirchian, B.; Lorenz, K.T.; Pabon Jr., R.A.; Reynolds, D.W.; Wirth, D.D.; Chion, H.S.; Marsh, B.K. *Acc. Chem. Res.*, *1987*, *20*, 371.

SECTION 210: ALKENES FROM MISCELLANEOUS COMPOUNDS

Cleary, D.G.; Paquette, L.A. *Synth. Commun.*, *1987*, *17*, 497.

Kamimura, A.; Ono, N. *J. Chem. Soc., Chem. Commun.*, **1988**, 1278.

Opitz, G.; Ehlis, T.; Rieth, K. *Tetrahedron Lett.*, **1989**, *30*, 3131.

REVIEWS:

"Organozirconium Compounds in Organic Synthesis"

Negishi, E.; Takahashi, T. *Synthesis*, **1988**, 11.

"Use of Cyclopropanes and Their Derivatives in Organic Synthesis"

Wong, H.N.C.; Hon, M.-Y.; Tse, C.-W.; Yip, Y.-C.; Tanko, J.; Hudlicky, T. *Chem. Rev.*, **1989**, *89*, 431.

CHAPTER 15

PREPARATION OF OXIDES

This chapter contains reactions which prepare the oxides of nitrogen, sulfur and selenium. Included are N-oxides, nitroso and nitro compounds, nitrile oxides, sulfoxides, selenoxides and sulfones. Oximes are considered to be amines and appear in those sections. Preparation of sulfonic acid derivatives are found in Chapter Two and the preparation of sulfonates in Chapter Ten.

SECTION 211: OXIDES FROM ALKYNES

Kosugi, H.; Kitaoka, M.; Tagami, K.; Takahashi, A.; Uda, H. J. Org. Chem., **1987**, 52, 1078.

SECTION 212: OXIDES FROM ACID DERIVATIVES

$$Tol\text{-}SO_2Cl \xrightarrow[\text{60}^\circ\text{C , 10h}]{\text{Na}_2\text{Te , EtI , THF}} Tol\text{-}SO_2Et \qquad 71\%$$

Suzuki, H.; Nishioka, Y.; Padmanabhan, S.I.; Ogawa, T. Chem. Lett., **1988**, 727.

$$CH_3(CH_2)_{14}COOH \qquad CH_3(CH_2)_{16}SO_2Ph \qquad 95\%$$

Barton, D.H.R.; Boivin, J.; Sarma, J.; da Silva, E.; Zard, S.Z. Tetrahedron Lett., **1989**, 30, 4237.

SECTION 213: OXIDES FROM ALCOHOLS AND THIOLS

Klunder, J.M.; <u>Sharpless, K.B.</u> *J. Org. Chem.*, **1987**, *52*, 2598.

Rebiere, F.; <u>Kagan, H.B.</u> *Tetrahedron Lett.*, **1989**, *30*, 3659.

SECTION 214: OXIDES FROM ALDEHYDES

NO ADDITIONAL EXAMPLES

SECTION 215: OXIDES FROM ALKYLS, METHYLENES AND ARYLS

<u>Olah, G.A.</u>; Rochin, C. *J. Org. Chem.*, **1987**, *52*, 701.

claycop = montmorillonite K-10 clay
CuNO$_2$(H$_2$O)$_3$

(87   :   12) **53%**

improved para selectivity

<u>Laszlo, P.</u>; Pennetreau, P. *J. Org. Chem., **1987**, 52,* 2407.

SECTION 216: OXIDES FROM AMIDES

NO ADDITIONAL EXAMPLES

SECTION 217: OXIDES FROM AMINES

88%

<u>Murahashi, S.</u>; Shiota, T. *Tetrahedron Lett., **1987**, 28,* 2383.

90%

<u>Pandey, G.</u>; Kumaraswamy, G.; Krishna, A. *Tetrahedron Lett., **1987**, 28,* 2649.

73%

<u>Zabrowski, D.L.</u>; Moorman, A.E.; Beck Jr., K.R. *Tetrahedron Lett., **1988**, 29,* 4501.

Murray, P.W.; Singh, M. *Tetrahedron Lett.*, *1988*, *29*, 4677.

Zajac Jr., W.W.; Walters, T.R.; Darcy, M.G. *J. Org. Chem.*, *1988*, *53*, 5856.

Christensen, D.; Jørgensen, K.A. *J. Org. Chem.*, *1989*, *54*, 126.

SECTION 218: OXIDES FROM ESTERS

NO ADDITIONAL EXAMPLES

SECTION 219: OXIDES FROM ETHERS, EPOXIDES AND THIOETHERS

Tiecco, M.; Tingoli, M.; Testaferri, L.; Bartoli,.D. *Tetrahedron Lett.*, *1987*, *28*, 3849.

Bu-S-Bu

$\xrightarrow{\text{H}_2\text{O}_2 - \text{CH}_2\text{Cl}_2 , 1\text{h}}$

(reagent: benzotriazole with N-Ac group)

$\underset{\overset{\parallel}{O}}{\text{Bu}-\text{S}-\text{Bu}}$

82%

Torrini, I.; Paradisi, M.P.; Zecchini, G.P.; Agrosi, F. *Synth. Commun., 1987, 17,* 515.

$\underset{\text{Ph}}{\overset{\text{S}}{\diagdown}}\text{Me}$

$\xrightarrow[\text{KO}_2 , \text{MeCN} , -25^\circ\text{C} , 4\text{h}]{\text{(2-nitrophenyl)sulfinyl chloride}}$

$\underset{\text{Ph}}{\overset{\overset{\parallel}{O}}{\diagdown}}\text{Me}$ 92%

Kim, Y.H.; Yoon, D.C. *Tetrahedron Lett., 1988, 29,* 6453.

(cyclohexenyl-SPh)

$\xrightarrow[\text{CH}_2\text{Cl}_2 , \text{RT}]{\text{oxone} , \text{Bu}_4\text{N}^+}$

(cyclohexenyl-SO$_2$Ph)   79%

Trost, B.M.; Braslau, R. *J. Org. Chem., 1988, 53,* 532.

$\underset{\text{Ph}}{\diagup}\overset{\text{S}}{\diagdown}\text{Ph}$

$\xrightarrow[\substack{\text{H}_2\text{O/CH}_2\text{Cl}_2 \\ 20 \text{ min}}]{\text{CAN} , \text{Bu}_4\text{NBr} , \text{RT}}$

$\underset{\text{Ph}}{\diagup}\overset{\overset{\parallel}{O}}{\underset{\diagdown}{S}}\text{Ph}$ 99%

Baciocchi, E.; Piermatteri, A.; Ruzziconi, R. *Synth. Commun., 1988, 18,* 2167.

Ph-S-Me

$\xrightarrow[\substack{\text{PhSO}_2\text{N=CHPh}(p\text{-NO}_2) \\ 15 \text{ min}}]{\text{Oxone} , \text{K}_2\text{CO}_3 , \text{CH}_2\text{Cl}_2 , 75^\circ\text{C}}$

$\underset{\text{Ph}}{\overset{\overset{\parallel}{O}}{\diagdown}}\text{Me}$

95%

Davis, F.A.; Lal, S.G.; Durst, H.P. *J. Org. Chem., 1988, 53,* 5004.

$\underset{\text{MeS}}{\diagdown}\overset{\text{Ph}}{\diagup}\text{OSiPh}_2t\text{-Bu}$

$\xrightarrow{\text{cumyl-OOH}}$

$\underset{\text{Me}}{\overset{\overset{\parallel}{O}}{\diagdown}}\text{S}\diagdown\overset{\text{Ph}}{\diagup}\text{OSiPh}_2t\text{-Bu}$

90%

87:13 (74% ee:76% ee)

Conte, V.; DiFuria, F.; Licini, G.; Modena, G. *Tetrahedron Lett., 1989, 30,* 4859.

56%

79% ee

Colonna, S.; Gaggero, N. *Tetrahedron Lett.*, *1989*, *30*, 6233.

REVIEW:

"Chiral Sulfoxidation by Biotransformation of Organic Sulfides"

Holland , H.L. *Chem. Rev.*, *1988*, *88*, 473.

SECTION 220: OXIDES FROM HALIDES AND SULFONATES

48%

Harris, A.R. *Synth. Commun.*, *1988*, *18*, 659.

61%

Sane, P.V.; Sharma, M.M. *Org. Prep. Proceed. Int.*, *1988*, *20*, 598.

85%

Ballini, R.; Marcantoni, E.; Petrini, M. *Tetrahedron*, *1989*, *45*, 6791.

SECTION 221: OXIDES FROM HYDRIDES

Sura, T.P.; Ramana, M.M.V.; Kudav, N.A. *Synth. Commun.*, *1988*, 18, 2161.

Poirier, J.-M.; Vottero, C. *Tetrahedron*, *1989*, *45*, 1415.

SECTION 222: OXIDES FROM KETONES

NO ADDITIONAL EXAMPLES

SECTION 223: OXIDES FROM NITRILES

NO ADDITIONAL EXAMPLES

SECTION 224: OXIDES FROM ALKENES

NO ADDITIONAL EXAMPLES

SECTION 225:   OXIDES FROM MISCELLANEOUS COMPOUNDS

1. n-Pr-Li

2. n-Pr-Li

$(EtO)_3P$ → 3. MeI → $(EtO)_3P$ 72%

Teulade, M.-P.; Savignac, P. *Tetrahedron Lett.*, *1987*, *28*, 405.

Me···P (Ph), , Zn(Cu) , RT

aq. EtOH ⟩⟩⟩⟩ → Me···P (Ph)

63% , S

Pietrusiewicz, K.M.; Zablocka, M. *Tetrahedron Lett.*, *1988*, *29*, 937.

O···P–H (Me) cat. Pd(PPh₃)₄ , NEt₃

PhH → O···P–Ph (Me)

90%

>97% ee , S

Zhang, J.; Xu, Y.; Huang, G.; Guo, H. *Tetrahedron Lett.*, *1988*, *29*, 1955.

=N⁺ (O⁻, OSiMe₃) 1. PhCHO , TBAF

THF , -78°C

2. NaH , reflux , THF → O₂N⟍⟋Ph

90%

Lee, K.; Oh, D.Y. *Synth. Commun.*, *1989*, *19*, 3055.

REVIEWS:

"Synthesis and Selected Reductions of Conjugated Nitroalkanes: A Review"

Kabalka, G.W.; Varma, R.S. *Org. Prep. Proceed. Int.*, *1987*, *19*, 285.

"Sultone Chemistry"

Roberts, D.W.; Williams, D.L. *Tetrahedron*, *1987*, *43*, 1027.

"Applications of Oxaziridines in Organic Synthesis"

Davis, F.A.; Sheppard, A.C. *Tetrahedron*, *1989*, *45*, 5703.

CHAPTER 16

PREPARATION OF DIFUNCTIONAL COMPOUNDS

SECTION 300: ALKYNE - ALKYNE

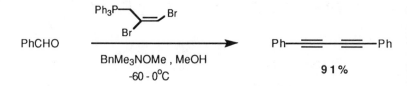

PhCHO $\xrightarrow{\text{BnMe}_3\text{NOMe , MeOH} \atop -60 - 0^{\circ}\text{C}}$ Ph—≡—≡—Ph

91%

Ziegler Jr., C.B.; Harris, S.M.; Baldwin, J.E. *J. Org. Chem.*, **1987**, *52*, 443.

Cu-C≡C-H $\xrightarrow[\text{3. Cu(NO}_3)_2 \cdot 3\text{ H}_2\text{O,} \atop \text{THF , 23}^{\circ}\text{C , 10 min}]{\text{1. BuLi , THF} \atop \text{2. Bu}_3\text{SnCl , THF , -20}^{\circ}\text{C}}$ Bu-C≡C-C≡C-Bu

60%

Ghosal, S.; Luke, G.P.; Kyler, K.S. *J. Org. Chem.*, **1987**, *52*, 4296.

Ph—≡—H $\xrightarrow[\text{2. TBAF , THF}]{\text{1. BuNH}_2\text{ , CuI , } \atop \text{Pd(PPh}_3)_4\text{ , 16h}}$ Ph—≡—≡—H

68%

Kende, A.S.; Smith, C.A. *J. Org. Chem.*, **1988**, *53*, 2655.

SECTION 301: ALKYNE - ACID DERIVATIVES

NO ADDITIONAL EXAMPLES

SECTION 302: ALKYNE - ALCOHOL, THIOL

$CH_2ClCH(OEt)_2$ → (1. $LiNEt_2$, THF / 2. acetone / 3. aq. NH_4Cl) → product **60%**

Raucher, S.; Bray, B.L. *J. Org. Chem.*, **1987**, *52*, 2332.

(95 : 5) **95%**

Mead, K.T. *Tetrahedron Lett.*, **1987**, *28*, 1019.

(25 : 75) **83%**

Tabuchi, T.; Inanaga, J.; Yamaguchi, M. *Chem. Lett.*, **1987**, 2273.

LTMP, THF/hexane, -78 - 23°C

12%

Marshall, J.A.; Nelson, D.J. *Tetrahedron Lett.*, **1988**, *29*, 741.

Yadav, J.S.; Chander, M.C.; Joshi, B.V. *Tetrahedron Lett., 1988, 29*, 2737.
Yadav, J.S.; Chander, M.C.; Srinivas Rao, C. *Tetrahedron Lett., 1989, 30*, 5455.

Jeffery, T. *Tetrahedron Lett., 1989, 30*, 2225.

Michelot, D. *Synth. Commun., 1989, 19*, 1705.

SECTION 303: ALKYNE - ALDEHYDE

NO ADDITIONAL EXAMPLES

SECTION 304: ALKYNE - AMIDE

CSI = chlorosulfonyl isocyanate

Page, P.C.B.; Rosenthal, S.; Williams, R.V. *Synthesis, 1988*, 621.

H—C≡C—C(=O)—OEt

$\xrightarrow[\text{PhNH}_2\text{ , 25}^{\circ}\text{C , 96h}]{\textit{Candida cylindracea}}$

H—C≡C—C(=O)—NHPh

80%

Rebolledo, F.; Brieva, R.; Gotor, V. *Tetrahedron Lett.*, *1989*, *30*, 5345.

SECTION 305: ALKYNE - AMINE

(95   :   5)   63%

Yamaguchi, R.; Hata, E.; Utimoto, K. *Tetrahedron Lett.*, *1988*, *29*, 1785.

Ph-C≡C-Li

$\xrightarrow[\text{2. H}_2\text{O   3. aq. NaOH}]{\text{1.}\qquad\qquad\text{, THF-hexane RT}}$

Ph-C≡C-CH$_2$NBn$_2$

76%

Katritzky, A.R.; Gallos, J.K.; Yannakopoulou, K. *Synthesis*, *1989*, 31.

$\xrightarrow[\text{3. LiAlH}_4]{\substack{\text{1. MeI , THF}\\\text{2. Li-C≡C-Ph}}}$

70%

Takahata, H.; Yakahashi, K.; Wang, E.C.; Yamazaki, T. *J. Chem. Soc., Perkin Trans. I,* *1989*, 1211.

Ph-C≡C-H

$\xrightarrow[\text{3. Ph-C≡C-Li , ether , 0}^{\circ}\text{C}]{\substack{\text{1. BuLi , ether}\\\text{2. PhNHCH}_2\text{OMe , ether ,0}^{\circ}\text{C}}}$

PhNHCH$_2$C≡C-H

90%

Barluenga, J.; Campos, P.J.; Canal, G. *Synthesis*, *1989*, 33.

SECTION 306: ALKYNE - ESTER

Onaka, M.; Mimura, T.; Ohno, R.; Ilzumi, Y. *Tetrahedron Lett., 1989, 30,* 6341.

SECTION 307: ALKYNE - ETHER, EPOXIDE, THIOETHER

Moyano, A.; Charbonnier, F.; Greene, A.E. *J. Org. Chem., 1987, 52,* 2919.

Hayashi, M.; Inumushi, A.; Mukaiyama, T. *Bull. Chem. Soc., Jpn., 1988, 61,* 4037.

SECTION 308: ALKYNE - HALIDE

$$Bu\text{-}C\equiv C\text{-}H \xrightarrow[\text{2. } IPy_2^+ BF_4^-, MeOH]{\text{1. NaOMe , MeOH}} Bu\text{-}C\equiv C\text{-}I$$

88%

Barluenga, J.; González, J.M.; Rodríquez, M.A.; Campos, P.J.; Asensio, G. *Synthesis, 1987,* 661.

$$H\!\!=\!\!\equiv\!\!-C_6H_{13} \xrightarrow[K_2ClO_3]{Cul\,,\,Bu_4NCl\,,\,DMF} I\!\!=\!\!\equiv\!\!-C_6H_{13}$$

90%

Jeffery, T. *J. Chem. Soc., Chem. Commun., 1988*, 909.

$$C_6H_{13}\text{-}C\!\equiv\!C\text{-}H \xrightarrow[\substack{2.\,Me_3Si\text{-}O\text{-}O\text{-}SiMe_3\\ZnI_2\,,\,ether\,,\,-78^oC\,\to\,RT}]{1.\,BuLi\,,\,ether\,,\,-78^oC} C_6H_{13}C\!\equiv\!C\text{-}I$$

80%

Ricci, A.; Taddei, M.; Dembechm, P.; Guerrini, A.; Seconi, G. *Synthesis, 1989*, 461.

SECTION 309: ALKYNE - KETONE

1.5 eq. PDC , RT

CH_2Cl_2

62%

Liotta, D.; Brown, D.; Hoekstra, W.; Monahan III, R. *Tetrahedron Lett., 1987, 28*, 1069.

TrClO$_4$, 6h

CH_2Cl_2 , -78°C

Ph$_{\diagdown}$OSiMe$_2t$-Bu

(99   :   1)   90%

Kobayashi, S.; Matsui, S.; Mukaiyama, T. *Chem. Lett., 1988*, 1491.

.05 CrO$_3$, RT , PH

7 eq t-BuOOH

.05 p-TsOH , H$_2$O

44h   62%

Muzart, J.; Piva, O. *Tetrahedron Lett., 1988, 29*, 2321.

erythro:threo = 91:9

Mukai, C.; Nagami, K.; Hanaoka, M. *Tetrahedron Lett.*, **1989**, *30*, 5623, 5627.

Haruta, J.; Nishi, K.; Matsuda, S.; Tamura, Y.; Kita, Y. *J. Chem. Soc., Chem. Commun.*, **1989**, 1065.

SECTION 310: ALKYNE - NITRILE

Yadla, R.; Rao, J.M. *Indian J. Chem.*, **1988**, *27B*, 1125.

SECTION 311: ALKYNE - ALKENE

Ratovelomanana, V.; Hammoud, A.; Linstrumelle, G. *Tetrahedron Lett.*, **1987**, *28*, 1649.

Me$_3$Sn-C≡C-H

$\xrightarrow{\text{Pd(PPh}_3)_4 \text{ , THF}}$
25°C , 24h

C≡C-H

90%

Stille, J.K.; Simpson, J.H. *J. Am. Chem. Soc., 1987, 109,* 2138.

n-Bu-C≡C-I-Ph $^+$ OTs$^-$

$\xrightarrow[\text{Ph-C≡C-H}]{\textit{n}\text{-BuCu}}$

Ph
|
Bu

78%   Bu

Stang, P.J.; Kitamura, T. *J. Am. Chem. Soc., 1987, 109,* 7561.

C$_6$H$_{13}$-C≡CH

$\xrightarrow[\text{2\% Pd(OAc)}_2 \text{ , 15h}]{2.2 \text{ eq } [\text{(OMe)}_3\text{P}] \text{ , PhH , RT}}$

C$_6$H$_{13}$C≡C   C$_6$H$_{13}$

63%

Trost, B.M.; Chan, C.; Ruhter, G. *J. Am. Chem. Soc., 1987, 109,* 3486.

C$_6$H$_{13}$——≡——I

$\xrightarrow[\substack{\text{Pd(OAc)}_2 \text{ , Bu}_4\text{NCl} \\ \text{DMF , RT}}]{\text{CO}_2\text{Et , K}_2\text{CO}_3}$

C$_6$H$_{13}$——≡————CO$_2$Et

50%

Jeffery, T. *Synthesis, 1987,* 70.

C$_5$H$_{11}$——≡——ZnCl

$\xrightarrow[\substack{\text{Pd(PPh}_3)_4 \text{ , } \\ \text{-20 - 0}^\circ\text{C}}]{\substack{\text{Br——=——SiMe}_3 \text{ , THF} \\ \text{Br——=——SiMe}_3}}$

C$_5$H$_{11}$——≡——=——SiMe$_3$

78%

99.5% E

Andreini, B.P.; Carpita, A.; Rossi, R. *Tetrahedron Lett., 1988, 29,* 2239.

TDMPP = Tris-(2,6-dimethoxyphenyl)
phosphine

Trost, B.M.; Matsubara, S.; Caringi, J.J. *J. Am. Chem. Soc.*, *1989*, *111*, 8745.

Hatanaka, Y.; Matsui, K.; Hiyama, T. *Tetrahedron Lett.*, *1989*, *30*, 2403.

SECTION 312: CARBOXYLIC ACID - CARBOXYLIC ACID

'good yield'

Belletire, J.L.; Fry, D.F. *J. Org. Chem.*, *1987*, *52*, 2549.

Osowska-Pacewicka, K.; Alper, H. *J. Org. Chem.*, *1988*, *53*, 808.

Depres, J.-P.; Greene, A.E. *Tetrahedron Lett.*, *1989*, *30*, 7065.

SECTION 313: CARBOXYLIC ACID - ALCOHOL, THIOL

OH
MeO_2C
CO_2Me
OH
THF
H H
N
N
N
N
S
MeO_2C
CO_2Me
O
O
S
n-Bu_3SnH
THF
OH
MeO_2C
CO_2Me
67%

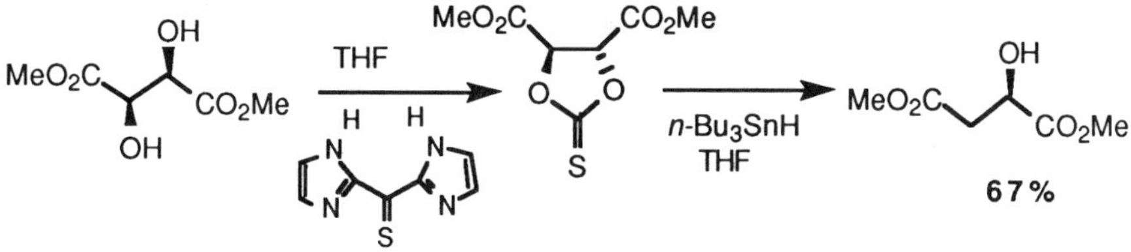

Alpegiani, M.; Hanessian, S. *J. Org. Chem.*, ***1987***, *52*, 278.

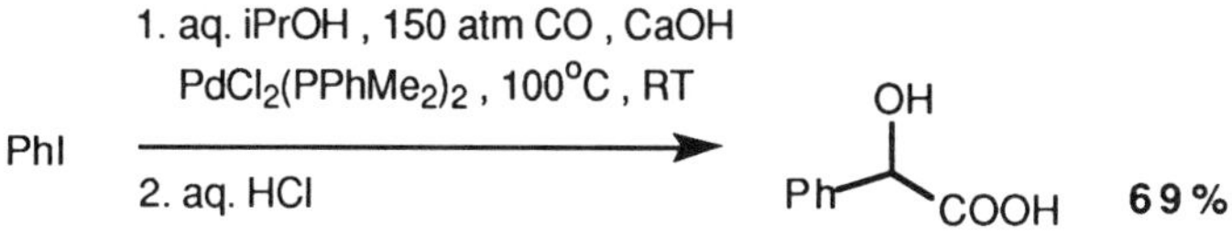

1. aq. iPrOH , 150 atm CO , CaOH
$PdCl_2(PPhMe_2)_2$, 100°C , RT
PhI
2. aq. HCl
OH
Ph
COOH
69%

Kobayashi, T.; Sakakura, T.; Tanaka, M. *Tetrahedron Lett.*, ***1987***, *28*, 2721.

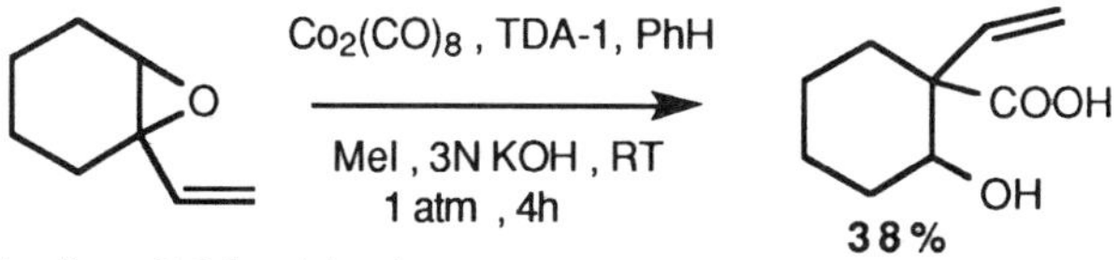

O
$Co_2(CO)_8$, TDA-1, PhH
MeI , 3N KOH , RT
1 atm , 4h
COOH
OH
38%

TDA-1 - tris-dioxa-3,6-heptylamine

Alper, H.; Calet, S. *Tetrahedron Lett.*, ***1988***, *29*, 1763.

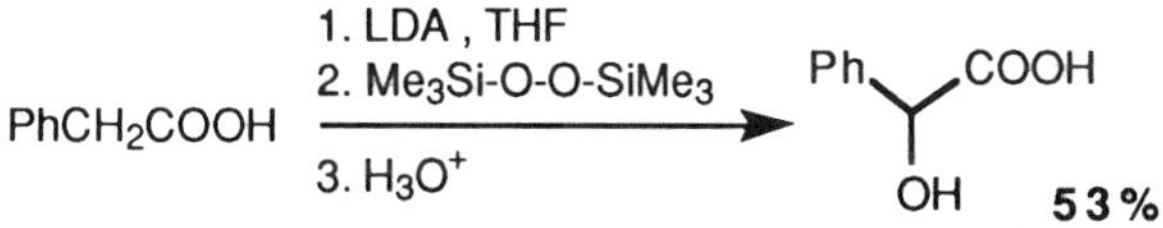

1. LDA , THF
2. Me_3Si-O-O-$SiMe_3$
$PhCH_2COOH$
3. H_3O^+
Ph
COOH
OH
53%

Pohmakotr, M.; Winotai, C. *Synth. Commun.*, ***1988***, *18*, 2141.

SECTION 314: CARBOXYLIC ACID - ALDEHYDE

NO ADDITIONAL EXAMPLES

SECTION 315: CARBOXYLIC ACID - AMIDE

1. LTMP , THF , -70°C
2. HgCl$_2$, THF
3. MeLi, ether , -70°C
4. CO$_2$
5. 10% aq. HCl

68%

Eaton, P.E.; Daniels, R.G.; Casucci, D.; Cunkle, G.T.; Engel, P. *J. Org. Chem.*, *1987, 52*, 2100.

1. PhNHSiMe$_3$, CH$_2$Cl$_2$
   1h , 20°C
2. MeOH

94%

Rigo, B.; Fasseur, D.; Cauliez, P.; Couturier, D. *Tetrahedron Lett.*, *1989, 30*, 3073.

SECTION 316: CARBOXYLIC ACID - AMINE

1. *n*-BuLi
2. BrCH$_2$CD$_2$OTf
3. 0.25 N HCl
4. 6N HCl

34% overall

46% ee

Subramanian, P.K.; Woodard, R.W. *J. Org. Chem.*, *1987, 52*, 15.

Zn^{+2} , pH 10.23 24h , RT , aq. MeOH

, CH$_3$CHO

73%

allo threo:threo = 1.7:1
% ee = 88% S : 74%S

Kuzuhara, H.; Watanabe, N.; Ando, M. *J. Chem. Soc., Chem. Commun.*, *1987*, 95.

1. $Ph_3P=CHCO_2Et$, PhMe , 2h
reflux

2. $(HCHO)_n$, PhH , reflux, 7h

3. 3M HCl, dioxane , reflux
18h

24%

Baldwin, J.E.; Adlington, R.M.; Robinson, N.G. *J. Chem. Soc., Chem. Commun.,* **1987,** 153.

1. Bu_2BOTf , CH_2Cl_2 , NEt_3
2. NBS , CH_2Cl_2 , -78°C
3. 3 eq   tetramethyl
guanidinium azide
CH_2Cl_2 , 0°C , 3h
4. H_2 , 10% Pd/C
5. (+)-MTPA-Cl , NEt_3 , 0°C
CH_2Cl_2 , 1h
6. $Ti(OBn)_4$

NH_3^+ MTPA

<78%

S:R = 82:18

Evans, D.A.; Ellman, J.A.; Dorow, R.L. *Tetrahedron Lett.,* **1987,** *28,* 1123.

1. LDA , THF , -70°C
2. THF , HMPA , -40°C

OSO_2OMe

3. H_3O^+

60%

40% ee , S

Duhamel, P.; Eddine, J.J.; Valnot, J.-Y. *Tetrahedron Lett.,* **1987,** *28,* 3801.

1. *t*-BuLi , THF , -50°C
30 min

2. NH_4Cl , H^+

BzONH   COOH   **69%**

Kolasa, T.; Sharma, S.; Miller, M.J. *Tetrahedron Lett.,* **1987,** *28,* 4973.

1. Ph$_3$BiCO$_3$, DMF
reflux

2. H$_3$O$^+$

54%

O'Donnell, M.J.; Bennett, W.D.; Jacobsen, W.N.; Ma, Y. *Tetrahedron Lett.*, *1989*, 30, 3913.

1. Et$_2$NH , CHCl$_3$
25°C , 17h

2. MeOH , 25°C
4h

99%

Kwiatkowski, S.; Jeganathan, A.; Tobin, T.; Watt, D.S. *Synthesis*, *1989*, 946.

REVIEW:

A series of articles and reviews concerning amino acid synthesis

O'Donnell, M.J. (Ed.), *Tetrahedron*, *1988*, 44, 5253-5614.

Related Methods: Section 315 (Carboxylic Acid - Amide).
                          Section 344 (Amide - Ester).
                          Section 351 (Amine - Ester).

SECTION 317: CARBOXYLIC ACID - ESTER

PPL , pH 7.2 , RT

KH$_2$PO$_4$

(R,S)

93% R
95% ee

76% S
73% ee

PPL = porcine pancreatic lipase

Guibé-Jampel, E.; Rousseau, G.; Salaün, J. *J. Chem. Soc., Chem. Commun.*, *1987*, 1080.

Yamamoto, K.; Nishioka, T.; Oda, J. *Tetrahedron Lett.*, **1988**, *29*, 1717.

Martín, V.S.; Nuñez, M.TY.; Tonn, C.E. *Tetrahedron Lett.*, **1988**, *29*, 2701.

Pseudomonas fluorescens (Amano P)	(92	:	8) **quant**
without lipase	(57	:	43)

Hiratake, J.; Yamamoto, K.; Yamamoto, Y.; Oda, J. *Tetrahedron Lett.*, **1989**, *30*, 1555.

SECTION 318: CARBOXYLIC ACID - ETHER, EPOXIDE, THIOETHER

NO ADDITIONAL EXAMPLES

SECTION 319: CARBOXYLIC ACID - HALIDE, SULFONATE

$$FCIC\text{-}CF_2 \xrightarrow[\substack{\text{3. KF, ethylene glycol} \\ \text{60 torr, } 100^\circ C \\ \text{4. aq. NaOH 5. conc. } H_2SO_4}]{\substack{\text{1. pyrrolidine, ether, } 10^\circ C, 2h \\ \text{2. } H_2O/\text{ice}}} F_2CHCO_2H$$

43%

Luz, M.C.; <u>Dailey, W.P.</u> *Org. Prep. Proceed. Int.,* **1987,** *19,* 468.

NiOOH/anode , $30^\circ C$

aq. KOH , 0.1 N NaOH

94%

Ruholl, H.; Schäfer, H.-J. *Synthesis,* **1987,** 408.

PhCHO

1. (Et, OSiMe₃ / H, OSiMe₃) , 2 TiCl₄

CH_2Cl_2 -, -60 - +20°C

2. H_2O

80%

Bellassoued, M.; Dubois, J.-E.; Bertounesque, E. *Tetrahedron Lett.,* **1988,** *29,* 1275.

$$ClCF_2CO_2CH=CH_2 \xrightarrow[\substack{\text{2. } SiO_2 \cdot H_2O, 12h \\ \text{3. EtOH , } CHCl_3, \text{ reflux , 12h}}]{\substack{\text{1. } Zn^\circ, Me_3SiCl, MeCN, \\ 100^\circ C, 2h}} CH_2=CHCH_2CF_2CO_2H$$

78%

Greuter, H.; <u>Lang, R.W.</u>; Romann, A.J. *Tetrahedron Lett.,* **1988,** *29,* 3291.

SECTION 320: CARBOXYLIC ACID - KETONE

5.5 NaIO₄ , RuCl₃•3H₂O

MeCN , CCl₄ , 5h

60%

Webster, F.X.; Rivas-Enterrios, J.; <u>Silverstein, R.M.</u> *J. Org. Chem.,* **1987,** *52,* 689.

SiO$_2$, KMnO$_4$

(chromatography
column - PhH eluant)

84%

Ferreira. J.T.B.; Cruz, W.O.; Vieira, P.C.; Yonashiro, M. *J. Org. Chem.*, *1987*, 52, 3698.

1. *t*-BuLi, THF , -78°C
2. CuI , 0°C
3. ClCH$_2$P(=O)(OEt)$_2$
   reflux

65%

Poindexter, M.K.; Katz. T.J. *Tetrahedron Lett.*, *1988*, 29, 1513.

MeCN , 60°C , O$_2$

H$_5$[PMo$_{10}$V$_2$O$_{40}$]•30-36 H$_2$O

94%

El Ali, B.; Brégeault. J.-M.; Mercier, J.; Martin, J.; Martin, C.; Convert, O. *J. Chem. Soc., Chem. Commun.*, *1989*, 825.

Also via: Section 360 (Ketone - Ester).

SECTION 321: CARBOXYLIC ACID - NITRILE

NO ADDITIONAL EXAMPLES

Also via: Section 361 (Nitrile - Ester).

SECTION 322: CARBOXYLIC ACID - ALKENE

3 CH$_2$(COOH)$_2$
reflux , 3 h

Py-H$^+$ OAc$^-$, xylene

E   60%

Ragoussis. N. *Tetrahedron Lett.,* **1987**,*28*, 93.

1. HCl

2. Ph—N(Li)—CH$_2$-adamantyl
   |
   Me

84%

82% ee , S

Duhamel. L.; Ravard, A.; Plaquevent, J.,-C.; Davoust, D. *Tetrahedron Lett.,* **1987**, *28*,
5517.

1. Me$_3$Si—C(OSiMe$_3$)=CH—OSiMe$_3$
    ZnBr$_2$, THF , RT , 24h

Ph-CHO

2. H$_2$O

Ph⌒COOH

95%

Bellassoued, M.; Gaudemar, M. *Tetrahedron Lett.,* **1988**, *29*, 4551.

Bu$_2$CuMgI , THF-Me$_2$S

-78°C , 1h

Me
|
Bu⌒COOH

83%

E:Z = 82:18

Kawashima, M.; Sato, T.; Fujiwsawa. T. *Bull. Chem. Soc., Jpn.,* **1988**, *61*, 3255.

1. PhCH(Li)CO$_2$Li

2. PhSO$_2$Cl , Py

3. MgBr$_2$

Ph
|
⌒—CH—COOH

56%

Black. T.H.; Maluleka, S.L. *Tetrahedron Lett.,* **1989**, *30*, 531.

Also via: Section 313 (Alcohol - Carboxylic Acids).
                   Section 349 (Amide - Alkene).
                   Section 362 (Ester - Alkene).
                   Section 376 (Nitrile - Alkene).

SECTION 323: ALCOHOL, THIOL - ALCOHOL, THIOL

1. Et_2BOMe , MeOH/THF
   -70 , 15 min

2. $NaBH_4$, 3 h
3. AcOH

99%

(syn:anti = 99:1)

Chen, K.-M.; Hardtmann, G.E.; Prasad. K.; Repič, O.; Shapiro, M.J. *Tetrahedron Lett.,* *1987,.28,* 155.

1. $(dba)_3Pd_2 \cdot CHCl_3$
   dppp , THF , 40 psi CO_2

2. Dibal-H , CH_2Cl_2 , -78°C

3. $Ph_3P=CMeCO_2Et$,
   CH_2Cl_2

4. NaOMe , MeOH , RT

30% overall

Trost. B.M.; Lynch, J.K.; Angle, S.R. *Tetrahedron Lett., 1987, 28,* 375.

82%   18%

Canonne. P.; Boulanger, R.; Bernatchez, M. *Tetrahedron Lett., 1987, 28,* 4997.

PNB = *p*-nitrobenzoate   (2.3   :   1) 66%

Ko, S.Y.; Masamune, H.; <u>Sharpless, K.B.</u> *J. Org. Chem.*, *1987*, *52*, 667.

Tamao, K.; Nakajo, E.; <u>Ito, Y.</u> *J. Org. Chem.*, *1987*, *52*, 957.

<u>Nakayama, J.</u>; Yamaoka, S.; Hoshino, M. *Tetrahedron Lett.*, *1987*, *28*, 1799.

anti:syn = 87:13

Murphy, P.J.; <u>Procter, G.</u>; Russell, A.T. *Tetrahedron Lett.*, *1987*, *28*, 2037.

(100   :   1) 98%

Handa, Y.; <u>Inanaga, J.</u> *Tetrahedron Lett.*, *1987*, *28*, 5717.

$$\text{5 eq. MeTi(OiPr)}_3 \xrightarrow{\quad \text{CH}_2\text{Cl}_2 \quad}_{-78^\circ\text{C - RT}}$$

93%

syn:anti = 7.3:1

Tomooka, K.; Okinaga, T.; Suzuki, K.; Tsuchihashi, G. Tetrahedron Lett., *1987, 28*, 6335.

1.

OsO$_4$, THF, -110°C, 6h

2. add LiAlH$_4$, RT, overnight

85%

97% ee

Tomioka, K.; Nakajima, M.; Iitaka, Y.; Koga, K. Tetrahedron Lett., *1988, 29*, 573.

$$\xrightarrow{\quad \text{TiCl}_4,\ \text{Mg}^\circ \quad}$$

+

(1   :   1) 60%

Rao, S.A.; Periasamy, M. Tetrahedron Lett., *1988, 29*, 1583.

$$\xrightarrow[\text{H}_2,\ \text{MeOH},\ 20\text{h (50 kgw/cm}^2)]{0.2\%\ \text{Ru}_2\text{Cl}_4[(\text{R})\text{-DINAP}]_2\text{·NEt}_3}$$

98%

BINAP = 2,2'-bis-(diphenylphosphino)-
1,1'-binaphthyl

anti:syn = 99:1
(>99% ee)

Kawano, H.; Ishii, Y.; Saburi, M.; Uchida, Y. J. Chem. Soc., Chem. Commun., *1988*, 87.

90%

Hou, Z.; Takamine, K.; Aoki, O.; Shiraishi, H.; Fujiwara, Y.; Tankguchi, H. *J. Chem. Soc., Chem. Commun.*, *1988*, 668.

72%

(200:1)

Molander, G.A.; Kenny, C. *J. Org. Chem.*, *1988*, *53*, 2132.

80%

Horiuchi, C.A.; Satoh, J.Y. *Chem. Lett.*, *1988*, 1209.

Anwar, S.; Davis, A.P. *Tetrahedron*, *1988*, *44*, 3761.

67%

Zhang, Y.; Liu, T. *Synth. Commun.*, *1988*, *18*, 2173.

60%   18%

Hou, Z.; Takamine, K.; Aoki, O.; Shiraishi, H.; Fujiwara, Y.; Taniguchi, H. *J. Org. Chem.*, **1988**, *53*, 6077.

(4   :   1)   67%

Fruedenberger, J.H.; Jonradi, A.W.; Pedersen, S.F. *J. Am. Chem. Soc.*, **1989**, *111*, 8014.

(7.5   :   1)

Burgess, K.; Ohlmeyer, M.J. *Tetrahedron Lett.*, **1989**, *30*, 395.

(100   :   0)   88%

1,12-dodecanedicarboxaldehyde gave 75% of cis:trans diols (25:75).

McMurry, J.E.; Rico, J.G. *Tetrahedron Lett.*, **1989**, *30*, 1169.

82%

syn:anti = 1:1

Takahara, P.M.; Freudenberger, J.H.; Konradi, A.W.; Pedersen, S.F. *Tetrahedron Lett.*, **1989**, *30*, 7177.

syn:anti = >200:1

Harada, T.; Matsuda, Y.; Uchimura, J.; Oku, A. J. Chem. Soc., Chem. Commun., **1989**, 1429.

Delair, P.; Luche, J.-L. J. Chem. Soc., Chem. Commun., **1989**, 398.

Léonard, E.; Duñach, E.; Périchon, J. J. Chem. Soc., Chem. Commun., **1989**, 274.

Venturello, C.; Gambaro, M. Synthesis, **1989**, 295.

100% ee , SS   **96%**

Oishi, T.; Hirama, M. J. Org. Chem., **1989**, 54, 5834.

REVIEWS:

"Polyketide-Derived Natural Products Containing Three or More Consecutive Stereogenic Centers"

Hoffmann, R.W. *Angew. Chem. Int. Ed., Engl,* **1987**, *26*, 489.

"Carbonyl Coupling Reactions Using Transition Metals, Lanthanides and Actinides"

Kahn, B.E.; Rieke, R.D. *Chem. Rev.,* **1988**, *88*, 733.

"Enzyme Catalyzed Synthesis of Carbohydrates"

Toone, E.J.; Simon, E.S.; Bednarshi, M.D.; Whitesides, G.M. *Tetrahedron,* **1989**, *45*, 5365.

"The Applications of Low-Valent Titanium Reagents in Organic Synthesis"

Lenior, D. *Synthesis,* **1989**, 883.

Also via: Section 327 (Alcohol - Ester).
                   Section 357 (Ester - Ester).

SECTION 324: ALCOHOL, THIOL - ALDEHYDE

1. $LiC_{10}H_8$, -78°C
2. $n\text{-}C_3H_7CHO$
3. aq. HCl

54%

Barluenga, J.; Rubiera, C.; Fernández, J.R.; Yus, M. *J. Chem. Soc., Chem. Commun.,* **1987**, 425.

1. sBuLi , THF , -78°C
2. PhCHO , THF , -78°C
15 min

77%

Tius, M.A.; Astrab, D.P.; Gu, X. *J. Org. Chem.,* **1987**, *52*, 2625.

Matsukawa, M.; Inanaga, J.; Yamaguchi, M. *Tetrahedron Lett., 1987, 28,* 5877.

Chambert, P.; Ousset, J.B.; Mioskowski, C. *Tetrahedron Lett., 1989, 30,* 179.

Related Methods: Section 330 (Alcohol - Ketone).

SECTION 325: ALCOHOL, THIOL - AMIDE

Shono, T.; Matsumura, Y.; Onomura, O.; Ogaki, M.; Kanazawa, T. *J. Org. Chem., 1987, 52,* 536.

Parker, K.A.; Koziski, K.A. *J. Org. Chem., 1987, 52,* 674.

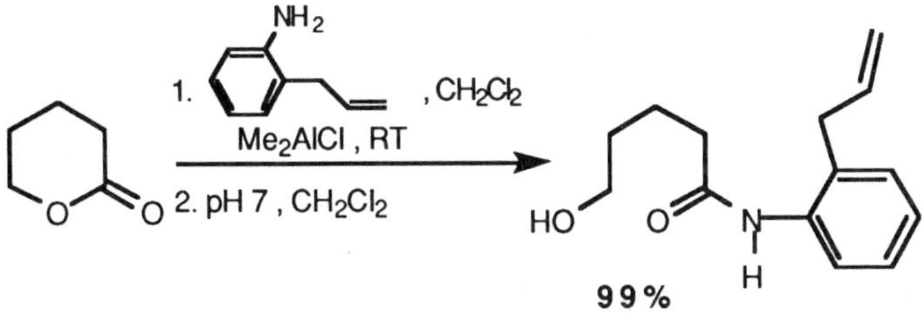

1.
NH2
, CH2Cl2
Me2AlCl , RT
2. pH 7 , CH2Cl2
O
O
HO
O
N
H
99%

Barrett, A.G.M.; Dhanak, D. *Tetrahedron Lett.*, ***1987***, *28*, 3327.

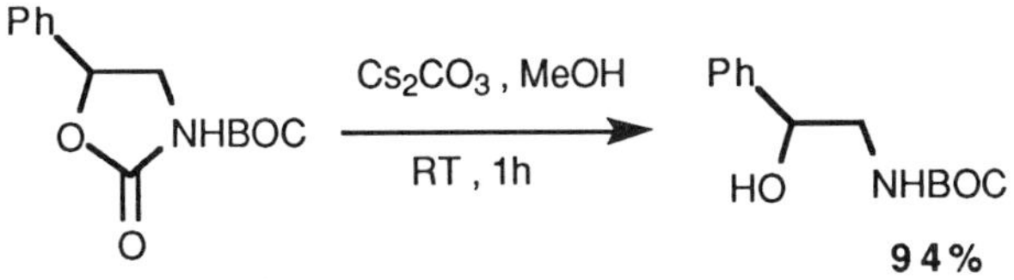

Ph
O
NHBOC
O
Cs2CO3 , MeOH
RT , 1h
Ph
HO
NHBOC
94%

Ishizuka, T.; Kunieda, T. *Tetrahedron Lett.*, ***1987***, *28*, 4185.

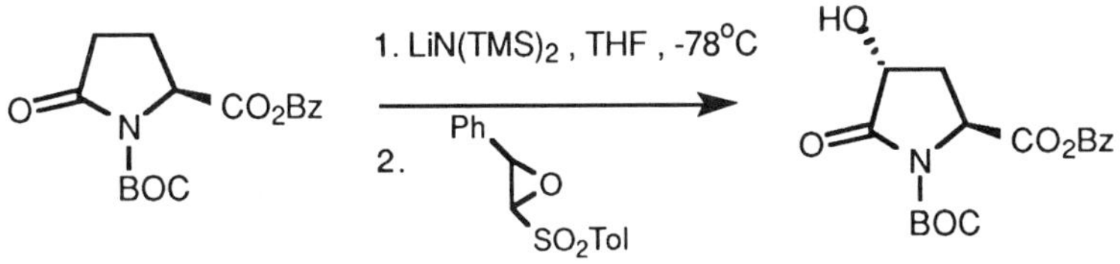

O
N
CO2Bz
BOC
1. LiN(TMS)2 , THF , -78°C
2.
Ph
O
SO2Tol
HO
O
N
CO2Bz
BOC

Ohta, T.; Hosoi, A.; Nozoe, S. *Tetrahedron Lett.*, ***1988***, *29*, 329.

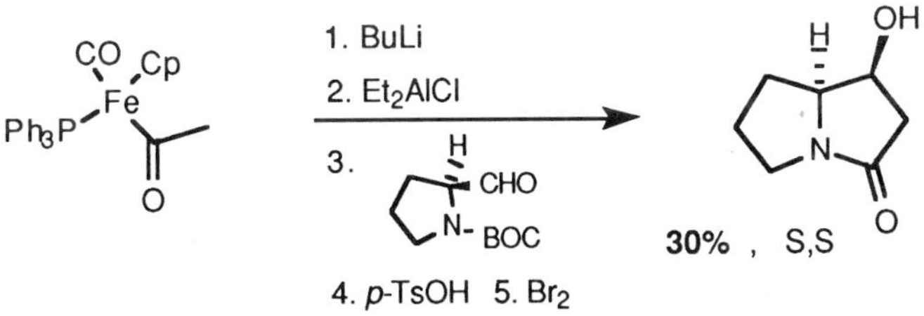

CO
Cp
Fe
Ph3P
O
1. BuLi
2. Et2AlCl
3.
H
CHO
N- BOC
4. p-TsOH 5. Br2
H
OH
N
O
30% , S,S

Beckett, R.P.; Davies, S.G. *J. Chem. Soc., Chem. Commun.*, ***1988***, 160.

60%   15%

Azzouzi, A.; Dufour, M.; Gramain, J.-C.; Remuson, R. *Heterocycles, 1988, 27,* 133.

(>99   :   1) 87%

Fujita, M.; Hiyama, T. *J. Org. Chem., 1988, 53,* 5415.

87%

Keusenkothen, P.F.; Smith, M.B. *Synth. Commun., 1989, 19,* 2859.

(80   :   20) 95%

Isayama, S.; Mukaiyama, T. *Chem. Lett., 1989,* 2005.

99%

Matsumoto, K.; Hashimoto, S.; Uchida, T.; Okamoto, T.; Otani, S. *Bull. Chem. Soc., Jpn.,*
1989, 62, 3138.
Matsumoto, K.; Hashimoto, S.; Okamoto, T.; Otani, S.; Hayami, J. *Chem. Lett., 1987,*
803.

SECTION 326: ALCOHOL, THIOL - AMINE

Marsella, J.A. *J. Org. Chem.*, *1987*, *52*, 467.

also with aldehydes

Roskamp, E.J.; Pedersen, S.F. *J. Am. Chem. Soc.*, *1987*, *109*, 6551.

Tanis, S.P.; Raggon, J.W. *J. Org. Chem.*, *1987*, *52*, 819.

Soai, K.; Niwa, S.; Kobayashi, T. *J. Chem. Soc., Chem. Commun.*, *1987*, 801.

Goralski, C.T.; Singaram, B.; Brown, H.C. *J. Org. Chem.*, *1987*, *52*, 4014.

36%

Guy, A.; Lemaire, M.; Graillot, Y.; Negre, M.; Guette, J.P. *Tetrahedron Lett.*, *1987*, *28*, 2969.

1. $(EtO)_2P(=O)Cl$
2. $BH_3 \cdot SMe_2$, $0°C$
3. $NaOH$, H_2O_2
4. HCl , reflux

65%

Dicko, A.; Montury, M.; Baboulene, M. *Tetrahedron Lett.*, *1987*, *28*, 6041.

1. Pd^0 , THF , reflux , 2h
2. 20% KOH , MeOH

90%

67% ee, S

Hayashi, T.; Yamamoto, A.; Ito, Y. *Tetrahedron Lett.*, *1988*, *29*, 99.

1. Et_2NH , DMF , 2h
2. $LiAlH_4$, THF

59%

Feringa, B.L.; de Lange, B. *Tetrahedron Lett.*, *1988*, *29*, 1303.

1. mesityl-Li , 0°C - RT
2. PhCHO
3. H₂O

66%

Comins, D.L.; LaMunyon, D.H. *Tetrahedron Lett.*, **1988**, *29*, 773.

Na / NH₃

74%

Kozikowski, A.P.; Musgrage, B.B. *J. Chem. Soc., Chem. Commun.*, **1988**, 198.

Mg(ClO)₄ , MeCN

, 25°C , 7d

72%

87% ee

Meyers, A.I.; Brown, J.D. *Tetrahedron Lett.*, **1988**, *29*, 5617.

¹DC≡N* , hv (>280 nm)

MeCN
4h

85%

Pandey, G.; Kumaraswamy, G. *Tetrahedron Lett.*, **1988**, *29*, 4153.

HNEt₂ , CH₂Cl₂ , 20h

Ph₄SbOTf

75%

Fujiwara, M.; Imada, M.; Baba, A.; Matsuda, H. *Tetrahedron Lett.*, **1989**, *30*, 739.

1. 1.5 eq. Hg(OAc)$_2$
2. 2 NaHCO$_3$, MeCN
   25°C , 10 min
2. 4 eq., LiBH$_4$,
   THF , -78°C
3. 4 eq NH$_2$NH$_2$ •H$_2$O
   cat. TsOH , EtOH , reflux

64%

Takacs, J.M.; Helle, M.A.; Yang, L. *Tetrahedron Lett.*, *1989*, *30*, 1777.

1. *t*-BuLi
2. PhCHO , BF$_3$ •OEt$_2$
3. NH$_2$NH$_2$

84%

syn: anti = 95:5

Sanner, M.A. *Tetrahedron Lett.*, *1989*, *30*, 1909.

1. Na° , iPrOH , 25°C
2. 4N HCl

(61   :   39)   95%

Barluenga, J.; Joglar, J.; González, F.J; Fustero, S. *Tetrahedron Lett.*, *1989*, *30*, 2001.

1. Bu$_3$PbNEt$_2$, ether
   RT , 30 min
2. 6N NaOH

(6   :   94)   84%

Yamada, J.; Yumoto, M.; Yamamoto, Y. *Tetrahedron Lett.*, *1989*, *30*, 4255.

LiBH$_4$, Me$_3$SiCl
THF , 24h

88%

Giannis, A.; Sandhoff, K. *Angew. Chem. Int. Ed., Engl.*, *1989*, *28*, 218.

SECTION 327: ALCOHOL, THIOL - ESTER

78:22 S:R

Brooks, D.W.; Kellogg, R.P.; Cooper, C.S. *J. Org. Chem.*, **1987**, *52*, 192.

Molander, G.A.; Etter, J.B. *J. Am. Chem. Soc.*, **1987**, *109*, 6556.

>99% ee , R

Noyori, R.; Ohkuma, T.; Kitamura, M.; Takaya, H.; Sayo, N.; Kumobayashi, H.; Akutagawa, S. *J. Am. Chem. Soc.*, **1987**, *109*, 5856.

syn:anti = 100:1

Bernardi, A.; Cardani, S.; Colombo, L.; Poli, G.; Schimperna, G.; Scolastico, C. *J. Org. Chem.*, **1987**, *52*, 888.

Bakers yeast , glucose

KH$_2$PO$_4$, NH$_4^+$H$_2$PO$_4^-$

MgSO$_4$, CaCO$_3$, H$_2$O

74%

threo   : erythro   = 2:3
(76% ee RS : 88% ee RR)

Tsuboi, S.; Nishiyama, E.; Furutani, H.; Utaka, M.; Takeda, A. *J. Org. Chem.*, **1987, 52,** 1359.

SmI$_2$•HMPA•DMAE

1 min , THF , RT

50%

DMAE = N,N-dimethylamino ethanol   β:α OH = 17:1

Otsubo, K.; Inanaga, J.; Yamaguchi, M. *Tetrahedron Lett.*, **1987, 28,** 4437.

1. BrCH$_2$CO$_2$Et , Zn$^\circ$,
   Me$_3$SiCl , ether

2. 2M HCl

improved Reformatsky reaction

70%

Picotin, G.; Miginiac, P. *J. Org. Chem.*, **1987, 52,** 4796.

mCPBA

53%

Russell, A.T.; Procter, G. *Tetrahedron Lett.*, **1987, 28,** 2041, 2045.

Me\
Me — ⟨ring⟩, O, O   →   50 atm H$_2$, [Rh(COD)Cl]$_2$\
THF , 50°C , 45h\
Me, O—H, PCy$_2$\
Me, O, PCy$_2$, H\
72% ee , R

Me\
Me — HO — ⟨ring⟩, O\
O   quant

Chiba, M.; Takahashi, H.; Takahashi, H.; Morimoto, T.; Achiwa, K. *Tetrahedron Lett.*, **1987**, *28*, 3675.

Ph — C(=O) — CO$_2$Et   →   ═CN , DABCO\
5d   EtO$_2$C — C(OH)(Ph)(═CH$_2$)CN   65%

Basavaiah, D.; Bharathi, T.K.; Gowriswari, V.V.L. *Tetrahedron Lett.*, **1987**, *28*, 4351.

Me — C(=O) — C(SMe)═(SMe)   →   OEt, ═C(O$^-$ ZnBr$^+$)\
1. ether/PhH , 40 h, reflux\
2. 5% H$_2$SO$_4$

OH\
CO$_2$Et\
Me ⟨ring⟩ SMe

Datta, A.; Ila, H.; Junjappa, H. *Tetrahedron Lett.*, **1988**, *29*, 497.

⟨structure⟩ — CO$_2$Et   →   GGDH , NAD$^+$, iPrOH   →   OH — CO$_2$Et   60%

GGDH = glycerol dehydrogenase   >99% ee

Nakamura, K.; Yoneda, T.; Miyai, T.; Ushio, K.; Oka, S.; Ohno, A. *Tetrahedron Lett.*, **1988**, *29*, 2453.

Kernan, M.R.; Faulkner, D.J. *J. Org. Chem.*, **1988**, *53*, 2773.

Gupton, J.T.; Dureanceau, S.J.; Miller, J.F.; Kosiba, M.L. *Synth. Commun.*, **1988**, *18*, 937.

Abd ElSamii, Z.K.M.; Al Ashmawy, M.I.; Mellor, J.M. *J. Chem. Soc., Perkin Trans. I*, **1988**, 2509.

RajanBabu, T.V.; Nugent, W.A. *J. Am. Chem. Soc.*, **1989**, *111*, 4525.

SAM-II = microbial ester hydrolase from
    Psuedomonas sp.

>95% ee

Ader, U.; Breitgoff, D.; Klein, P.; Laumen, K.E.; Schneider, M.P. *Tetrahedron Lett.*, **1989**, *30*, 1793.

* = immobilized Bakers Yeast on
    magnesium alginate

58% (81% ee)

Nakamura, K.; Kawai, Y.; Oka, S.; Ohno, A. *Tetrahedron Lett., 1989, 30*, 2245.

(58   :   42)   81%

Enholm, E.J.; Prasad, G. *Tetrahedron Lett., 1989, 30*, 4939.

83%

Deardorff, D.R.; Shambayati, S.; Linde II, R.G.; Dunn, M.M. *J. Org. Chem., 1988, 53*, 189.

97%

Hara, R.; Mukaiyama, T. *Chem. Lett., 1989*, 1909.

REVIEW:

"Recent Advances in the Reformatsky Reaction"

Fürstner, A. *Synthesis, 1989*, 571.

Also via: Section 313 (Alcohol - Carboxylic Acid).

SECTION 328: ALCOHOL, THIOL - ETHER, EPOXIDE, THIOETHER

(50 : 1) 65%

Keck. G.E.; Abbott, D.E.; Wiley, M.R. *Tetrahedron Lett., 1987,28*, 139.

(14 : 1) 87%

Koreeda. M.; Tanaka, Y. *Tetrahedron Lett., 1987,28*, 143.

95%

McCombie. S.W.; Metz, W.A. *Tetrahedron Lett., 1987, 28,* 383.

76%   19%

Yamaguchi, Y.; Yamada, H.; Hayakawa, K.; Kanematsu. K. *J. Org. Chem., 1987, 52,* 2040.

1. Et-S-Et , SO_2Cl_2

2. NEt_3

67%

Sato, K.; Inoue, S.; Miyamoto, O.; Ikeda, H.; Ota, T. *Bull. Chem. Soc., Jpn.,* ***1987***, *60*, 4184.
Sato, K.; Inoue, S.; Ozawa, K.; Kobayashi, T.; Ota, T.; Tazaki, M. *J. Chem. Soc., Perkin Trans. I,* ***1987***, 1753.

Me_2BBr , Bu_2CuLi

CH_2Cl_2

HO   Bu $OC_{12}H_{25}$ 89%

a variety of nucleophiles can be used

Guindon, Y.; Bernstein, M.A.; Anderson, P.C. *Tetrahedron Lett.,* ***1987***, *28*, 2225.

1. NBS , =   OH

   CH_2Cl_2 , -40°C

2. Bu_3SnCl ,$NaCNBH_3$

   AIBN , *t*-BuOH , 80°C

3. *p*-TsOH , PhH , RT

HO   Me   51%

Srikrishna, A.; Pullaiah, K.C. *Tetrahedron Lett.,* ***1987***, *28*, 5203.

0.1 $SmCl_3$, PhSH

CH_2Cl_2 , RT

OH   ''SPh   84%

Vougioukas, A.E.; Kagan, H.B. *Tetrahedron Lett.,* ***1987***, *28*, 6065.

0.3% $[Pt(PPh_3)_2(CH_2=CH_2)]$

THF , 60°C , 20h

64%   OH   OH

syn:anti = 94:1

Tamao, K.; Nakagawa, Y.; Arai, H.; Higuchi, N.; Ito, Y. *J. Am. Chem. Soc.,* ***1988***, *110*, 3712.

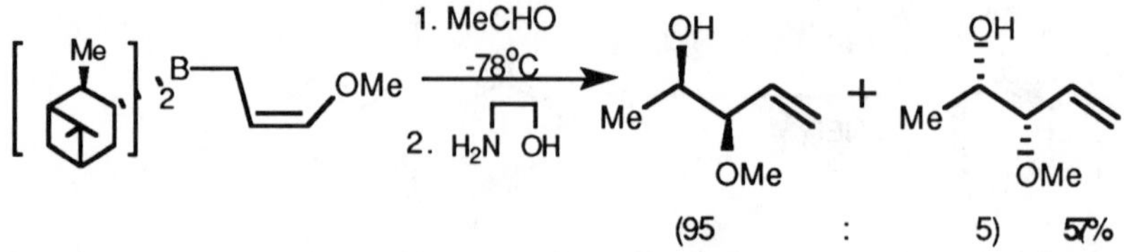

(95   :   5)   57%

Brown, H.C.; Jadhav, P.K.; Bhat, K.S. *J. Am. Chem. Soc.*, *1988*, *110*, 1535.

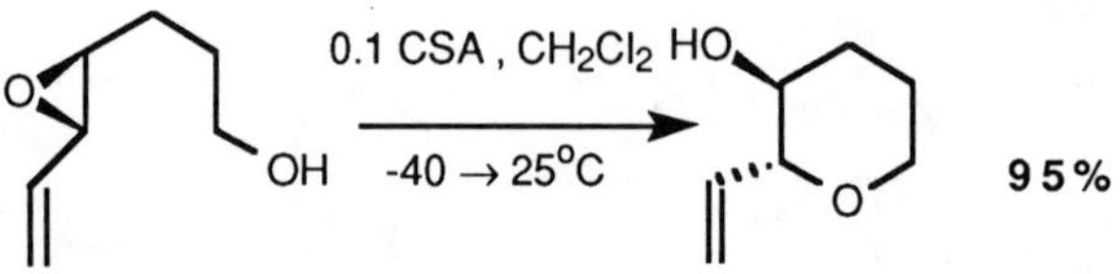

0.1 CSA , CH$_2$Cl$_2$

-40 → 25°C

95%

Nicolaou, K.C.; Prasad, C.V.C.; Somers, P.K.; Hwang, C.-K. *J. Am. Chem. Soc.*, *1989*, *111*, 5330, 5335.

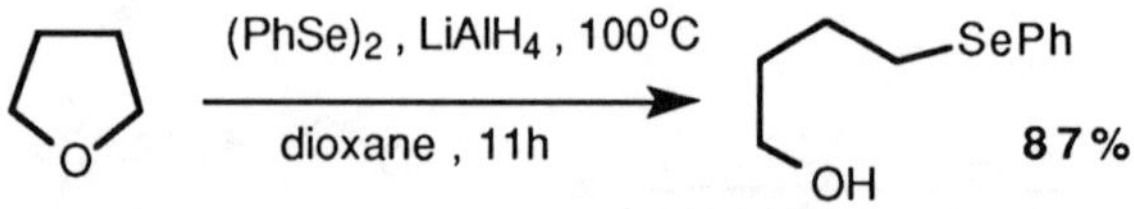

(PhSe)$_2$, LiAlH$_4$, 100°C

dioxane , 11h

87%

Haraguchi, K.; Tanaka, H.; Hayakawa, H.; Miyasaka, T. *Chem. Lett.*, *1988*, 931.

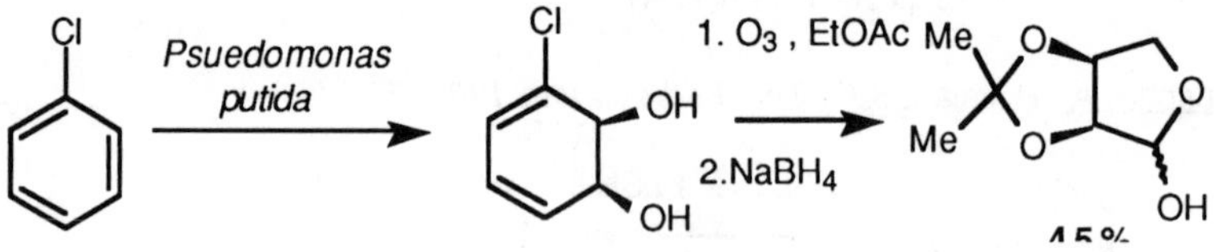

Psuedomonas putida

1. O$_3$, EtOAc

2.NaBH$_4$

45%

Hudlicky, T.; Luna, H.; Price, J.D.; Rulin, F. *Tetrahedron Lett.*, *1989*, *30*, 4053.
Hudlicky, T.; Luna, H.; Barbieri, G.; Kwart, L.D. *J. Am. Chem. Soc.*, *1988*, *110*, 4735.

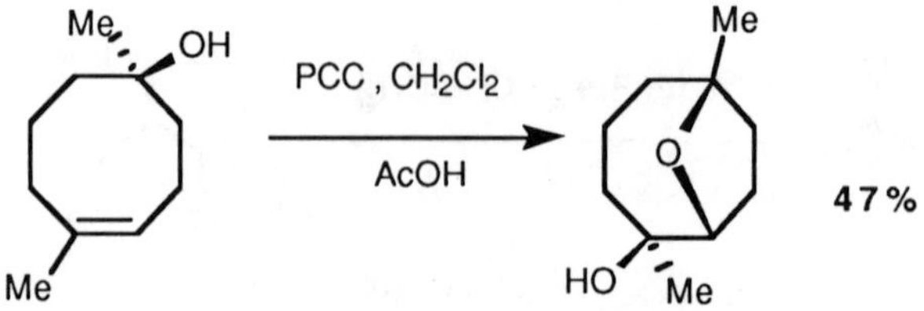

PCC , CH$_2$Cl$_2$

AcOH

47%

Schlecht, M.F.; Kim, H. *J. Org. Chem.*, *1989*, *54*, 583.

(PhSe)₂CHCH₂SiMe₃ →

1. BuLi , ether/THF
-80°C

2. PhCHO
-80°C - RT

Me₃Si–CH₂–CH(OH)–CH(SePH)–Ph

85%

Sarkar, T.K.; Satapathi, T.K. *Tetrahedron Lett., 1989, 30,* 3333.

Me₂CuLi , BF₃ •OEt₂
ether
→

67% de

95%

Normant, J.F.; Alexakis, A.; Ghribi, A.; Mangeney, P. *Tetrahedron, 1989, 45,* 507.

REVIEW:
"Selective Dealkylations of Aryl Alkyl Ethers, Thioethers and Selenoethers"

Tiecco, M. *Synthesis, 1988,* 749.

SECTION 329: ALCOHOL, THIOL - HALIDE, SULFONATE

1. Bu₂BOTf , NEt₃

2. (i-Pr)–CHO

→

75%

95:5 (major isomer:all others)

Evans, D.A.; Sjogren, E.B.; Weber, A.E.; Conn, R.E. *Tetrahedron Lett., 1987,.28,* 39.

Me₂B-Br , NEt₃ , CH₂Cl₂
0°C , 2h
→

Br–CH₂CH₂CH₂CH₂–OH

quant.

Guindon, Y.; Therien, M.; Girard, Y.; Yoakim, C. *J. Org. Chem., 1987, 52,* 1680.

Gassman, P.G.; O'Reilly, N.J. *J. Org. Chem.*, *1987*, *52*, 2481.

Kitazume, T.; Kobayashi, T.; Yamamoto, T.; Yamazaki, T. *J. Org. Chem.*, *1987*, *52*, 3218.

Otsubo, K.; Inanaga, J.; Yamaguchi, M. *Tetrahedron Lett.*, *1987*, *28*, 4435.

Kuroboshi, M.; Ishihara, T. *Tetrahedron Lett.*, *1987*, *28*, 6481.

Iqbal, J.; Khan, M.A. *Chem. Lett.*, *1988*, 1157.

Shimizu, M.; Yoshioka, H. *Tetrahedron Lett.*, *1989*, *30*, 967.
Shimizu, M.; Yoshioka, H. *Tetrahedron Lett.*, *1988*, *29*, 4101.

(19   :   9) 90%

Iqbal, J.; Khan, M.A.; Ahmad, S. *Synth. Commun., 1989, 19*, 641.

SECTION 330: ALCOHOL, THIOL - KETONE

68%

92% ee

Silverman, I.R.; Edington, C.; Elliott, J.D.; Johnson, W.S. *J. Org. Chem., 1987, 52,* 180.

54%

Miller, J.A. *J. Org. Chem., 1987, 52*, 322.

50%

Fleming, I.; Reddy, N.L.; Takaki, K.; Ware, A.C. *J. Chem. Soc., Chem. Commun., 1987*, 1472.

45% ee , SS

Ando, A.; <u>Shioiri, T.</u> J. Chem. Soc., Chem. Commun., **1987**, 1620.

<u>Fukuzawa, S.</u>; Tsuruta, T.; Fujinami, T.; Sakai, S. J. Chem. Soc., Perkin Trans. I, **1987**, 1473.

<u>Moriarty, R.M.</u>; Duncan, M.P.; Prakash, O. J. Chem. Soc., Perkin Trans. I, **1987**, 1781.

α:β = 80:20

<u>Clive, D.L.J.</u>; Cheshire, D.R.; Set, L. J. Chem. Soc., Chem. Commun., **1987**, 353.

<u>Davis, F.A.</u>; Sheppard, A.C. J. Org. Chem., **1987**, 52, 954.

n-C5H11CO2Me

1. THF, $\overset{\curvearrowright Li}{\underset{SO_2}{O}}$, -78°C

2. SiO2 , CH2Cl2 ,
   30 min

n-C5H11 ——— OH

64%

Fuji, K.; Node, M.; Usami, Y.; Kirya, Y. *J. Chem. Soc., Chem. Commun.*, *1987*, 449.

Ph—CH(Me)—C(=O)—Br

1. e⁻ , iPrCHO , LaBr3 , THF

LiClO4 , C cathode

2. aq. HCl

Ph—C(=O)—CH(Me)—CH(OH)—CH(Me)—Me

84%

erythro:threo = 35:65

Fry, A.J.; Susla, M.; Welta, M. *J. Org. Chem.*, *1987*, *52*, 2496.

PhCHO , TiCl4

CH2Cl2

syn   :   anti   = 15:85
4.5:1   :   34:1
60%ee:94%ee : 10%ee : 64% ee

Gennari, C.; Colombo, L.; Bertolini, G.; Schimperna, G. *J. Org. Chem.*, *1987*, *52*, 2754.

Me3SiCH2OH

1. *n*-BuLi
2. CO2
3. sBuLi , THF , -25°C
4. Me—⟨⟩—CO2Et
5. H3O⁺

Me—⟨⟩—C(=O)—CH2—OH

68%

Katritzky, A.R.; Sengupta, S. *Tetrahedron Lett.*, *1987*, *28*, 1847.

1. Me3SiOTf , CH2Cl2
   -78 - -10°C

2. H2O

72%

Suzuki, K.; Miyazawa, M.; Tsuchihashi, G. *Tetrahedron Lett.*, *1987*, *28*, 3515.

Nokami, J.; Nishimura, A.; Sunami, M.; Wakabayashi, S. *Tetrahedron Lett., 1987, 28,* 649.

Basavaiah, D.; Bharathi, T.K.; Gowriswari, V.V.L. *Synth. Commun., 1987, 17,* 1893.

Kim, K.S.; Kim, S.J.; Song, Y.-H.; Hahn, C.S. *Synthesis, 1987,* 1017.

Iwasawa, N.; Mukaiyama, T. *Chem. Lett., 1987,* 463.
Mukaiyama, T.; Kobayashi, S. *Chem. Lett., 1987,* 491.

Baraldi, P.G.; Barco, A.; Benetti, S.; Manfredini, S.; Simoni, D. *Synthesis, 1987,* 276.

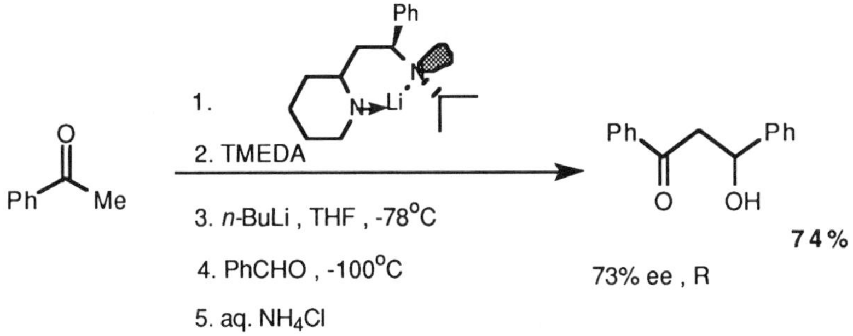

1.

2. TMEDA

3. *n*-BuLi , THF , -78°C

4. PhCHO , -100°C

5. aq. NH₄Cl

Ph-CO-CH₂-CH(OH)-Ph

73% ee , R

74%

Muraoka, M.; Kawasaki, H.; Koga, K. *Tetrahedron Lett., 1988, 29,* 337.

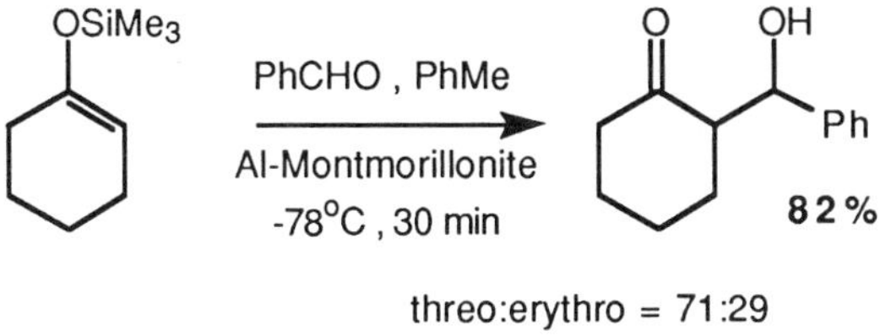

PhCHO , PhMe

Al-Montmorillonite

-78°C , 30 min

82%

threo:erythro = 71:29

Kawai, M.; Onaka, M.; Iizumi, Y. *Bull. Chem. Soc., Jpn., 1988, 61,* 1237.

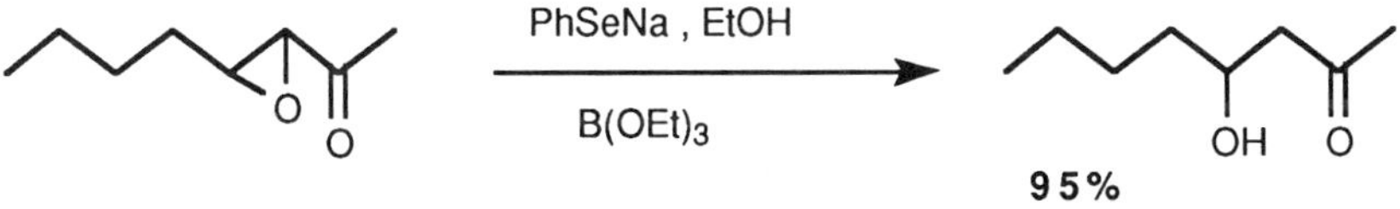

PhSeNa , EtOH

B(OEt)₃

95%

Miyashita, M.; Hoshino, M.; Yoshikoshi, A. *Tetrahedron Lett., 1988, 29,* 347.
Miyashita, M.; Suzuki, T.; Yoshikoshi, A. *Tetrahedron Lett., 1987, 28,* 4293.

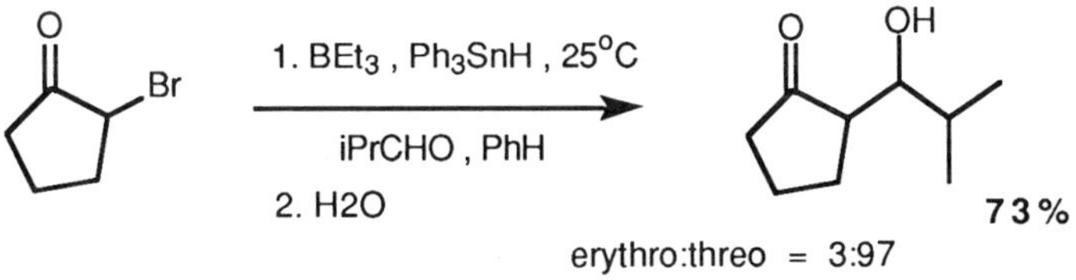

1. BEt₃ , Ph₃SnH , 25°C

iPrCHO , PhH

2. H2O

73%

erythro:threo = 3:97

Nozaki, K.; Oshima, K.; Utimoto, K. *Tetrahedron Lett., 1988, 29,* 1041.

Me OH   1. NaH , ether/hexane ,   Me   O

                          0 - 25°C , 2h

                          2. MeOH   71%

                                                                      OH

Carey, J.T.; Helquist, P. *Tetrahedron Lett.*, **1988**, *29*, 1243.

                          1. CF₃SO₃H , DMSO , 23°C   OH

              O

                          2. iPr₂NEt , CH₂Cl₂   61%
                          -78°C   O

Trost, B.M.; Fray, M.J. *Tetrahedron Lett.*, **1988**, *29*, 2163.

                                  O
                          Br
                              Me   , Bu₃Sb , cat. I₂
                     1.
                          Me   Me

              CHO   RT , 24h   Me

                          2. H₂O   OH   O

                                        erythro:threo = 57:43   60%

Huang, Y.-Z.; Chen, C.; Shen, Y. *J. Chem. Soc., Perkin Trans. I*, **1988**, 2855.

                              O₂N

                                      CO₃H

C₆H₁₃   Me   O₂N   C₆H₁₃   Me

         O—N   catalyst , CH₂Cl₂   OH   O
                       RT , 16h   56%
         catalyst = 4,4'-thiobis(2-*t*-butyl-6-methyl)phenol

Park, P.; Kozikowski, A.P. *Tetrahedron Lett.*, **1988**, *29*, 6703.

              O   1. LDA , THF   O   OH   O   OH
                        2. Me₃GeCl
                        3. PhCHO , LiBr   Ph   Ph + Ph   Ph
      Ph
                        -78 → -40°C   Me   Me
                                                  (76   :   24)   82%

Yamamoto, Y.; Yamada, J. *J. Chem. Soc., Chem. Commun.*, **1988**, 802.

Smith III, A.B.; Dorsey, B.D.; Ohba, M.; Lupo Jr., A.J.; Malamas, M.S. *J. Org. Chem.*, *1988*, *53*, 4314.

DeShong, P.; Sidler, D.R. *J. Org. Chem.*, *1988*, *53*, 4892.

(76 : 1) 88%

Siegel, C.; Thornton, E.R. *J. Am. Chem. Soc.*, *1989*, *111*, 5722.

68% ee

Paterson, I.; Goodman, J.M. *Tetrahedron Lett.*, *1989*, *30*, 997.

Horiguchi, Y.; Nakamura, E.; Kuwajima, I. *Tetrahedron Lett.*, *1989*, *30*, 3323.

Mucha, B.; Hoffman, H.M.R. *Tetrahedron Lett., 1989, 30*, 4489.

Crout, D.H.G.; Rathbone, D.L. *Synthesis, 1989*, 40.

Reydellet, V.; Helquist, P. *Tetrahedron Lett., 1989, 30*, 6837.

Kobayashi, S.; Sano, T.; Mukaiyama, T. *Chem. Lett., 1989*, 1319.

Kanemoto, S.; Saimoto, H.; Oshima, K.; Utimoto, K.; Nozaki, H. *Bull. Chem. Soc., Jpn., 1989, 62*, 519.

Ph———△——O

DMSO , TFA , MS 4Å
──────────────→
20h

Ph———C(=O)——CH2——OH

62%

Tsuji, T. *Bull. Chem. Soc., Jpn.*, *1989*, *62*, 645.

KMnO4 •CuSO4
5 H2O •Cu(OAc)2
──────────────→
H2O , t-BuOH
30 min

59%

Baskaran, S.; Das, J.; Chandrasekaran, S. *J. Org. Chem.*, *1989*, *54*, 5182.

REVIEWS:

"Stereoselective Carbon-Carbon Coupling"

Braun, M. *Angew. Chem. Int. Ed., Engl, 1987*, *26*, 24.

"Cross-Coupling Reactions Based on Acetals"

Mukaiyama, T.; Murakami, M. *Synthesis, 1987*, 1043.

SECTION 331: ALCOHOL, THIOL - NITRILE

NO ADDITIONAL EXAMPLES

SECTION 332: ALCOHOL, THIOL - ALKENE

Allylic and benzylic hydroxylation (C=C-C-H → C=C-C-OH, etc.) is listed in
Section 41 (Alcohols from Hydrides).

Me_3Si———OH

1. VO(acac)2 , 2 t-BuOOH
DCE , CH2Cl2 , 0°C , 8h
──────────────→
2. H3O+

3. SiO2

OH   OH

35%

97:3 erythro:threo

Mohr, P.; Tamm, C. *Tetrahedron Lett., 1987*, *28*, 395.

Roush, W.R.; Palkowitz, A.D.; Palmer, M.A.J. *J. Org. Chem.*, **1987**, *52*, 316.

Whitesell, J.K.; Carpenter, J.F. *J. Am. Chem. Soc.*, **1987**, *109*, 2839.

Molander, G.A.; Etter, J.B.; Zinke, P.W. *J. Am. Chem. Soc.*, **1987**, *109*, 453.

(96   :   4) **no yield**

Brown, H.C.; Bhat, K.S.; Randad, R.S. *J. Org. Chem.*, **1987**, *52*, 319.

1. -78°C , /\/PPh$_3$

Ph$_2$MeSiLi

2. -78°C , Ph/\(Me)CHO

syn:anti = 15:1
E:Z   = 30:1

Me
Ph/\/(OSiMe$_2$Ph)/\=/Me

79%

Tsukamoto, M.; Iio, H.; Tokoroyama, T. *Tetrahedron Lett.*, **1987**, *28*, 4561,

[furanyl-Me structure]

PhMgBr , PhH , RT

(Ph$_3$P)$_2$NiCl$_2$, 20 min

HO\/\/Ph=/Me

85%

Wadman, S.; Whitby, R.; Yeates, C.; Kocienski, P. *J. Chem. Soc., Chem. Commun.*, **1987**, 241.

Me\(Me)=/CO$_2$Et

1. BuLi , THF , -78°C
2. Bu$_3$SnCl
3. PhCHO , BF$_3$•OEt$_2$
   CH$_2$Cl$_2$

Me
Ph\(OH)\/CO$_2$Et

quant.

Yamamoto, Y.; Hatsuya, S.; Yamada, J. *J. Chem. Soc., Chem. Commun.*, **1987**, 561.

Me\/\Br

PhCHO , BiCl$_3$•Al

aq. THF

OH
Ph\/(Me)\/=

87%

erythro:threo: 84:16

Wada, M.; Ohki, H.; Akiba, K. *J. Chem. Soc., Chem. Commun.*, **1987**, 708.

68% overall

Vaultier, M.; Truchet, F.; Carboni, B.; Hoffmann, R.W.; Denne, I. *Tetrahedron Lett.*, *1987*, *28*, 4169.

82%

Gauthier, J.Y.; Guindon, Y. *Tetrahedron Lett.*, *1987*, *28*, 5985.

(1   :   1) 98%

Ziegler, F.E.; Zheng, Z. *Tetrahedron Lett.*, *1987*, *28*, 5973.

(12   :   88) 88%

Marshall, J.A.; Trometer, J.D.; Blough, B.E.; Crute, T.D. *Tetrahedron Lett.*, *1988*, *29*, 913.
Marshall, J.A.; Trometer, J.D. *Tetrahedron Lett.*, *1987*, *28*, 4985.

1. TiCl₄ , CH₂Cl₂ , 0°C
3.5h

2. aq. NH₄Cl

78%

Aubert, C.; Bégué, J.-P. *Tetrahedron Lett.*, *1988*, *29*, 1011.

NapthPhMeSi
(S)

1. 2 *t*-BuLi , THF , -78°C
2. PhCHO , THF
BF₃•OEt₂ , -78°C
3. TBAF , DMSO , 25°C

Ph—_/—Me + Ph—_/—Me
   OH   OH
(1   :   1)
(88% ee   :   76% ee)

Torres, E.; <u>Larson, G.L.</u>; <u>McGarvey, G.J.</u> *Tetrahedron Lett.*, *1988*, *29*, 1355.

1. (Ph₃P)₂NiCl₂ , PhMgBr , PhH

2. aq. NH₄Cl

HO Ph

70%

<u>Kocieński, P.</u>; Dixon, N.J.; Wadman, S. *Tetrahedron Lett.*, *1988*, *29*, 2353.

PhHgCl , 0.1 Li₂PdCl₄

0 → +25°C , 8h

Ph   OH

68%

E:Z = 79:21

<u>Larock, R.C.</u>; Stolz-Dunn, S.K. *Tetrahedron Lett.*, *1988*, *29*, 5069.

1. 2 eq. ClCH₂I , LiBr
2. 2 eq. MeLi , LiI , -78°C
3. -78 → +25°C
4. aq. NH₄Cl

OH

70%

<u>Barluenga, J.</u>; Concellón, J.M.; Fernández-Simón, J.; Yus, M. *J. Chem. Soc., Chem. Commun.*, *1988*, 536.

79%   **6%**

Brückner, R.; Priepke, H. *Angew. Chem. Int. Ed., Engl, 1988, 27,* 278.

(79   :   21) **90%**

Maruoka, K.; Banno, H.; Nonoshita, K.; Yamamoto, H. *Tetrahedron Lett., 1989, 30,* 1265.

78%

quant

5

Alexakis, A.; Marek, I.; Mangeney, P.; Normant, J.F. *Tetrahedron Lett., 1989, 30,* 2391.

70%

Avignon-Tropis, M.; Pougny, J.R. *Tetrahedron Lett., 1989, 30,* 4951.

Teuting, D.R.; Echavarren, A.M.; Stille, J.K. Tetrahedron, **1989**, 45, 979.

60%

Alexakis, A.; Jachiet, D. Tetrahedron, **1989**, 45, 6197.

59%

Liebeskind, L.S.; Gasdaska, J.R.; McCallum, J.S. J. Org. Chem., **1989**, 54, 669.

REVIEW:

"Intramolecular Addition Reactions of Allylic and Propargylic Silanes"

Schinzer, D. Synthesis, **1988**, 263.

Also via: Section 302 (Alkyne - Alcohol).

SECTION 333: ALDEHYDE - ALDEHYDE

NO ADDITIONAL EXAMPLES

SECTION 334: ALDEHYDE - AMIDE

150 psi H_2 , 1700 psi CO
$[Rh(dppb)(NBD)]^+ ClO_4^-$
autoclave (stainless steel)
THF , 100°C ,pyrex vessel

B87%

Ojima, I.; Zhang, Z. *J. Org. Chem.*, *1988*, *53*, 4422.

SECTION 335: ALDEHYDE - AMINE

REVIEW:

"Optically Active N-Protected α-Amino Aldehydes in Organic Synthesis"

Jurczak, J.; Gułębiowski, A. *Chem. Rev.*, *1989*, *89*, 149.

SECTION 336: ALDEHYDE - ESTER

$POCl_3$, DMF , RT
CH_2Cl_2

52%

Reddy, C.P.; Tanimoto, S. *Synthesis, 1987*, 575.

SECTION 337: ALDEHYDE - ETHER, EPOXIDE, THIOETHER

1. PhSCHLiOMe
2. $SOCl_2$, Py

86%

Jansen, B.J.M.; Peperzak, R.M.; de Groot, Ae. *Recl. Trav. Chim., Pays-Bas,*
1987, *106*, 489.

Benati, L.; Montevecchi, P.C.; Spagnoli, P. *Tetrahedron Lett.*, *1988*, *29*, 2381.

SECTION 338: ALDEHYDE - HALIDE, SULFONATE

NO ADDITIONAL EXAMPLES

SECTION 339: ALDEHYDE - KETONE

Klix, R.C.; Bach, R.D. *J. Org. Chem.*, *1987*, *52*, 580.

Kitahara, H.; Tozawa, Y.; Fujita, S.; Tajira, A.; Morita, N.; Asao, T. *Bull. Chem. Soc., Jpn.*, *1988*, *61*, 3362.

SECTION 340: ALDEHYDE - NITRILE

NO ADDITIONAL EXAMPLES

SECTION 341: ALDEHYDE - ALKENE

For the oxidation of allylic alcohols to alkene aldehydes, also see Section 48 (Aldehydes from Alcohols).

E   72%

Yoshida, J.; Murata, T.; Isoe, S. *Tetrahedron Lett.*, *1987*,.28, 211.

66%

Hagen, J.P.; Harris, J.J.; Lakin, D. *J. Org. Chem.*, *1987*, 52, 782.

55%

Campi, E.; Fitzmaurice, N.J.; Jackson, W.R.; Perlmutter, P.; Smallridge, A.J. *Synthesis*, *1987*, 1032.

87%

E:Z = >100:1

Desmond, R.; Mills, S.G.; Volante, R.P.; Shinkai, I. *Tetrahedron Lett.*, *1988*, 29, 3895.

C_3H_7CHO , $TiCl_4$

EtO—〜〜—SiMe₃ → C_3H_7—〜〜—SiMe₃

CH_2Cl_2 , -70 → 0°C

CHO   62%

Pornet, J.; Rayadh, A.; Miginiac, L. *Tetrahedron Lett.*, *1988*, *29*, 4717.

cat. Me—〜—COOH , PhCO2Me

Me—C(Me)(OH)—≡≡

$Ti(OiPr)_4$, cat. CuCl, 130°C
1h

Me₂C=CH—CHO   89%

Chabardes. P. *Tetrahedron Lett.*, *1988*, *29*, 6253.

1. Ph—〜〜—MgBr , THF , 0 →+35°C

〜〜—ZnBr

2. Me_3SnCl , 0°C , 15 min

3. dry air , Me_3SiCl , -10 → 0°C

Ph—CH(CHO)—CH₂—CH=CH₂   89%

Knochel. P.; Xiao, C.; Yeh, M.C.P. *Tetrahedron Lett.*, *1988*, *29*, 6697.

2% $RuCl_2(PPh_3)_3$, $PiPr_3$

C_6H_{13}—≡≡—〜OH

PhMe , reflux , 32h

C_6H_{13}—〜〜—CHO   80%

Ma, D.; Lu. X. *J. Chem. Soc., Chem. Commun.*, *1989*, 890.

1. $(COCl)_2$, DMSO

Ph—〜〜—OH

NEt₃ , -70°C

2. $Me_2N^+=CH_2$ Cl⁻
RT

Ph—CH₂—C(=CH₂)—CHO   72%

Takano. S.; Inomata, K.; Samizu,K.; Tomita, S.; Yanase, M.; Suzuki, M.; Iwabuchi, Y.; Sugihara, T.; Ogasawara, K. *Chem. Lett.*, *1989*, 1283.

53%

84% ee

Furuta, K.; Shimizu, S.; Miwa, Y.; Yamamoto, H. *J. Org. Chem.*, *1989*, *54*, 1481.

Also via β-Hydroxy aldehydes: Section 324 (Alcohols - Aldehyde).

SECTION 342: AMIDE - AMIDE

69%

Ruppin, C.; Dixneuf, P.H.; Lecolier, S. *Tetrahedron Lett.*, *1988*, *29*, 5365.

Also via Dicarboxylic Acids: Section 312 (Carboxylic Acid - Carboxylic Acid)

Diamines   Section 350 (Amines - Amines)

SECTION 343: AMIDE - AMINE

80%

cis:trans = 70:30

Van der Steen, F.H.; Kleijn, H.; Jastrebski, J.T.B.H.; van Koten, G. *Tetrahedron Lett.*, *1989*, *30*, 765.

SECTION 344: AMIDE - ESTER

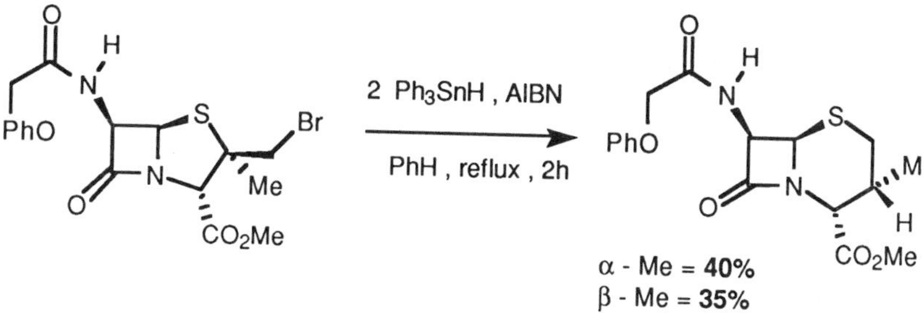

2 Ph₃SnH , AIBN

PhH , reflux , 2h

α - Me = **40%**
β - Me = **35%**

<u>Baldwin, J.E.</u>; Adlington, R.M.; Kang, T.W.; Lee, E.; Schofield, C.J. *J. Chem. Soc., Chem. Commun., **1987**, 104.*

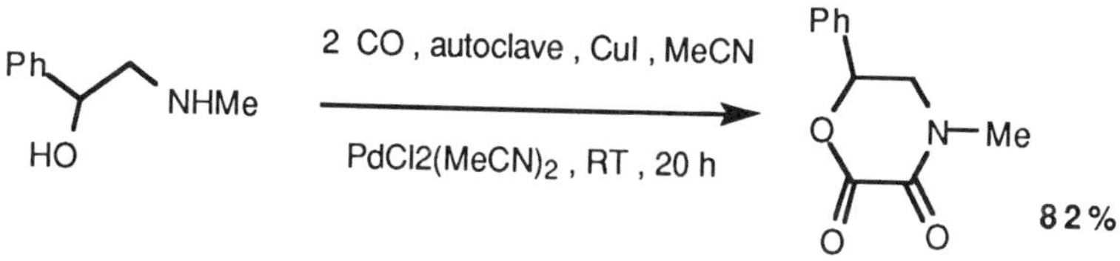

2 CO , autoclave , CuI , MeCN

PdCl2(MeCN)₂ , RT , 20 h

82%

<u>Murahashi, S.</u>; Mitsue, Y.; Ike, K. *J. Chem. Soc., Chem. Commun., **1987**, 125.*

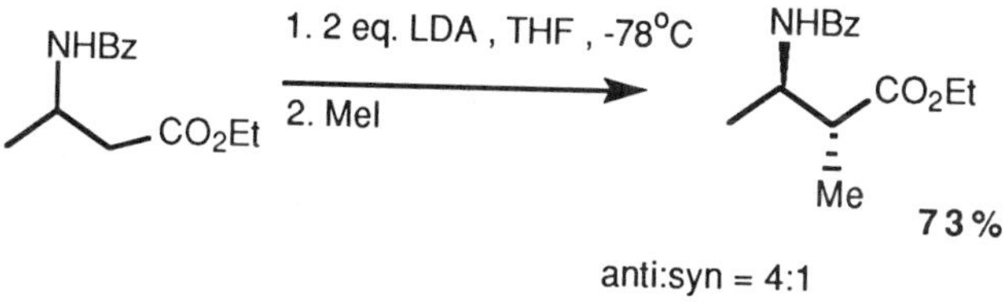

1. 2 eq. LDA , THF , -78°C
2. MeI

73%

anti:syn = 4:1

<u>Seebach, D.</u>; Estermann, H. *Tetrahedron Lett., **1987**, 28, 3103.*

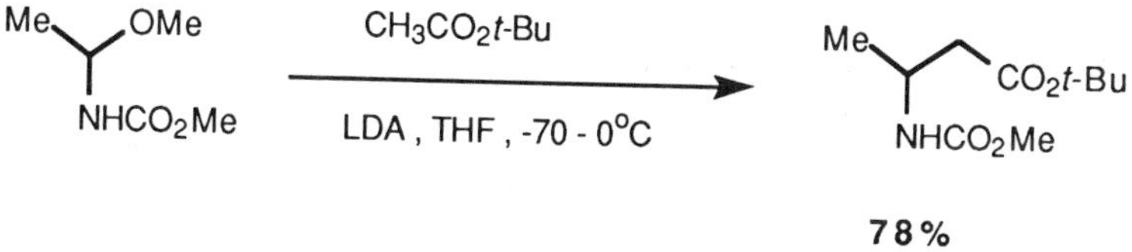

CH₃CO₂t-Bu

LDA , THF , -70 - 0°C

78%

<u>Shono, T.</u>; Kise, N.; Sanda, F.; Ohi, S.; Tsubata, K. *Tetrahedron Lett., **1988**, 29, 231.*

Zezza, C.A.; Smith, M.B. *J. Org. Chem.*, **1988**, *53*, 1161.

Ciattini, P.G.; Morena, E.; Ortar, G. *Synthesis*, **1988**, 140.

Mori, M.; Kagechika, K.; Tohjima, K.; Shibasaki, M. *Tetrahedron Lett.*, **1988**, *29*, 1409.

Bertenshaw, S.; Kahn, M. *Tetrahedron Lett.*, **1989**, *30*, 2731.

Baldwin, J.E.; Adlington, R.M.; O'Neil, I.A.; Schofield, C.; Spivey, A.C.; Sweeney, J.B. *J. Chem. Soc., Chem. Commun.*, **1989**, 1852.

(37:63)

Mooiweer, H.H.; Hiemstra,H.; Speckamp, W.N. *Tetrahedron, 1989, 45,* 4627.

Related Methods: Section 315 (Carboxylic Acid - Amide)
                              Section 316 (Carboxylic Acid - Amine)
                              Section 351 (Amine - Ester)

SECTION 345: AMIDE - ETHER, EPOXIDE, THIOETHER

Beak, P.; Wilson, K.D. *J. Org. Chem., 1987, 52,* 218.

Kogen, H.; Nishi, T. *J. Chem. Soc., Chem. Commun., 1987,* 311.

Yoshida, J.; Isoe, S. *Tetrahedron Lett., 1987, 28,* 6621.

(20   :   80)   60%

Harding. K.E.; Hollingsworth, D.R. *Tetrahedron Lett.*, *1988*, *29*, 3789.
Harding , K.E.; Nam, D. *Tetrahedron Lett.*, *1988*, *29*, 3793.

86%

Moeller. K.D.; Tarazi, S.; Marzabadi, M.R. *Tetrahedron Lett.*, *1989*, *30*, 1213.

SECTION 346: AMIDE - HALIDE, SULFONATE

65%

Shimizu. M.; Tanaka, E.; Yoshioka. H. *J. Chem. Soc., Chem. Commun.*, *1987*, 136.

72%

Takahata. H.; Takamatsu, T.; Mozumi, M.; Chen, Y.-S.; Yamazaki. T.; Aoe, K. *J. Chem. Soc., Chem. Commun.*, *1987*, 1627.

Knapp, S.; Levorse, A.T. *J. Org. Chem.*, *1988*, *53*, 4006.

SECTION 347: AMIDE - KETONE

Shimizu, M.; Yoshioka, H. *J. Chem. Soc., Chem. Commun.*, *1987*, 689.

Ley, S.V.; Woodward, P.R. *Tetrahedron Lett.*, *1987*, *28*, 3019.

Sasaki, Y.; Dixneuf, P.H. *J. Org. Chem.*, *1987*, *52*, 4389.

1. NaH , RT
2. DMF , 80°C , 4h

3. mesitylene , reflux , 12h

33%

Flitsch, W.; Hampel, K.; Hohenhorst, M. *Tetrahedron Lett., 1987, 28,* 4395.

1. LDA , ether , -78°C
2. Ph-N=C=O
3. aq. NH₄Cl

98%

Hendi, S.B.; Hendi, M.S.; Wolfe, J.F. *Synth. Commun., 1987, 17,* 13.

CO , MeI , Co₂(CO)₈ , CH₂Cl₂

aq. KOH , PEG-400 , 60°C
1 atm

76%

Vasapollo, G.; Alper, H. *Tetrahedron Lett., 1988, 29,* 5113.

1% Rh₂(OAc)₄ , CH₂Cl₂

5°C , RT

49%

Saba, A.; Selva, A. *Heterocycles, 1988, 27,* 867.

1. CH₃C≡O⁺ BF₄⁻ , ClCH₂CN
2. aq. NaOH

Ph-C≡C-H

41%

Gridnev, I.D.; Balenkova, E.S. *J. Org. Chem., U.S.S.R., 1988, 24,* 1447.

Vedejs. E.; Stults, J.S. *J. Org. Chem.*, **1988**, *53*, 2226.

(9   :   1) 52%

Shono. T.; Kise, N.; Sanda, F.; Ohi, S.; Yoshioka, K. *Tetrahedron Lett.*, **1989**, *30*, 1253.

82%

Pouilhès, A.; Thomas. S.E. *Tetrahedron Lett.*, **1989**, *30*, 2285.

79%

Middleton, D.S.; Simpkins. N.S.; Terrett, N.K. *Tetrahedron Lett.*, **1989**, *30*, 3865.

Giovannini, A.; Savoia, D.; Umana-Ronchi, A. *J. Org. Chem.*, **1989**, *54*, 228.

SECTION 348: AMIDE - NITRILE

Ognyanov, V.I.; Hesse, M. *Helv. Chim. Acta*, **1989**, *72*, 1522.

SECTION 349: AMIDE - ALKENE

Sasaki, Y.; Dixneuf, P.H. *J. Org. Chem.*, **1987**, *52*, 314.

Alper, H.; Hamel, N. *Tetrahedron Lett.*, **1987**, *28*, 3237.

Baldwin, J.E.; Li, C.-S. *J. Chem. Soc., Chem. Commun.*, **1987**, 166.

Hayashi, T.; Yamamoto, A.; Ito, Y. *Tetrahedron Lett.*, **1987**, *28*, 4837.

Sakaitani, M.; Ohfune, Y. *Tetrahedron Lett.*, **1987**, *28*, 3987.

Cossy, J.; Belotti, D.; Pete, J.P. *Tetrahedron Lett.*, **1987**, *28*, 4545, 4547.

$$\equiv\!\!-Bu \xrightarrow[\text{Ru(COD)(COT)}]{Et_2NH\,,\,CO_2\,,\,PCy_3}$$

62%

Z:E = 92:8

Mitsudo, T.; Hori, Y.; Yamakawa, Y.; Watanabe, Y. *Tetrahedron Lett., 1987, 28,* 4417.

$$\xrightarrow[\substack{\text{cat. } p\text{-TsOH , reflux} \\ \text{3h}}]{C_5H_{11}CHO\,,\,PhMe}$$

66%

Zezza, C.A.; Smith, M.B. *Synth. Commun., 1987, 17,* 729.

$$\xrightarrow[\text{NaH , PhH}]{(EtO)_2P(=O)CH_2CO_2Et}$$

40%

Fillion, H.; Hseine, A.; Péra, M.-H. *Synth. Commun., 1987, 17,* 929.

$$\xrightarrow[\text{80°C , 06.h}]{4\%\text{ Pd(PPh}_3)_4\,,\,AcOH}$$

(88   :   12)   76%

Oppolzer, W.; Gaudin, J.-M.; Bedoya-Zurita, M.; Hueso-Rodriguez, J.; Raynham, T.M.; Robyr, C. *Tetrahedron Lett., 1988, 29,* 4709.

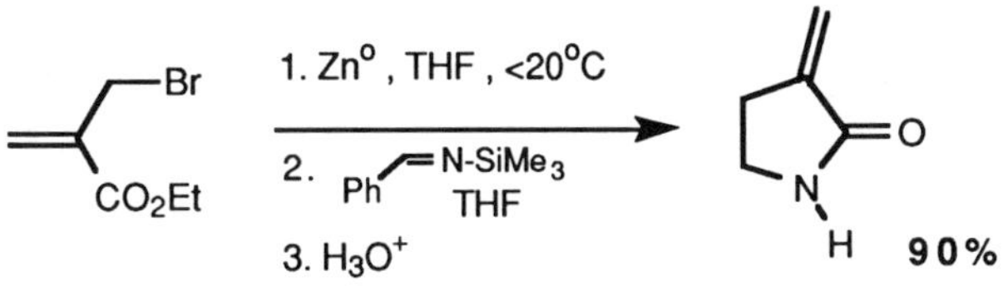

1. Zn°, THF, <20°C

2. Ph—⟨=N-SiMe₃ THF

3. H₃O⁺

90%

El Alami, N.; Belaud, C.; Villieras, J. *Synth. Commun.*, *1988*, *18*, 2073.

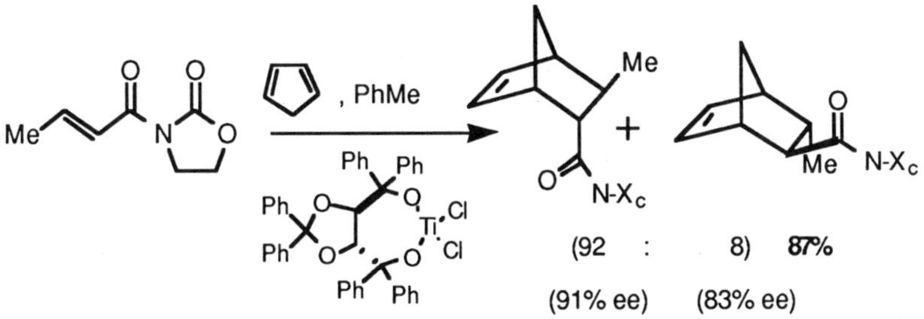

(92 : 8) 87%

(91% ee) (83% ee)

Narasaka, K.; Iwasawa, N.; Inoue, M.; Yamada, T.; Nakashima, M.; Sugimori, J. *J. Am. Chem. Soc.*, *1989*, *111*, 5340.

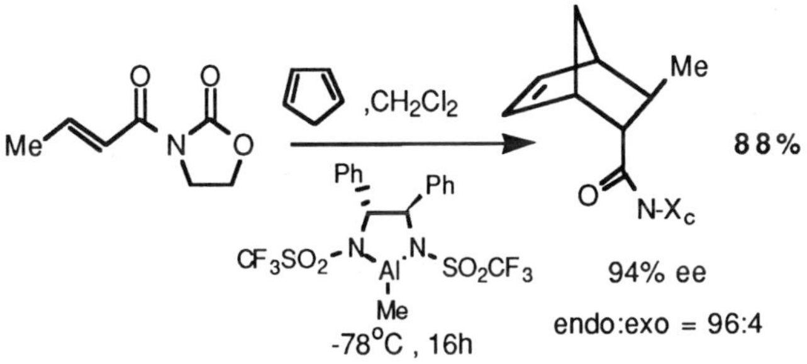

, CH₂Cl₂

88%

94% ee

endo:exo = 96:4

CF_3SO_2-N-Al-N-SO_2CF_3

Me

-78°C, 16h

Corey, E.J.; Imwinkelried, R.; Pikul, S.; Xiang,Y.B. *J. Am. Chem. Soc.*, *1989*, *111*, 5493.

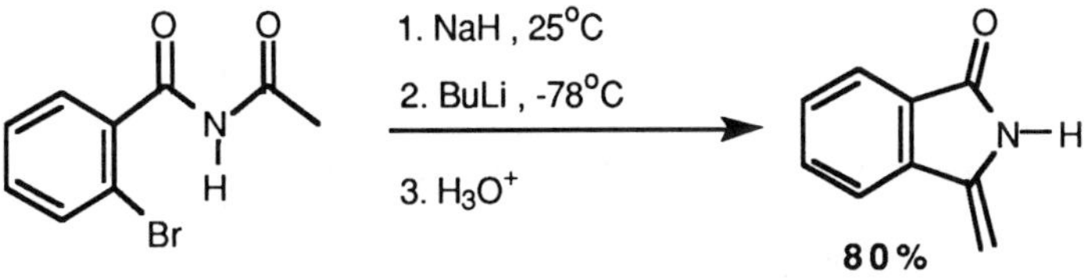

1. NaH, 25°C

2. BuLi, -78°C

3. H₃O⁺

80%

Hendi, M.S.; Natalie Jr., K.J.; Hendi, S.B.; Campbell, J.A.; Greenwood, T.D.; Wolfe, J.F. *Tetrahedron Lett.*, *1989*, *30*, 275.

PhHgCl , Li$_2$PdCl$_4$

THF , 0°C - RT , 3h

60%

E:Z = 74:26

Larock, R.C.; Ding, S. *Tetrahedron Lett.*, **1989**, *30*, 1897.

1.

Me$_3$SiCN , AlCl$_3$

2. SiO$_2$

3. PhH , reflux , 2h

54%

(>25:1)

Teng, M.; Fowler, F.W. *Tetrahedron Lett.*, **1989**, *30*, 2481.

, aq. EtOH

reflux , 3h

73%

Menezes, R.F.; Zezza, C.A.; Sheu, J.; Smith, M.B. *Tetrahedron Lett.*, **1989**, *30*, 3295.

Mn(OAc)$_3$, EtOH

RT , 1h

60%

Cossy, J.; LeBlanc, C. *Tetrahedron Lett.*, **1989**, *30*, 4531.

TsO~~~~~SPh
1. chloramine-T , CH$_2$Cl$_2$
   C$_{16}$H$_{33}$Bu$_3$P$^+$Br$^-$, P(OEt)$_3$
2. aq. NaOH

→ [pyrrolidine with N-Ts and vinyl group] 85%

Dolle, R.E.; Li, C.-S.; Shaw, A.N. *Tetrahedron Lett.,* **1989,** *30,* 4723.

TsHN~Ph, Me (with terminal alkene)

1. 9-BBN , THF
   -78°C→RT
2. catechol borane
   2% [Rh(COD)Cl]$_2$
   8% PPh$_3$
3. NaOH , H$_2$O$_2$

TsHN Ph~~OH, Me   +   TsHN Ph~~OH, Me

(7   :   1)   90%

Burgess, K.; Ohlmeyer, M.J. *Tetrahedron Lett.,* **1989,** *30,* 5857.

[bicyclic isoxazoline with Me, CO$_2$Me]

[isoprene] , 24h
sealed tube
60°C

→ [product with Me, CO$_2$Me, Me]   +   [product with Me, CO$_2$Me, Me]

(1   :   2)   74%

Meyers, A.I.; Busacca, C.A. *Tetrahedron Lett.,* **1989,** *30,* 6973, 6971.

(EtO)$_2$P(=O)–C(Me)(H)–C(=O)–N(pyrrolidine)

1. BuLi , THF , -70°C
2. PhCHO

→ Ph~C(Me)=CH–C(=O)–N(pyrrolidine)   83%

E:Z = 100:0

Tay, M.K.; About-Jaudet, E.; Collignon, N.; Savignac, P. *Tetrahedron,* **1989,** *45,* 4415.

[triene with oxazolidinone acyl group]

[Ph, Me dioxolane with CPh$_2$OH groups]

TiCl$_2$(OiPr)$_4$, RT
161h

→ [bicyclic product with oxazolidinone]   78%

73% ee

Iwasawa, N.; Sugimori, J.; Kawase, Y.; Narasaka, K. *Chem. Lett.,* **1989,** 1947.
Also via Alkenyl Acids: Section 322 (Carboxylic Acid -Alkene)

SECTION 350: AMINE - AMINE

Ni/Al , H₂O , KOH
reflux , 19h

69%

Lunn, G. *J. Org. Chem.*, **1987**, *52*, 1043.

1. NbCl₄(THF)₂
DME , 4h
2. KOH

60%

d,l:meso = 19:1

Roskamp, E.J.; Redersen, S.F. *J. Am. Chem. Soc.*, **1987**, *109*, 3152.

1. PhNH₂ , ZnCl₂ , PhMe
36h
2. Ph₂CHNH₂ , TiCl₄
PhH , 0°C - RT (4d)

42%

Armesto, D.; Bosch, P.; Gallego, M.G.; Martin, J.F.; Ortiz, M.J.; Perez-Ossorio, R.; Ramos, A. *Org. Prep. Proceed. Int.*, **1987**, *19*, 181.

1. LDA , THF , 0°C , 2h
2. 0.5

75%

Sulmon, P.; DeKimpe, N.; Schamp, N. *Org. Prep. Proceed. Int.*, **1987**, *19*, 17.

1. NaN$_3$, NaH, DMF, 23°C, 6h
2. H$_2$, Pd/C
3. CH$_2$=CHCN
4. BH$_3$·SMe$_2$
5. HCl

36%

Knapp, S.; Levorse, A.T. *Tetrahedron Lett.*, **1987**, *28*, 3213.

THF, RT, 3h

75%

Furukawa, N.; Shibutani, T.; Fujihara, H. *Tetrahedron Lett.*, **1987**, *28*, 5845.

1. TFA, PbBr$_2$, RT, 6h
2. H$_3$O$^+$

NHBn   NHBn   **82%**

Tanaka, H.; Dhimane, H.; Fujita, H.; Ikemoto, Y.; Torii, S. *Tetrahedron Lett.*, **1988**, *29*, 3811.

TiCl$_4$, Mg^0
THF

meso:d,l = 1:3

NH$_2$ NH$_2$ **52%**

Betschart, C.; Schmidt, B.; Seebach, D. *Helv. Chim. Acta*, **1988**, *71*, 1999.

1. HgCl$_2$, Mg^0, THF
2. TiCl$_4$, THF

(80   :   20)   **63%**

Mangeney, P.; Tejero, T.; Alexakis, A.; Grosjean, F.; Normant, J. *Synthesis*, **1988**, 255.

Barluenga, J.; Aguilar, E.; Joglar, J.; Olano, B.; Fustero, S. *J. Chem. Soc., Chem. Commun.*, *1989*, 1132.

Fraenkel, G.; Gallucci, J.; Rosenzweig, H.S. *J. Org. Chem.*, *1989*, *54*, 677.

SECTION 351: AMINE - ESTER

Padwa, A.; Dent, W. *J. Org. Chem.*, *1987*, *52*, 235.

Baldwin, J.E.; North, M.; Flinn, A. *Tetrahedron Lett.*, *1987*, *28*, 3167.

Guanti, G.; Narisano, E.; Banfi, L. *Tetrahedron Lett., 1987, 28*, 4331, 4335.

Celerier, J.P.; Haddad, M.; Jacoby, D.; Lhommet, G. *Tetrahedron Lett., 1987, 28*, 6597.

Tarnchampoo, B.; Thebtaranonth, C.; Thebtaranonth, Y. *Tetrahedron Lett., 1987, 28*, 6675.

Gajda, T. *Synthesis, 1987*, 1108.

(4.1   :   1) 98%

Perlmutter, P.; Mabone, M. *Tetrahedron Lett.*, *1988*, *29*, 949.

89%

70% ee , S

Ito, Y.; Sawamura, M.; Shirakawa, E.; Hayashizaki, K.; Hayashi, T. *Tetrahedron Lett.*, *1988*, *29*, 239.

1. Hg(NO₃)₂ , MeCN
2. KBr
3. O₂ , NaBH₄
4. Jones'
5. HBr , AcOH

54%

Harding, K.E.; Marman, T.H.; Nam, D. *Tetrahedron Lett.*, *1988*, *29*, 1627.

1. NaH
2. Pd(dppe)₂
3. MeI
4. H₃O⁺

88%

van der Werf, A.; Kellogg, R.M. *Tetrahedron Lett.*, *1988*, *29*, 4981.

Murahashi, S.; Kodera, Y.; Hosomi, T. *Tetrahedron Lett.*, *1988*, *29*, 5949.

Genet, J.P.; Juge, S.; Mallart, S. *Tetrahedron Lett.*, *1988*, *29*, 6765.

Menezes, R.; Smith, M.B. *Synth. Commun.*, *1988*, *18*, 1625.

Wasserman, H.H.; Cook, J.D.; Fukuyama, J.M.; Rotello, V.M. *Tetrahedron Lett.*, *1989*, *30*, 1721.

Mukaiyama, T.; Kashiwagi, K.; Matsui, S. *Chem. Lett.*, *1989*, 1397.

1. TiCl$_3$, NaBH$_4$, aq. MeOH
L-tartaric acid

2. HCl

N-OH
Me   CO$_2$Me

→

NH$_2$ • HCl
Me   CO$_2$Me

73%

Hoffman, C.; Tanke, R.S.; Miller, M.J. *J. Org. Chem.*, **1989**, *54*, 3750.

Related Methods: Section 315 (Carboxylic Acid - Amide)
                          Section 316 (Carboxylic Acid - Amine)
                          Section 344 (Amide - Ester)

SECTION 352: AMINE - ETHER, EPOXIDE, THIOETHER

H$_2$N-CN , *p*-TsOH , 25°C

N-PSP , CH$_2$Cl$_2$, 25°C

→

SePh
NHCN

71%

Hernández, R.; León, E.I.; Salazar, J.A.; Suárez, E. *J. Chem. Soc., Chem. Commun.*, **1987**, 312.

BOCNH   CHO

1. KHMDS , THF/DMSO
MePh$_3$P$^+$ Br$^-$, -78→ -40°C

2. mCPBA

→

BOCNH

+

BOCNH

(15   :   1) 35%

Luly, J.R.; Dellaria, J.F.; Plattner, J.J.; Soderquist, J.L. *J. Org. Chem.*, **1987**, *52*, 1487.

Me
Me   =NPh

PhCHO , Eu(fod)$_3$

CCl$_4$, 40°C

→

Me
Me   NPh
              O
Ph

85%

Barbaro, G.; Battaglia, A.; Giorgianni, P. *Tetrahedron Lett.*, **1987**, *28*, 2995.

88%

Tominaga, Y.; Matsuoka, Y.; Hayashida, H.; Kohra, S.; Hosomi, A. *Tetrahedron Lett.,* **1988**, *29*, 5771.

REVIEW:

"Chemistry of Lactam Acetals"

Anand, N.; Singh, J. *Tetrahedron,* **1988**, *44*, 5975.

SECTION 353: AMINE - HALIDE, SULFONATE

$CF_3CF_2CHFCO_2R^1$

R^1 = cholestanyl-3β

78%

Portella, C.; Iznaden, M. *Tetrahedron Lett.,* **1987**, *28*, 1655.

1. Bu_3SnH , AIBN
   PhH , reflux , 3h

2. 35% aq. HCl

75%

Watanabe, Y.; Ueno, Y.; Tanaka, C.; Okawara, M.; Endu, T. *Tetrahedron Lett.,* **1987**, *28*, 3953.

I_2 , MeCN , NaHCO$_3$

87%

trans:cis = 67:33

Williams, D.R.; Osterhout, M.H.; McGill, J.M. *Tetrahedron Lett.,* **1989**, *30*, 1331.

SECTION 354: AMINE - KETONE

Differding, E.; Vandevelde, O.; Roekens, B.; Van, T.T.; Ghosez, L. *Tetrahedron Lett.*, *1987*, *28*, 397.

McNab, H.; Monahan, L.C.; Gray, T. *J. Chem. Soc., Chem. Commun.*, *1987*, 140.

Westling, M.; Livinghouse, T. *J. Am. Chem. Soc.*, *1987*, *109*, 590.

Pilli, R.A.; Russowsky, D. *J. Chem. Soc., Chem. Commun.*, *1987*, 1053.

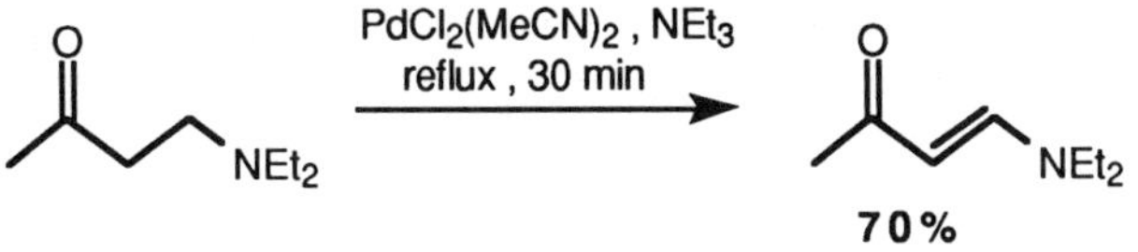

PdCl$_2$(MeCN)$_2$, NEt$_3$
reflux , 30 min

70%

Murahashi, S.; Mitsue, Y.; Tsumiyama, T. *Bull. Chem. Soc., Jpn.,* **1987,** *60,* 3285.

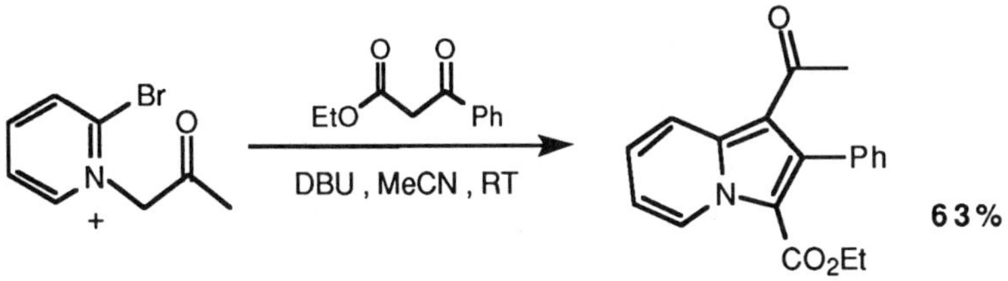

EtO ⟶ Ph

DBU , MeCN , RT

Ph

CO$_2$Et

63%

Nugent, R.A.; Murphy, M. *J. Org. Chem.,* **1987,** *52,* 2206.

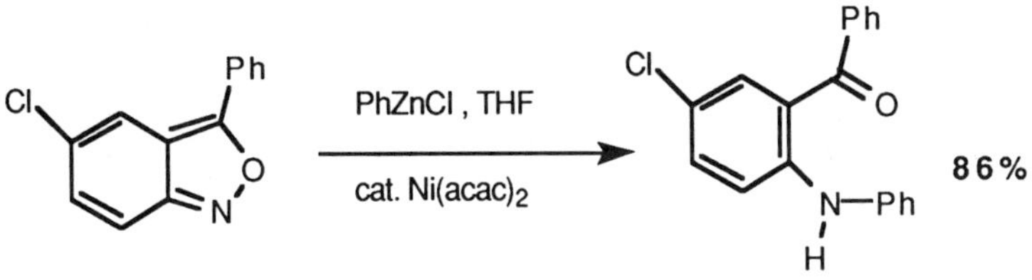

PhZnCl , THF

cat. Ni(acac)$_2$

86%

Baum, J.S.; Condon, M.E.; Shook, D.A. *J. Org. Chem.,* **1987,** *52,* 2983.

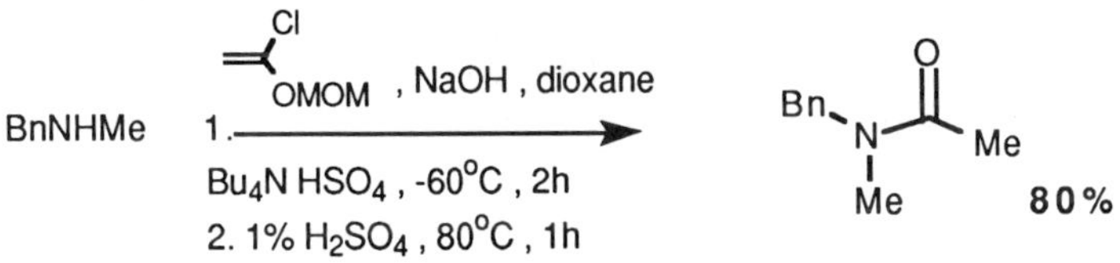

Cl
OMOM , NaOH , dioxane

BnNHMe   1.

Bu$_4$N HSO$_4$, -60°C , 2h

2. 1% H$_2$SO$_4$, 80°C , 1h

80%

Gu, X.-P.; Nishida, N.; Ikeda, I.; Okahara, M. *J. Org. Chem.,* **1987,** *52,* 3192.

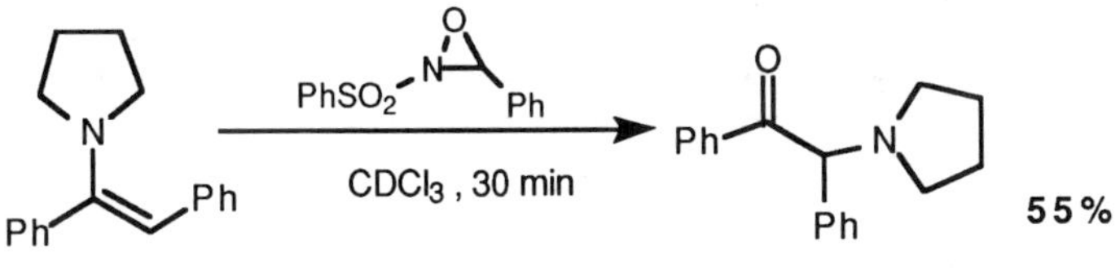

PhSO$_2$—N

Ph

CDCl$_3$, 30 min

55%

Davis, F.A.; Sheppard, A.C. *Tetrahedron Lett.,* **1988,** *29,* 4365.

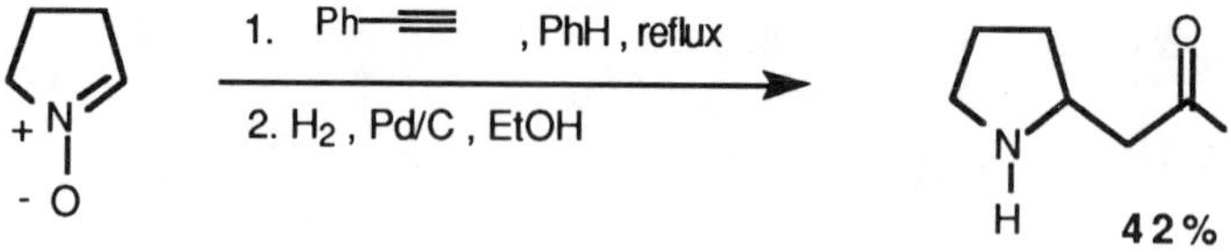

1. Ph—≡ , PhH , reflux

2. H₂ , Pd/C , EtOH

42%

Mancuso, V.; Hootelé. C. Tetrahedron Lett., **1988**, 29, 5917.

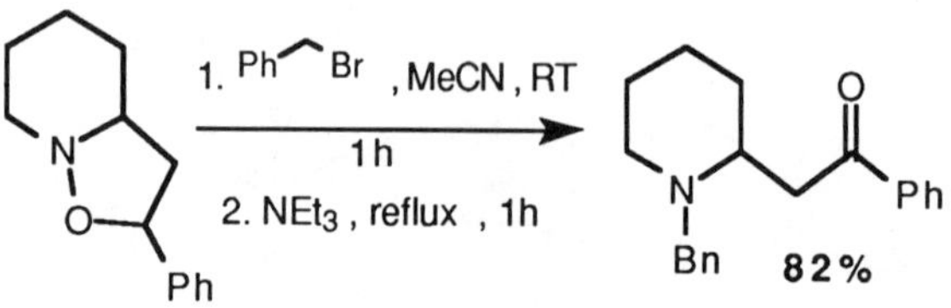

1. Ph⌒Br , MeCN , RT
1h

2. NEt₃ , reflux , 1h

Bn   82%

Murahashi. S.; Kodera, Y.; Hosomi, T. Tetrahedron Lett., **1988**, 29, 5949.

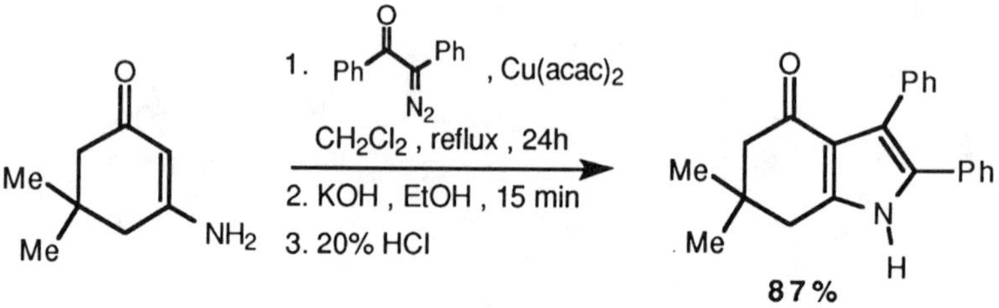

1. Ph⨍Ph , Cu(acac)₂
   N₂
   CH₂Cl₂ , reflux , 24h

2. KOH , EtOH , 15 min

3. 20% HCl

87%

Eberlin, M.N.; Kascheres. C. J. Org. Chem., **1988**, 53, 2084.

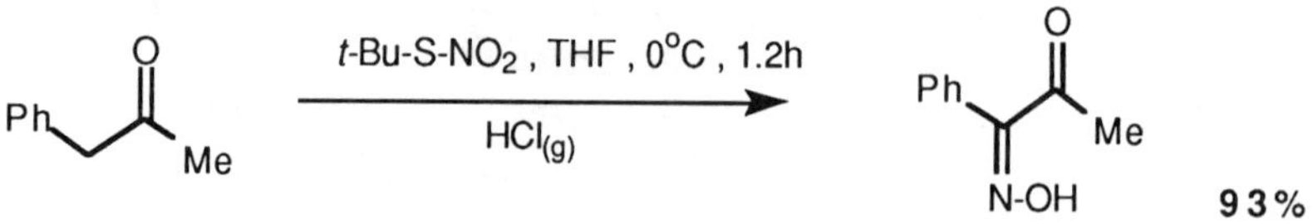

t-Bu-S-NO₂ , THF , 0°C , 1.2h

HCl₍g₎

N-OH   93%

Kim. Y.H.; Park, J.; Kim, K. Tetrahedron Lett., **1989**, 30, 2833.

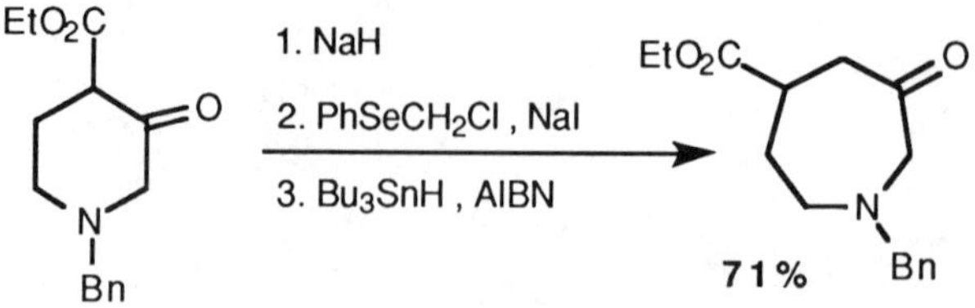

1. NaH

2. PhSeCH₂Cl , NaI

3. Bu₃SnH , AIBN

71%   Bn

Dowd. P.; Choi, S.-C. Tetrahedron Lett., **1989**, 30, 6129.

Tian, W.-S.; Livinghouse, T. *J. Chem. Soc., Chem. Commun.,* **1989**, 819.

SECTION 355: AMINE - NITRILE

Murahashi, S.; Shioto, T. *Tetrahedron Lett.,* **1987**, 28, 6469.

Hanafusa, T.; Ichihara, J.; Ashida, T. *Chem. Lett.,* **1987**, 687.

Sulmon, P.; DeKimpe, N.; Schamp, N. *Tetrahedron,* **1989**, 45, 3907.

SECTION 356: AMINE - ALKENE

Barluenga, J.; Tomás, M.; Ballesteros, A.; Gotor, V. *J. Chem. Soc., Chem. Commun.,* **1987**, 1195.

Grieco, P.A.; Fobare, W.F. *J. Chem. Soc., Chem. Commun.,* **1987**, 185.

Kinsman, R.; Lathbury, D.; Vernon, P.; Gallagher, T. *J. Chem. Soc., Chem. Commun.,* **1987**, 245.

Suzuki, K.; Ohkuma, T.; Tsuchihashi, G. *J. Org. Chem.,* **1987**, *52*, 2929.

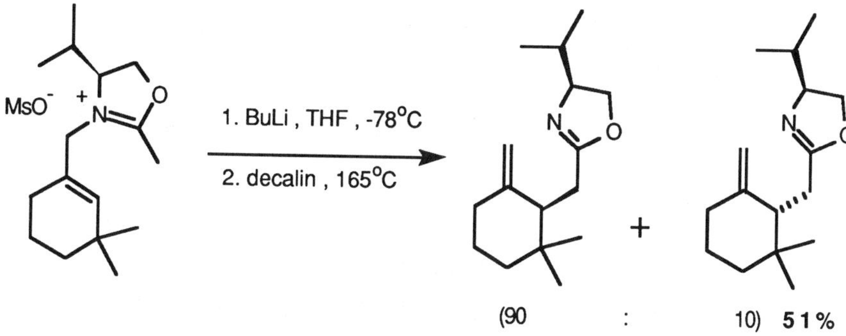

MsO⁻
N
O
1. BuLi, THF, -78°C
2. decalin, 165°C
N
O
+
N
O
(90 : 10) 51%

Kurth, M.J.; Soares, C.J. *Tetrahedron Lett.*, ***1987***, *28*, 1031.

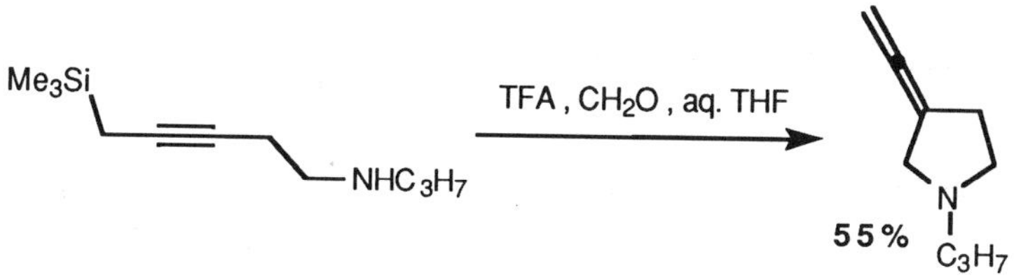

Me3Si
NHC3H7
TFA, CH2O, aq. THF
55%
N
C3H7

Damour, D.; Pornet, J.; Miginiac, L. *Tetrahedron Lett.*, ***1987***, *28*, 4689.

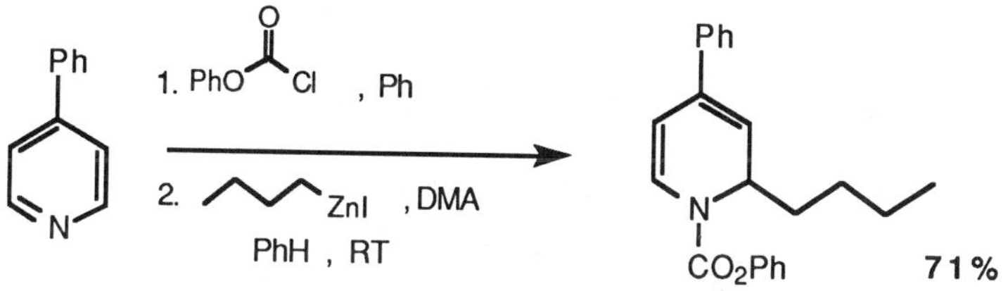

Ph
N
1. PhO Cl, Ph
2. ZnI, DMA
PhH, RT
Ph
N
CO2Ph
71%

Comins, D.L.; O'Connor, S. *Tetrahedron Lett.*, ***1987***, *28*, 1843.

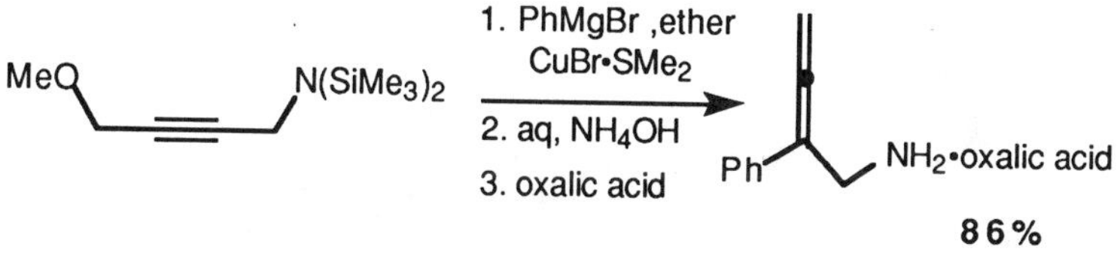

MeO
N(SiMe3)2
1. PhMgBr, ether
CuBr•SMe2
2. aq, NH4OH
3. oxalic acid
Ph
NH2•oxalic acid
86%

McCarthy, J.R.; Barney, C.L.; Matthews, D.P.; Bargar, T.M. *Tetrahedron Lett.*, ***1987***, *28*, 207.

Sant, J.Y.; South, M.S. *Tetrahedron Lett.*, *1987*, *28*, 6019.

Overman, L.E.; Sharp, M.J. *J. Am. Chem. Soc.*, *1988*, *110*, 612.
Overman, L.E.; Sharp, M.J. *Tetrahedron Lett.*, *1988*, *29*, 901.

Tamao, K.; Kobayashi, K.; Ito, Y. *J. Am. Chem. Soc.*, *1988*, *110*, 1286.

Barluenga, J.; Ferrero, M.; Palacios, F. *Tetrahedron Lett.*, *1988*, *29*, 4863.

Grieco, P.A.; Bahsas, A. *Tetrahedron Lett.*, *1988*, *29*, 5855.

Grimaldi, J.; Cormons, A. *Tetrahedron Lett.*, *1988*, *29*, 6609.

Gupta, A.K.; Ila, H.; Junjappa, H. *Tetrahedron Lett.*, *1988*, *29*, 6633.

Comins, D.L.; Foley, M.A. *Tetrahedron Lett.*, *1988*, *29*, 6711.

EE:EZ = 6:1

Chen, S.-F.; Ho, E.; Mariano, P.S. *Tetrahedron*, *1988*, *44*, 7013.

Murahashi, S.; Makabe, Y.; Kunita, K. *J. Org. Chem.*, *1988*, *53*, 4489.

$$E:Z = 98:2$$

<u>Takai, K.</u>; Fujimura, O.; Kataoka, Y.; Utimoto, K. *Tetrahedron Lett.*, *1989*, *30*, 211.

1. 2.2 eq. sBuLi , THF

2. PhCH$_2$Br

3. H$_2$O

<u>Kempf, D.J.</u> *Tetrahedron Lett.*, *1989*, *30*, 2029.

1. Ph—⟨   ⟩—OTf   , Pd(PPh$_3$)$_4$

CuI , Et$_2$NH , RT , DMF ,
RT , 4h

2. PdCl$_2$, MeCN , 75°C , 3h

Arcadi, A.; <u>Cacchi, S.</u>; Marinelli, F. *Tetrahedron Lett.*, *1989*, *30*, 2581.

, e$^-$, PbBr$_2$, Bu$_4$NBr

THF , Al anode , Pt cathode

Tanaka, H.; Nakahara, T.; Dhimane, H.; <u>Torii, S.</u> *Tetrahedron Lett.*, *1989*, *30*, 4161.

1. BuLi

2. PhCHO

3. H$_2$O

<u>Barluenga, J.;</u> Merino, I.; Palacios, F. *Tetrahedron Lett.*, *1989*, *30*, 5493.

Oppolzer, W.; Keller, T.H.; Bedoya-Zurita, M.; Stone, C. *Tetrahedron Lett.; 1989, 30,* 5883.

Freeman, F.; Kim, D.S.H.L. *Synthesis, 1989,* 698.

Tamao, K.; Kobayashi, K.; Ito, Y. *J. Org. Chem., 1989, 54,* 3517.

REVIEW:

"Synthesis of 1,4-Dihydropyridines by Cyclocondensation Reactions"
"Reactions of 1,4-Dihydropyridines"

Sausiňs, A.; Duburs, G. *Heterocycles, 1988, 27,* 269, 291.

SECTION 357: ESTER - ESTER

Tamaru, Y.; Hojo, M.; Yoshida, Z. *Tetrahedron Lett., 1987, 28,* 325.

1. 2 LDA , THF , HMPA , -78°C

2. THF , 0°C - RT , 14 h

22%

Garratt, P.J.; Porter, J.R. *Tetrahedron Lett.*, *1987*, *28*, 351.

1. $Ph_3P=CHC_3H_7$, 0°C
   30 min

2. HI

75%

Bazureau, J.P.; LeCorre, M. *Tetrahedron Lett.*, *1988*, *29*, 1919.
Bazureau, J.P.; LeRoux, J.; LeCorre, M. *Tetrahedron Lett.*, *1988*, *29*, 1921.

PhI ,4% $Pd(dppe)_2$

NaH , DMSO , 85°C

4% $Pd(dba)_2$

75%

Fournet, G.; Balme, G.; Gore, J. *Tetrahedron Lett.*, *1989*, *30*, 69.

5% $Pd(OAc)_2$

AcOH , acetone
benzoquinone , 20°C

+ 2 eq. LiOAc , 0.5 LiCl

(99	:	1) 77%
(24	:	76) 57%

Bäckvall, J.-E.; Andersson, P.G.; Vägberg, J.O. *Tetrahedron Lett.*, *1989*, *30*, 137.

Oumar-Mahamat, H.; Moustrou, C.; Surzur, J.-M.; Bertrand, M.P. *J. Org. Chem.*, *1989*, *54*, 5684.

REVIEW:

"The Chemistry of Methane Tricarboxylic Esters: A Review"

Newkome, G.R.; Baker, G.R. *Org. Prep. Proceed. Int.*, *1986*, *18*, 119.

Also via Dicarboxylic Acids:	Section 312 (Carboxylic Acids - Carboxylic Acids)
Hydroxy-esters	Section 327 (Alcohol - Ester)
Diols	Section 323 (Alcohol - Alcohol)

SECTION 358: ESTER - ETHER, EPOXIDE, THIOETHER

Walkup, R.D.; Park, G. *Tetrahedron Lett.*, *1987, 28*, 1023.

Larchevêque, M.; Petit, Y. *Tetrahedron Lett.*, *1987, 28*, 1993.

$$E:Z = 95:5$$

98%

Ricci, A.; Degl'Innocenti, A.; Borselli, G.; Reginato, G. *Tetrahedron Lett., 1987, 28,* 4093.

BHT = 2,6-di-*t*-butyl-4-methylphenol

91%

Revis, A.; Hilty, T.K. *Tetrahedron Lett., 1987, 28,* 4809.

71%

Konstantinović, S.; Vukićević, R.; Mihailović, M.L. *Tetrahedron Lett., 1987, 28,* 6511.

(83   :   17) 95%

Doyle, M.P.; Bagheri, V.; Harn, N.K. *Tetrahedron Lett., 1988, 29,* 5119.

70%

Maruoka, K.; Shinoda, K.; Yamamoto, H. *Synth. Commun., 1988, 18,* 1029.

PdCl$_2$(PPh$_3$)$_2$, Ac$_2$O

NEt$_3$, PhH , CO , 1.5h
170°C

85%

Iwasaki, M.; Li, J.; Kobayashi, Y.; Matsuzaka, H.; Ishii, Y.; Hidai, M. *Tetrahedron Lett.,* *1989, 30,* 95.

CH$_2$Cl$_2$ / acetone
-20°C , 3.5h

95%

Adam, W.; Hadjiarapoglou, L.; Jäger, V.; Seidel, B. *Tetrahedron Lett., 1989, 30,* 4223.

C$_6$H$_{13}$CH=CH$_2$

1. PhSeBr , AcOH
   Ac$_2$O , KOAc , 15h
2. BF$_3$ •OEt$_2$
   CHCl$_3$, RT
   24h

C$_6$H$_{13}$

(96   :   4)   79%

Engman, L. *J. Org. Chem., 1989, 54,* 884.

SECTION 359: ESTER - HALIDE, SULFONATE

MeO$_2$C

Me$_6$Sn$_2$, AIBN

PhH , 24h , reflux

CO$_2$Me

+

CO$_2$Me

cis : trans   (53:40   2.4:4.6)   83%

Curran, D.P.; Chang, C.-T. *Tetrahedron Lett., 1987, 28,* 2477.

BzCl , MeCN , 24h

10 KBar , 62°C

OBz

Cl

57%

Kotsuki, H.; Ichikawa, Y.; Nishizawa, H. *Chem. Lett., 1988,* 673.

Iqbal, J.; Aminkhan, M.; Srivastava, P.R. *Tetrahedron Lett., 1988, 29*, 4985.
Ahmad, S.; Iqbal, J. *Chem. Lett., 1987*, 953.

Cambie, R.C.; Rutledge, P.S.; Somerville, R.F.; Woodgate, P.D. *Synthesis, 1988*, 1009.

Coghlan, M.J.; Caley, B.A. *Tetrahedron Lett., 1989, 30*, 2033.

Balko, T.W.; Brinkmeyer, R.S.; Terando, N.H. *Tetrahedron Lett., 1989, 30*, 2045.

1. Bu$_3$P=CFCO$_2$Et , 78h

n-C$_7$H$_{15}$-Br $\xrightarrow{\text{2. aq. NaHCO}_3}$ n-C$_7$H$_{15}$CHFCO$_2$Et

52%

Thenappan, A.; Burton, D.J. *Tetrahedron Lett., 1989, 30*, 3641.

SECTION 360: ESTER - KETONE

syn:anti = 81:19

Berrada, S.; Metzner, P. *Tetrahedron Lett., 1987, 28,* 409.

75%   12%

Dowd, P.; Choi, S.-C. *J. Am. Chem. Soc., 1987, 109,* 6548.
Dowd, P.; Choi, S.-C. *J. Am. Chem. Soc., 1987, 109,* 3493.

(87   :   13) 60%

Taber, D.F.; Raman, K.; Gaul, M.D. *J. Org. Chem., 1987, 52,* 28.

87%

Ahmad, S.; Iqbal, J. *J. Chem. Soc., Chem. Commun., 1987,* 114.

90%

Miller, R.D.; Theis, W. *Tetrahedron Lett.*, *1987*, *28*, 1039.

97%

Henegar, K.E.; Winkler, J.D. *Tetrahedron Lett.*, *1987*, *28*, 1051.

41%

Moody, C.J.; Taylor, R.J. *Tetrahedron Lett.*, *1987*, *28*, 5351.

77%

Black, T.H.; Arrivo, S.M.; Schumm, J.S.; Knobeloch, J.M. *J. Org. Chem.*, *1987*, *52*, 5425.

Short, R.P.; Masamune. S. *J. Org. Chem.*, *1987*, *52*, 2841.

Shono. T.; Kashimura, S.; Yamaguchi, Y.; Kuwata, F. *Tetrahedron Lett.*, *1987*, *28*, 4411.

Corey. E.J.; Su, W. *Tetrahedron Lett.*, *1987*, *28*, 5241.

Baskaran, S.; Islam, I.; Baghavan, M.; Chandrasekaran. S. *Chem. Lett.*, *1987*, 1175.

$$\text{CH}_2=\text{CHCO}_2\text{Me} \xrightarrow[\substack{\text{2. sealed tube , 10 Torr} \\ \text{MeCN , RT , 10h} \\ \text{hv (350 nm)}}]{\substack{\text{1. MeMn(CO)}_5 \text{ , 6 KBar} \\ \text{THF , 48h}}} \text{Me-CO-CH}_2\text{CH}_2\text{-CO}_2\text{Me}$$

72%

DeShong, P.; Sidler, D.R.; Rybczynski, P.J.; Slough, G.A.; Rheingold, A.L. *J. Am. Chem. Soc.*, *1988*, *110*, 2575.

$$\text{Ph-C(H)(OH)-C}\equiv\text{C-CO}_2\text{Et} \xrightarrow[\substack{60^\circ\text{C , 5h} \\ \text{2. 0.1N HCl , ether}}]{\substack{\text{1. Pd(OAc)}_2(\text{PPh}_3)_2 \\ \text{Bu}_3\text{N , HCO}_2\text{H , DMF}}} \text{Ph-CO-CH}_2\text{CH}_2\text{-CO}_2\text{Et}$$

65%

Arcadi, A.; Cacchi, S.; Marinelli, F.; Misiti, D. *Tetrahedron Lett.*, *1988*, *29*, 1457.

79%

Pirrung, M.C.; Nunn, D.S. *Tetrahedron Lett.*, *1988*, *29*, 163.

83%

Stork, G.; Nakatani, K. *Tetrahedron Lett.*, *1988*, *29*, 2283

Aoki, S.; Nakamura, E.; Kawajima, I. *Tetrahedron Lett.*, *1988*, *29*, 1541.

Ramaiah, P.; Rao, A.S. *Tetrahedron Lett.*, **1988**, *29*, 2119.

Camps, F.; Coll, J.; Llebaria, A.; Moretó, J.M. *Tetrahedron Lett.*, **1988**, *29*, 5811.

Devanne, D.; Ruppin, C.; Dixneuf, P.H. *J. Org. Chem.*, **1988**, *53*, 925.

Reed, K.L.; Gupton, J.T.; McFarlane, K.L *Synth. Commun.*, **1989**, *19*, 2595.

Nudelman, A.; Kelner, R.; Broida, N.; Gottlieb, H.E. *Synthesis*, **1989**, 387.

Turner, J.A.; Jacks, W.S. *J. Org. Chem.*, *1989*, *54*, 4229.

Mandal, A.K.; Jawalkar, D.G. *J. Org. Chem.*, *1989*, *54*, 2364.

Holmquist, C.R.; Roskamp, E.J. *J. Org. Chem.*, *1989*, *54*, 3258.

Hirao, T.; Mori, M.; Oshiro, Y. *Bull. Chem. Soc., Jpn.*, *1989*, *62*, 2399.

REVIEW:

"Preparation and Reactions of 4-Oxocarbonyl Compounds"

Miyakoshi, T. *Org. Prep. Proceed. Int.*, *1989*, *21*, 661.

Also via Ketoacids   Section 320 (Carboxylic Acid - Ketone)
          Hydroxyketones   Section 330 (Alcohol - Ketone)

SECTION 361: ESTER - NITRILE

NO ADDITIONAL EXAMPLES

SECTION 362: ESTER - ALKENE

This section contains syntheses of enol esters and esters of unsaturated acids as well as ester molecules bearing a remote alkenyl unit.

Ley, S.V.; Woodward, P.R. *Tetrahedron Lett.*, *1987, 28*, 345.

Stille, J.K.; Tanaka, M. *J. Am. Chem. Soc.*, *1987, 109*, 3785.

Chan, D.M.T.; Marder, T.B.; Milstein, D.; Taylor, N.J. *J. Am. Chem. Soc.*, *1987, 109*, 6385.

Masuyama, Y.; Yamada, K.; Kurusu, Y. *Tetrahedron Lett.*, *1987, 28*, 443.

Marino, J.P.; Laborde, E. *J. Org. Chem.*, *1987*, *52*, 1.

Xu, Y.; Zhou, B. *J. Org. Chem.*, *1987*, *52*, 974.

81:19 mixture

Shing, T.K.M.; Lloyd-Williams, P. *J. Chem. Soc., Chem. Commun.*, *1987*, 423.

allylic acetates react to form substituted 1,3-dienes

Mitsudo, T.; Kadokura, M.; Watanabe, Y. *J. Org. Chem.*, *1987*, *52*, 1695.

Mitsudo, T.; Hori, Y.; Yamakawa, Y.; Watanabe, Y. *J. Org. Chem.*, *1987*, *52*, 2230.

Rizzacasa, M.A.; Sargent, M.V. *Aust. J. Chem.*, *1987*, *40*, 1737.

Srikrishna, A. *J. Chem. Soc., Chem. Commun.*, *1987*, 587.

Burgess, K. *J. Org. Chem.*, *1987*, *52*, 2046.

Larock, R.C.; Leuck, D.J.; Harrison, L.W. *Tetrahedron Lett.*, *1987*, *28*, 4977.

syn:anti = 27:73

Kawai, M.; Onaka, M.; Izumi, Y. *J. Chem. Soc., Chem. Commun.*, *1987*, 1203.

trans:cis = 91:9

Bäckvall, J.-E.; Gogoll, A. *J. Chem. Soc., Chem. Commun., 1987*, 1236.

Suemune, H.; Oda, K.; Sakai, K. *Tetrahedron Lett., 1987, 28*, 3373.

Tamaru, Y.; Bando, T.; Hojo, M.; Yoshida, Z. *Tetrahedron Lett., 1987, 28*, 3497.

Harvey, R.G.; Cortez, C.; Ananthanarayan, T.P.; Schmolka, S. *Tetrahedron Lett., 1987, 28*, 6137.

Michaelis, R.; Müller, U.; Schäfer, H.J. *Angew. Chem. Int. Ed., Engl, 1987, 26*, 1026.

Ph—CH=CH—TePh $\xrightarrow[\text{NEt}_3\text{, MeOH}]{\text{1 atm CO , PdCl}_2}$ Ph—CH=CH—CO$_2$Me

96%

Ohe, K.; Takahashi, H.; <u>Uemura, S.</u>; Sugita, N. *J. Org. Chem.*, *1987*, *52*, 4859.

$\xrightarrow[\substack{\text{PhMe , MeCN} \\ 50°\text{C , 15 min}}]{\text{Pd}_2(\text{dba})_3 \cdot \text{CHCl}_3}$

74%

<u>Tsuji, J.</u>; Nisar, M.; Minami, I. *Chem. Lett.*, *1987*, 23.

$\xrightarrow[\substack{\text{(dark) , 2d ,} \\ 12°\text{C}}]{\text{KI}_3\text{ , MeCN , NaHCO}_3}$

42%

Z:E = 8:1

Tsuboi, S.; Wada, H.; Mimura, S.; <u>Takeda, A.</u> *Chem. Lett.*, *1987*, 937.

Me$_3$Si—CH=CH—CH(Ph)—OCO$_2$Me + CH$_3$COCH$_2$CO$_2$Me

$\xrightarrow[\text{THF , 25°C , 2h}]{\text{PBu}_3\text{ ,Pd}_2(\text{dba})_3\cdot\text{CHCl}_3}$

83%

Tsuji, J.; Yuhara, M.; Minato, M.; Yamada, H.; Sato, F.; Kobayashi, Y. *Tetrahedron Lett.*, *1988*, *29*, 343.

$\xrightarrow{\text{NaI , NEt}_3}$

76%

Z:E = 98:2

<u>Shimagaki, M.</u>; Shiokawa, M.; Sugai, K.; Teranaka, T.; Nakata, T.; <u>Oishi, T.</u> *Tetrahedron Lett.*, *1988*, *29*, 659.

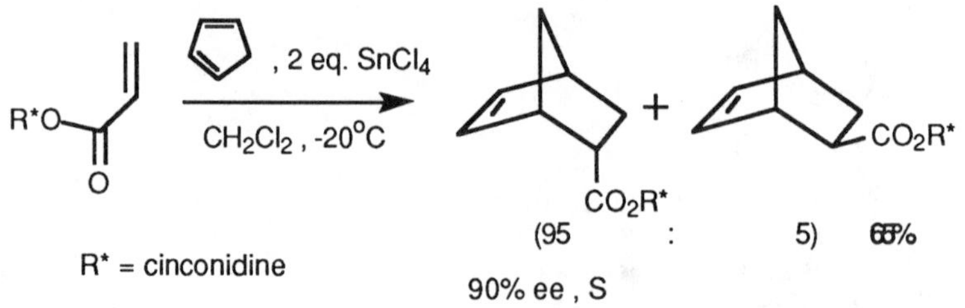

R*O— (acrylate), 2 eq. SnCl$_4$, CH$_2$Cl$_2$, -20°C

R* = cinconidine

(95 : 5) 65%

CO$_2$R*

90% ee , S

Suzuki, H.; Mochizuki, K.; Hattori, T.; Takahashi, N.; Tajima, O.; Takiguchi, T. *Bull. Chem. Soc., Jpn.,* **1988**, *61*, 1999.

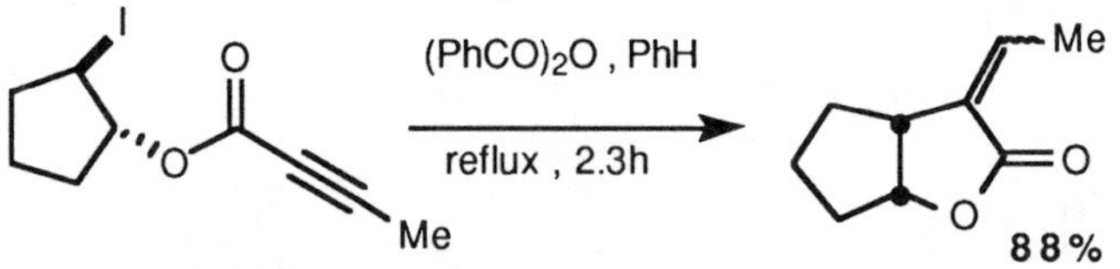

(PhCO)$_2$O , PhH

reflux , 2.3h

88%

Haaima, G.; Weavers, R.T. *Tetrahedron Lett.,* **1988**, *29*, 1085.

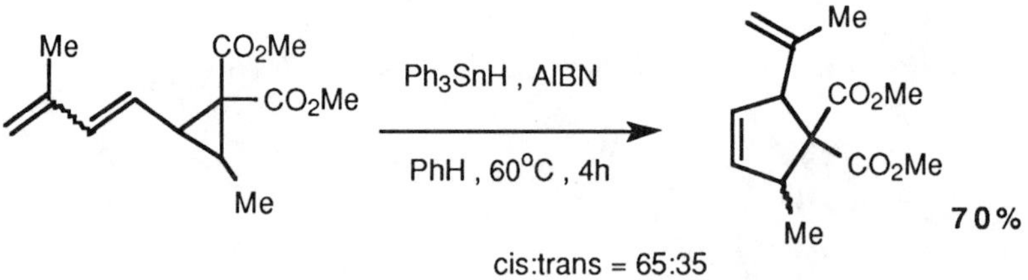

Ph$_3$SnH , AIBN

PhH , 60°C , 4h

cis:trans = 65:35

70%

Muira, K.; Fugami, K.; Oshima, K.; Utimoto, K. *Tetrahedron Lett.,* **1988**, *29*, 1543.

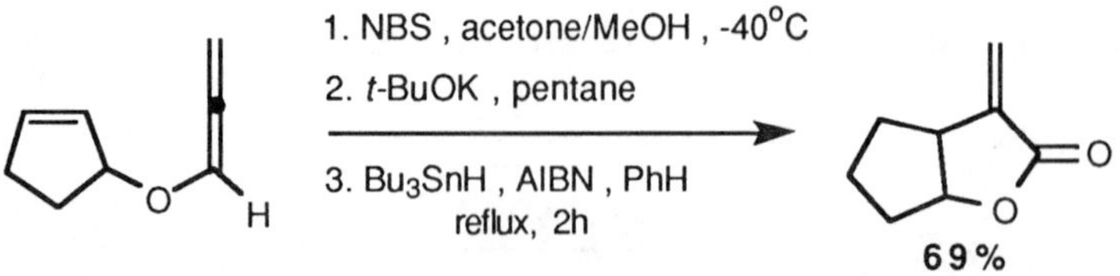

1. NBS , acetone/MeOH , -40°C
2. *t*-BuOK , pentane
3. Bu$_3$SnH , AIBN , PhH
   reflux, 2h

69%

Dulcere, J.P.; Mihoubi, M.N.; Rodriguez, J. *J. Chem. Soc., Chem. Commun.,* **1988**, 237.

1. COCl$_2$
2. PhSeH
3. Bu$_3$SnH , AIBN
PhMe , 110°C , 8h

55%

Bachi, M.D.; Bosch, E. *Tetrahedron Lett.*, **1988**, *29*, 2581.

Pd$_2$(dba)$_3$ •CHCl$_3$
PPh$_3$, NaBr , EtOH
iPr$_2$NEt , CO , 50°C
20h , autoclave

CO$_2$Et

80%

Murahashi, S.; Imada, Y.; Taniguchi, Y.; Higashiura, S. *Tetrahedron Lett.*, **1988**, *29*, 4945.

Ph

1. Bu$_3$SnH , BEt$_3$, PhMe
-78°C
2. 1N HCl

Ph

78%

Nozaki, K.; Oshima, K.; Utimoto, K. *Tetrahedron Lett.*, **1988**, *29*, 6127.

Bu —= I , Pd(OAc)$_2$

COOH

DMF , 80°C , Bu$_4$NCl
sealed tube , 20h

Bu

73%

Larock, R.C.; Leuck, D.J.; Harrison, L.W. *Tetrahedron Lett.*, **1988**, *29*, 6399.

1. NaH , DMSO , 4h
95°C
2. aq. H$_2$SO$_4$, heat

62%

Karim, M.R.; Sampson, P. *Tetrahedron Lett.*, **1988**, *29*, 6897.

Rigby, J.H.; Senanayaker, C. *J. Org. Chem.*, **1988**, *53*, 440.

Heumann, A.; Moberg, C. *J. Chem. Soc., Chem. Commun.*, **1988**, 1516.

Wu, T.-C.; Riecke, R.D. *J. Org. Chem.*, **1988**, *53*, 2381.

Tsuda, T.; Ohashi, Y.; Nagahama, N.; Sumiya, R.; Saegusa, T. *J. Org. Chem.*, **1988**, *53*, 2650.

Ma, D.; Lu, X. *Tetrahedron Lett.*, **1989**, *30*, 843.

Araki, S.; Katsumura, N.; Ito, H.; Butsugan, Y. *Tetrahedron Lett., 1989, 30*, 1581.

Chidambaram, N.; Satyanarayana, K.; Chandrasekaran, S. *Tetrahedron Lett., 1989, 30*, 2429.

E:Z = 98:2

Curran, D.P.; van Elburg, P.A. *Tetrahedron Lett., 1989, 30*, 2501.

Fournier, J.; Bruneau, C.; Dixneuf, P.H. *Tetrahedron Lett., 1989, 30*, 3981.

$R^* =$ (98   :   2)   **78%**

endo:exo = 54:1

$R^* =$ (2   :   98)   **70%**

endo:exo = 38:1

Linz, G.; Weetman, J.; Abdel-Hady, A.F.; Helmchen, G. *Tetrahedron Lett.*, *1989*, *30*, 5599.

67%

Tanino, K.; Sato, K.; Kumajima, I. *Tetrahedron Lett.*, *1989*, *30*, 6551.

1. MeZnCl ,
   5% PdCl$_2$(PPh$_3$)$_2$

2. NaOAc , LiCl , 3% PdCl$_2$
   benzoquinone

73%

Yamashina, N.; Hyuga, S.; Hara, S.; Suzuki, A. *Tetrahedron Lett.*, *1989*, *30*, 6555.

REVIEW:

"Nucleophilic Addition Reactions with Cationic Iron π-Alkyne and Related Complexes"

Reger, D.L. *Acc. Chem. Res.*, *1988*, *21*, 229.

Related Methods: Section 60A   (Protection of Aldehydes).
                          Section 180A (Protection of Ketones).

Also via Acetylenic Esters: Section 306 (Alkyne - Ester).
    Alkenyl Acids: Section 322 (Carboxylic Acid - Alkene).
    β-Hydroxy-esters: Section 327 (Alcohol - Ester).

SECTION 363:   ETHER, EPOXIDE, THIOETHER - ETHER, EPOXIDE, THIOETHER

See Section 60A (Protection of Aldehydes) and Section 180A (Protection of Ketones) for reactions involving formation of Acetals and Ketals.

$$\text{CH}_3\text{CH=CHCHO} \xrightarrow[\text{-50°C - RT}]{3.1 \ \text{PhSH} , 1.5 \ \text{SnCl}_4} \quad 55\%$$

Cossy, J.; Henin, F.; Leblanc, C. *Tetrahedron Lett.*, *1987*, *28*, 1417.

$$\xrightarrow[\text{-20°C , 1h}]{\text{MeNO}_2 , \text{BF}_3 \cdot \text{OEt}_2} \quad 50\%$$

Poirier, J.-M.; Dujardin, G. *Tetrahedron Lett.*, *1987*, *28*, 3337.

$$\xrightarrow[\text{2. } \text{H}_3\text{O}^+]{\text{1. MeSeH , NaH} \quad \text{DMF , 20°C , 1h}} \quad \textbf{quant}$$

Kreif, A.; Trabelsi, M. *Tetrahedron Lett.*, *1987*, *28*, 4225.

$$\xrightarrow[\text{30 min}]{\text{1. mCPBA} \quad \text{2. PhH , reflux}} \quad 82\%$$

Bhupathy, M.; Cohen, T. *Tetrahedron Lett.*, *1987*, *28*, 4793, 4797.

83%

Murata. S.; Suzuki, T. *Tetrahedron Lett., 1987, 28,* 4415.
Murata. S.; Suzuki, T. *Tetrahedron Lett., 1987, 28,* 4297.

79%

Mihailović, M.L.; Konstantinović, S.; Vukićević, R. *Tetrahedron Lett., 1987, 28,* 4343.

86%

Tuladhar, S.M.; Fallis. A.G. *Tetrahedron Lett., 1987, 28,* 523.

quant.

Bird. C.W.; Hormozi, N. *Tetrahedron Lett., 1988, 29,* 705.

84%

Mukaiyama, T.; Wariishi, K.; Saito, Y.; Hayashi, M.; Kobayashi, S. *Chem. Lett., 1988,* 1101.

92%

Nicolaou, K.C.; Prasad, C.V.C.; Hwang, C.-K.; Duggan, M.E.; Veale, C.A. *J. Am. Chem. Soc.*, *1989*, *111*, 5321.

89%

Tiecco, M.; Testaferri, L.; Tingoli, M.; Chianelli, D.; Bartoli, D. *Tetrahedron Lett.*, *1989*, *30*, 1417.

64%

Thomas, A.; Singh, G.; Ila, H.; Junjappa, H. *Tetrahedron Lett.*, *1989*, *30*, 3093.

60%

Lee, T.V.; Ellis, K.L. *Tetrahedron Lett.*, *1989*, *30*, 3555.

70%

Iqbal, J.; Srivastava, R.R.; Gupta, K.B.; Klan, M.A. *Synth. Commun.*, *1989*, *19*, 901.

94%

Hu, N.X.; Aso, Y.; Otsubo, T.; Ogura. F. *J. Org. Chem.*, *1989, 54*, 4391.

REVIEW:

"Chemistry of Spiroketals"

Perron, F.; Albizati. K.F. *Chem. Rev.*, *1989, 89*, 1617.

SECTION 364:   ETHER, EPOXIDE, THIOETHER - HALIDE, SULFONATE

86%

Beckwith. A.L.J.; Meijs. G.F. *J. Org. Chem.*, *1987, 52*, 1922.

77%

Ishibashi. H.; Nakatani, H.; Maruyama, K.; Minami, K.; Ikeda, M. *J. Chem. Soc., Chem. Commun.*, *1987*, 1443.

62%

Coppi, L.; Ricci, A.; Taddei. M. *Tetrahedron Lett.*, *1987, 28*, 973.

1. Hg(OAc)$_2$, MeOH

2. NaCl

3. AcOF , CHCl$_3$,
   -78°C , 5 min

90%

Hebel, D.; Rozen, S. *J. Org. Chem.*, *1987, 52*, 2588.

Ph⎯⎯Br , neat , 50°C

24 h

quant

Trujillo, D.A.; McMahon Jr., W.A.; Lyle, R.E. *J. Org. Chem.*, *1987, 52*, 2932.

I$_2$, NaHCO$_3$, 3h
aq. ether

0 → 20°C

+

(39   :   61)   94%

Tamaru, Y.; Hojo, M.; Kawamura, S.; Sawada, S.; Yoshida, Z. *J. Org. Chem.*,
1987, 52, 4062.

1. 0.4 C$_3$H$_7$CHO

2. 0.1 AlCl$_3$

3. 0.4 n-C$_6$H$_{13}$CHO

4. 0.6 AlCl$_3$

n-C$_6$H$_{13}$

77%

Wei, Z.Y.; Li, J.S.; Wang, D.; Chan, T.H.P. *Tetrahedron Lett.*, *1987, 28*, 3441.

HgO , I$_2$

hν

68%

Kraus, G.A.; Thurston, J. *Tetrahedron Lett.*, *1987, 28*, 4011.

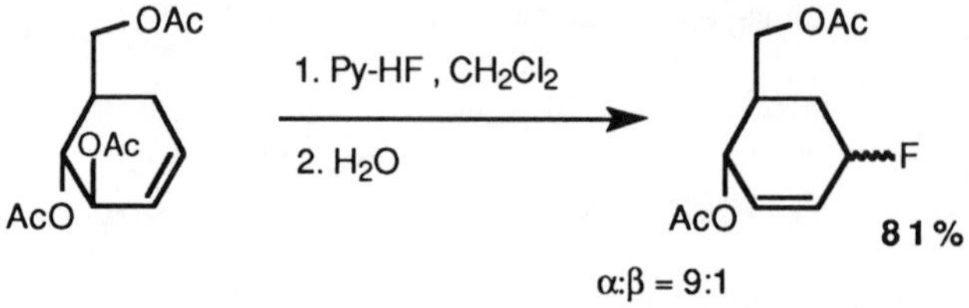

OAc
OAc
AcO
1. Py-HF , CH2Cl2
2. H2O
OAc
F
AcO
81%
α:β = 9:1

Macdonald, S.J.F.; McKenzie, T.C. *Tetrahedron Lett.*, ***1988***, *29*, 1363.

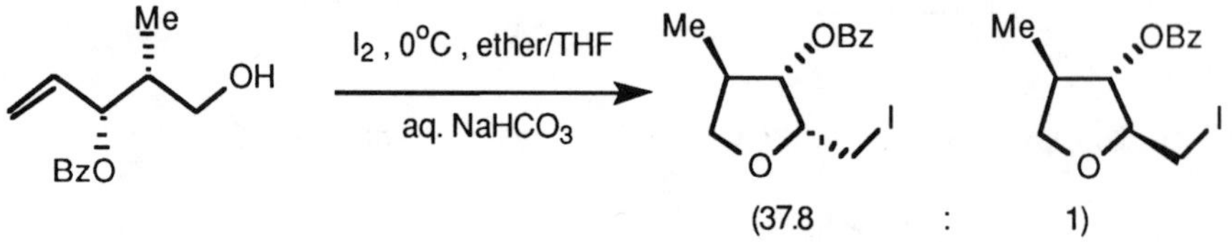

Me
OH
BzO
I2 , 0°C , ether/THF
aq. NaHCO3
Me
OBz
I
O
(37.8 : 1)

Kim, Y.G.; Cha, J.K. *Tetrahedron Lett.*, ***1988***, *29*, 2011.

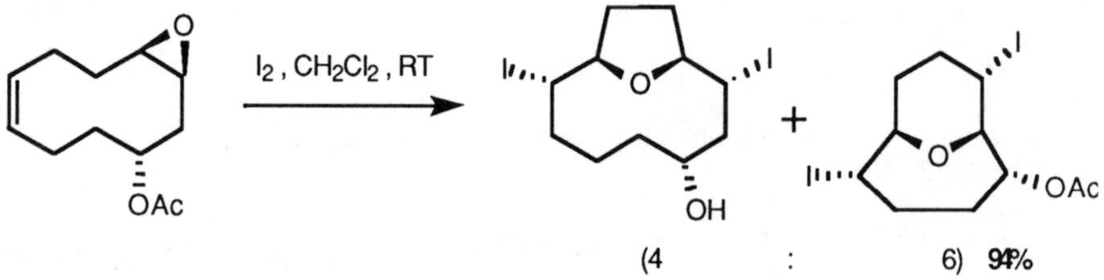

O
OAc
I2 , CH2Cl2 , RT
I
O
I
OH
+
I
O
I
OAc
(4 : 6) 94%

Alvarez, E.; Manta, E.; Martin, J.D.; Rodriguez, M.L.; Ruiz-Perez, C. *Tetrahedron Lett.*, ***1988***, *29*, 2093.

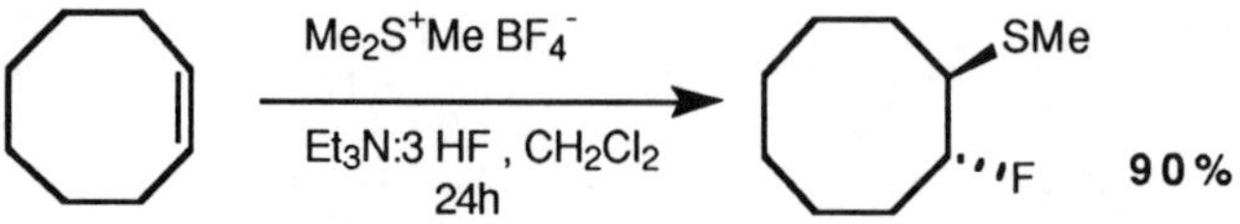

Me2S+Me BF4-
Et3N:3 HF , CH2Cl2
24h
SMe
F
90%

Haufe, G. *Tetrahedron Lett.*, ***1988***, *29*, 2311.

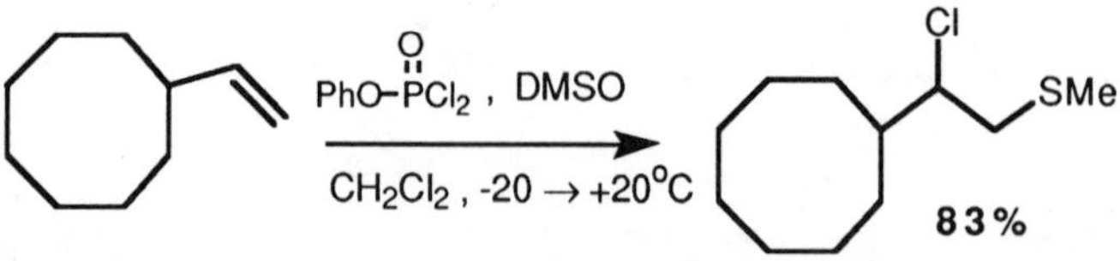

O
PhO-PCl2 , DMSO
CH2Cl2 , -20 → +20°C
Cl
SMe
83%

Liu, H.-J.; Nyangulu, J.M. *Tetrahedron Lett.*, ***1988***, *29*, 5467.

Coppi, L.; Ricci, A.; Taddei, M. *J. Org. Chem.*, *1988*, *53*, 911.

Wagner, A.; Heitz, M.-P.; Mioskowski, C. *Tetrahedron Lett.*, *1989*, *30*, 1971.

Chan, T.-H.; Arya, P. *Tetrahedron Lett.*, *1989*, *30*, 4065.

Joseph, S.P.; Keshavamurthy, K.S.; Dhar, D.N. *Synth. Commun.*, *1989*, *19*, 889.

Nikolic, N.A.; Gonda, E.; Desmond Longford, C.P.; Lane, N.T.; Thompson, D.W. *J. Org. Chem.*, *1989*, *54*, 2748.

Me   Ph   C₆H₁₃CHO , AlCl₃ , CH₂Cl₂
\Si   ────────────────────>
/ \   -78 → 0°C
Me   O   Me

Wei, Z.Y.; Wang, D.; Li, J.S.; Chan, T.H. *J. Org. Chem.*, *1989*, *54*, 5768.

SECTION 365: ETHER, EPOXIDE, THIOETHER - KETONE

1. Me₃SiCl , Et₃N•DMF
────────────────────>
2. (PhIO)ₙ , BF₃•OEt₂
   MeOH

78%

Moriarty, R.M.; Prakash, O.; Duncan, M.P.; Vaid, R.K.; Musllam, H.A. *J. Org. Chem.*, *1987*, *52*, 150.

2 PhSeCl , 30°C
────────────────────>
aq. MeCN , 15 min

70%

Mehta, G.; Prakash Rao, H.S.; Reddy, K.R. *J. Chem. Soc., Chem. Commun.*, *1987*, 78.

1. MeOCH₂SO₂Ph
   t-BuLi , DME , -78°C
────────────────────>
2. Et₂AlCl , CHCl₃ , 0°C

53%

Trost, B.M.; Mikhail, G.K. *J. Am. Chem. Soc.*, *1987*, *109*, 4124.

1. PhH , collidine
      reflux
────────────────────>
2. TFAA , reflux
      2h

70%

Brady, W.T.; Giang, Y.-S.F.; Weng, L.; Dad, M.M. *J. Org. Chem.*, *1987*, *52*, 2216.

Ether - Ketone

Et-C(OEt)$_3$

1. 80°C Me-C(=O)-CN
2. MeLi , ether
3. aq. MeOH
4. Py-H$^+$ OTs$^-$

→

70%

Babler, J.H.; Marcuccilli, C.J. *Tetrahedron Lett.*, **1987**, *28*, 4657.

C$_{10}$H$_{21}$OH

1. Cl-C(=)-OMOM , NaOH , dioxane
   Bu$_4$N HSO$_4$, -60°C , 2h
2. 1% H$_2$SO$_4$, 80°C , 1h

→ C$_{10}$H$_{12}$O-CH$_2$-C(=O)-Me

74%

Gu, X.-P.; Nishida, N.; Ikeda, I.; Okahara, M. *J. Org. Chem.*, **1987**, *52*, 3192.

PPTS , EtOH

reflux , 1h

83%

PPTS = *p*-toluenesulfonate

Satoh, T.; Iwamoto, K.; Yamakawa, K. *Tetrahedron Lett.*, **1987**, *28*, 2603.

1. PhSSPh , Se, CO , H$_2$O
   autoclave , 5h , 50°C
2. aq. HCl

→

Ogawa, A.; Nishiyama, Y.; Kambe, N.; Murai, S.; Sonoda, N. *Tetrahedron Lett.*, **1987**, *28*, 3271.

Ph-C(=O)-Cl

=-OBu , Pd(OAc)$_2$, NEt$_3$

sealed vessel , 70°C

→

69%

Andersson, C.-M.; Hallberg, A. *Tetrahedron Lett.*, **1987**, *28*, 4215.

Baciocchi, E.; Civitarese, G.; Ruzziconi, R. *Tetrahedron Lett.*, **1987**, *28*, 5357.

Herrinton, P.M.; Hopkins, M.H.; Mishra, P.; Brown, M.J.; Overman, L.E. *J. Org. Chem.*, **1987**, *52*, 3711.

Moriarty, R.M.; Vaid, R.K.; Duncan, M.P. *Synth. Commun.*, **1987**, *17*, 709.

chloreal = trichloroisocyanuric acid

de Groot, A.; Peperzaki, R.M.; Vader, J. *Synth. Commun.*, **1987**, *17*, 1607.

Miyashita, M.; Suzuki, T.; Yoshikoshi, A. *Chem. Lett.*, **1987**, 285.

77%

diastereomers = 69:31

Linderman, R.J.; Godfrey, A. *J. Am. Chem. Soc., 1988, 110*, 6249.

81%

81% ee

Maruoka, K.; Itoh, T.; Shirasaka, T.; Yamamoto, H. *J. Am. Chem. Soc., 1988, 110*, 310.

(18 : 1) 61%

Krafft, M.E. *J. Am. Chem. Soc., 1988, 110*, 968.

Lee, T.V.; Boucher, R.J.; Rockell, C.J.M. *Tetrahedron Lett., 1988, 29*, 689.

Middleton, D.S.; Simpkins, N.S.; Terrett, N.K. *Tetrahedron Lett., 1988, 29*, 1315.

Satoh, T.; Iwamoto, K.; Sugimoto, A.; Yamakawa, K. *Bull. Chem. Soc., Jpn., 1988, 61*, 2109.

endo:exo methyl = 1.7:1

Daude, N.; Eggert, U.; Hoffmann, H.M.R. *J. Chem. Soc., Chem. Commun., 1988*, 206.

Zhaishibekov, B.S.; Pat-saev, A.K.; Erzhanov, K.B. *J. Org. Chem., U.S.S.R.,* *1988,* *24,* 2219.

Heslin, J.C.; <u>Moody, C.J.</u> *J. Chem. Soc., Perkin Trans. I,* *1988,* 1417.
<u>Moody, C.J.</u>; Taylor, R.J. *J. Chem. Soc., Perkin Trans. I,* *1989,* 721.

Kim, H.; <u>Schlecht, M.F.</u> *Tetrahedron Lett.,* *1988,* *29,* 1771.

<u>Moody, C.J.</u>; Taylor, R.J. *Tetrahedron Lett.,* *1988,* *29,* 6005.

<u>Vinogradov, M.G.;</u> Kondorsky, A.E.; Nikishin, G.I. *Synthesis,* *1988,* 60.

26% ee

Sera. A.; Takagi, K.; Katayama, H.; Yamada, H.; Matsumoto, K. *J. Org. Chem.*, *1988*, *53*, 1157.

63%

Nicolaou. K.C.; Hwang, C.-K.; Nugiel, D.A. *Angew. Chem. Int. Ed., Engl*, *1988*, *27*, 1362.

88%

Antonioletti, R.; Bunadies, F.; Scettri. A. *J. Org. Chem.*, *1988*, *53*, 5540.

82%

α:β = 4:1

Hirst, G.C.; Howard, P.N.; Overman. L.E. *J. Am. Chem. Soc.*, *1989*, *111*, 1514.

Hoppe, D.; Krämer, T.; Erdbrügger, C.F.; Egert, E. *Tetrahedron Lett.*, *1989*, *30*, 1233.

Adams, J.; Poupart, M.-A.; Grenier, L.; Schaller, C.; Ouimet, N.; Frenette, R. *Tetrahedron Lett.*, *1989*, *30*, 1749.

erythro:threo = 78:22

Ishihara, K.; Yamamoto, H.; Heathcock, C.H. *Tetrahedron Lett.*, *1989*, *30*, 1825.

Molander, G.A.; Andrews, S.W. *Tetrahedron Lett.*, *1989*, *30*, 2351.

84%

Kim, S.; Park, J.H. *Tetrahedron Lett.*, *1989*, *30*, 6181.

94%

Hosokawa, T.; Shinohara, T.; Ooka, Y.; Murahashi, S. *Chem. Lett.*, *1989*, 2001.

84%

Brown, D.S.; Ley, S.V.; Bruno, M. *Heterocycles*, *1989*, *28*, 773.

51%

Perron, F.; Albizati, K.F. *J. Org. Chem.*, *1989*, *54*, 2044.

68%

Yoshida, J.; Nakajan, S.; Isoe, S. *J. Org. Chem.*, *1989*, *54*, 5655.

81%

Fukuzawa, S.; Fukushima, M.; Fujinami, T.; Sakai, S. *Bull. Chem. Soc., Jpn.*, *1989*, *62*, 2348.

63%

Hunter, R.; Carlton, L.; Cirillo, P.F.; Michael, J.P.; Simon, C.D.; Walter, D.S. *J. Chem. Soc., Perkin Trans. I*, *1989*, 1631.

SECTION 366: ETHER, EPOXIDE, THIOETHER - NITRILE

80%   **26%**

Parker, K.A.; Spero, D.M.; Koziski, K.A. *J. Org. Chem.*, *1987, 52*, 183.

86%

Kimura, M.; Koie, K.; Matsubara, S.; Sawaki, Y.; Iwamura, H. *J. Chem. Soc., Chem. Commun.*, *1987*, 122.

90%

Imi, K.; Yanagihara, N.; Utimoto, K. *J. Org. Chem.*, *1987*, *52*, 1013.

91%

Foucaud, A.; Bakouetila, M. *Synthesis, 1987*, 854.

SECTION 367: ETHER, EPOXIDE, THIOETHER - ALKENE

Enol ethers are found in this section as well as alkenyl ethers.

70%

Grimshaw, J.; Thompson, N. *J. Chem. Soc., Chem. Commun., 1987*, 240.

Molander, G.A.; Shubert, D.C. *J. Am. Chem. Soc., 1987*, *109*, 6877.

98%

E:Z = 2:1

Ishibashi, H.; Nakatani, H.; Sakashita, H.; Ikeda, M. *J. Chem. Soc., Chem. Commun.,* **1987**, 338.

15% ee , R

quant.

Quimpère, M.; Jankowski, K. *J. Chem. Soc., Chem. Commun.,* **1987**, 676.

Z:E = 10:1

49%

Charbonnier, F.; Moyano, A.; Greene, A.E. *J. Org. Chem.,* **1987**, *52*, 2303.

94%

Fugami, K.; Oshima, K.; Utimoto, K. *Tetrahedron Lett.,* **1987**, *28*, 809.

89%

Z:E = 95:5

Okazoe, T.; Takai, K.; Oshima, K.; Utimoto, K. *J. Org. Chem.,* **1987**, *52*, 4410.

79%

Dulcere, J.P.; Rodriguez, J.; Santelli, M.; Zahra, J.P. *Tetrahedron Lett., **1987**, 28,* 2009.

65%

Nicolaou, K.C.; Hwang, C.-K.; Duggan, M.E.; Bal Reddy, K. *Tetrahedron Lett., **1987**, 28,* 1501.

TMTHF = 2,2,5,5-tetramethyl tetrahydrofuran

52%   **8%**

Broka, C.A.; Reichert, D.E.C. *Tetrahedron Lett., **1987**, 28,* 1503.

91%

Bartels, B.; Hunter, R.; Simon, C.D.; Tomlinson, G.D. *Tetrahedron Lett., **1987**, 28,* 2985.

1. 2.5 eq. *t*-BuLi
   THF , -80°C

2. AcOH , MeOH
   -80°C

75%

Whitby, R.; Kocieński, P. *Tetrahedron Lett., **1987**, 28,* 3619.

Takacs, J.M.; Anderson, L.G.; Creswell, M.W.; Takacs, B.E. *Tetrahedron Lett., 1987, 28,* 5627.

Huang, Y.; Lu, X. *Tetrahedron Lett., 1987, 28,* 6219.

Sato, T.; Okura, S.; Otera, J.; Nozaki, H. *Tetrahedron Lett., 1987, 28,* 6299.

Pale, P.; Chuche, J. *Tetrahedron Lett., 1987, 28,* 6447.

Tanaka, H.; Yamashita, S.; Ikemoto, Y.; Torii, S. *Tetrahedron Lett., 1988, 29,* 1721.

1. Siam$_2$BH , -15°C

2. NaOAc , H$_2$O$_2$
40°C

76%

Zweifel, G.; Najafi, M.R.; Rajagopalan, S. *Tetrahedron Lett., 1988, 29,* 1895.

1. sBuLi

2. Me$_3$Si-O-O-SiMe$_3$

70%

E:Z = 92:8

Davis, F.A.; Sankar Lal, G.; Wei, J. *Tetrahedron Lett., 1988, 29,* 4269.

Bu$_2$(SPh)$_2$, BF$_3$ •OEt$_2$

PhMe , -78°C

93%

E:Z = 81:19

Sato, T.; Okazaki, H.; Otera,J.; Nozaki, H. *Tetrahedron Lett., 1988, 29,* 2979.

1.

, AlCl$_3$

PhMe , 0°C

2. pH 7 buffer

70%

Mann, A.; Ricci, A.; Taddei, M. *Tetrahedron Lett., 1988, 29,* 6175.

5% Pd(OAc)$_2$, MeCN
3h

80%

Andersson, C.-M.; Hallberg, A. *J. Org. Chem., 1988, 53,* 235.

iPr$_2$NEt , Me$_3$Si-OTf

CH$_2$Cl$_2$, -20°C , 2h

95%

Gassman, P.G.; Burns, S.J. *J. Org. Chem., 1988, 53,* 5574.

Ph—O—≡ $\xrightarrow[\text{reflux , 2h}]{\text{HgCl}_2\text{ , ether , MeOH}}$ Ph—O—/\ (79%, OMe)

79%

Barluenga. J.; Aznar, F.; Bayod, M. *Synthesis*, *1988*, 144.

$\xrightarrow[\substack{\text{benzoquinone ,}\\\text{EtOH , MeSO}_3\text{H}}]{\text{5\% Pd(OAc)}_2\text{ , 20}^\circ\text{C}}$ EtO—⬡—OEt

72%

>98% cis

Bäckvall. J.-E.; Vågberg, J.O. *J. Org. Chem.*, *1988*, *53*, 5695.

Me₂C=CH—CHO + Me₃Sn—C(=CH₂)—CH₂—OAc $\xrightarrow{\text{Pd(OAC)}_2\text{ , PPh}_3\text{ , THF}}$

73%

Trost. B.M.; King, S.A.; Schmidt, T. *J. Am. Chem. Soc.*, *1989*, *111*, 5902.

Ph—CH(OMe)₂ $\xrightarrow[\substack{\text{BuB(allyl)}_3^-\text{ Li}^+\\\text{2. H}_2\text{O}_2\text{ , KOH , MeOH , 0}^\circ\text{C}}]{\text{1. Me}_3\text{SiOTf , THF , -78}^\circ\text{C}}$ Ph—CH(OMe)—CH₂—CH=CH₂

94%

Hunter. R.; Tomlinson, G.D. *Tetrahedron Lett.*, *1989*, *30*, 2013.

(dihydropyran) $\xrightarrow[\substack{\text{9\% PPh}_3\text{ , Ag}_2\text{CO}_3\text{ , 80}^\circ\text{C}\\\text{48h}}]{\text{PhI , 4\% Pd(OAc)}_2\text{ , MeCN}}$ Ph—(pyran)

96%

Larock. R.C.; Gong, W.H.; Baker, B.E. *Tetrahedron Lett.*, *1989*, *30*, 2603.

Me₃Si—CH=CH—Br $\xrightarrow[\substack{\text{4\% Pd(PPh}_3)_4\text{ , 2h}\\\text{PhH}}]{\text{Me}_3\text{Sn-SPh , 40}^\circ\text{C}}$ Me₃Si—CH=CH—SPh

91%

>99% E

Carpita, A.; Rossi. R.; Scamuzzi, B. *Tetrahedron Lett.*, *1989*, *30*, 2699.

60%

Mukaiyama, T.; Wariishi, K.; Furuya, M.; Kobayashi, S. *Chem. Lett., 1989*, 1277.

94%

van der Louw, J.; Out, G.J.J.; van der Baan, J.L.; de Kanter, F.J.J.;
Bickelhaupt, F.; Klumpp, G.W. *Tetrahedron Lett., 1989, 30*, 4863.

53%

van der Louw, J.; Slagt, M.; van der Baan, J.L.; Bickelhaupt, F.; Klumpp,
G.W. *Tetrahedron Lett., 1989, 30*, 5497.
van der Louw, J.; van der Baan, J.L.; Stichter,H.; Out, G.J.J.; Bickelhaupt, F.; Klumpp,
G.W. *Tetrahedron Lett., 1988, 29*, 3579.

63%

E:Z = 70:30

Ikura, K.; Ryu, I.; Ogawa, A.; Kambe, N.; Sonoda, N. *Tetrahedron Lett., 1989, 30*, 6887.

82%

Fugami, K.; Oshima, K.; Utimoto, K. *Bull. Chem. Soc., Jpn., 1989, 62*, 2050.

1. Bu$_3$SnH , AIBN
2. BuLi , THF , -70°C
3. aq. NH$_4$Cl

66%

Biggs, K.R.; <u>Parsons, P.J.</u>; Tapolzcay, D.J.; Underwood, J.M. *Tetrahedron Lett.,* *1989, 30*, 7115.

1. NIS , BnOH
2. *t*-BuOK , THF

66%

Middleton, D.S.; <u>Simpkins, N.S.</u> *Synth. Commun., 1989, 19*, 21.

PhSH , PhMe , 6h
——————————
Montmorillonite
reflux

60%

Labiad, B.; <u>Villemin, D.</u> *Synthesis, 1989*, 143.

1. Me$_2$(MeO)Si—SiMe$_3$ / Li
   THF , -78→25°C
2. H$_2$O 3. 100°C , 2h

68%

Bates, T.F.; <u>Thomas, R.D.</u> *J. Org. Chem., 1989, 54*, 1784.

PhSnBu$_3$, HMPA
1% Pd(PPh$_3$)$_4$
——————————
100°C , 20h
sealed tube

74%

Kosugi, M.; Miyajima, Y.; Nakanishi, H.; Sano, H.; <u>Migita, T.</u> *Bull. Chem. Soc., Jpn., 1989, 62*, 3383.

Related Methods: Section 180A (Protection of Ketones)

SECTION 368:   ETHER, EPOXIDE, THIOETHER - HALIDE, SULFONATE

Halocyclopropanations are found in Section 74F (Alkyls from Alkenes).

Chi, D.Y.; Kibourn, M.R.; Katzenellenbogen, J.A.; Welch, M.J. *J. Org. Chem.*, *1987*, *52*, 658.

Hashimoto, T.; Surya Prakash, G.K.; Shih, J.G.; Olah, G.A. *J. Org. Chem.*, *1987*, *52*, 931.

Evans, R.D.; Schauble, J.H. *Synthesis*, *1987*, 551.

Alvernhe, G.; Laurent, A.; Haufe, G. *Synthesis*, *1987*, 562.

DBH = 1,3-dibromo-5,5-dimethyl hydantoin

Shimizu, M.; Nakahara, Y.; Yoihioka, H. *J. Chem. Soc., Chem. Commun.*, *1989*, 1881.

SECTION 369: HALIDE, SULFONATE - KETONE

Saraf, S.D.; Al-Omran, F. *Org. Prep. Proceed. Int.*, *1987*, *19*, 455.

Abraham, W.D.; Bhupathy, M.; Cohen, T. *Tetrahedron Lett.*, *1987*, *28*, 2203.

Sha, C.-K.; Young, J.-J.; Jean, T.-S. *J. Org. Chem.*, *1987*, *52*, 3919.

Kulinkovich, O.G.; Tishcenko, I.G.; Sviridov, S.V. *J. Org. Chem., U.S.S.R.*, *1987*, *23*, 885.

Horiuchi, C.A.; Kiji, S. *Chem. Lett.*, *1988*, 31.

Smith III, A.B.; Leenay, T.L. *Tetrahedron Lett., 1988, 29,* 49.

Guerrero, A.F.; Kim, H.; Schlecht, M.F. *Tetrahedron Lett., 1988, 29,* 6707.

Dauben, W.G.; Warshawsky, A.M. *Synth. Commun., 1988, 18,* 1323.

Kajigaeshi, S.; Kakinami, T.; Moriwaki, M.; Fujisaki, S.; Maeno, K.; Okamoto, T. *Synthesis, 1988,* 545.

Lee, J.G.; Ha, D.S. *Tetrahedron Lett., 1989, 30,* 193.

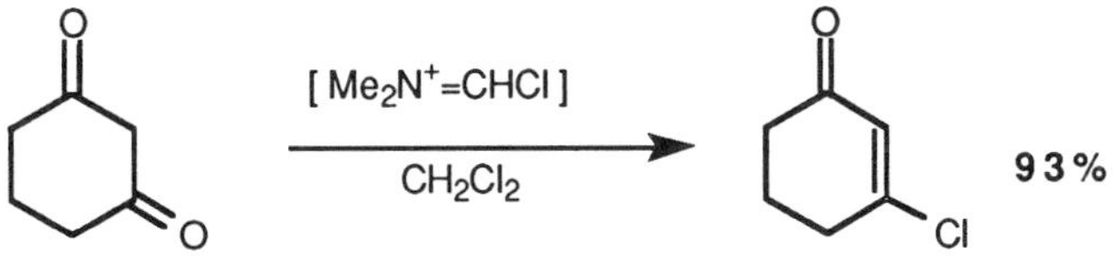

Mewshaw, R.E. *Tetrahedron Lett., **1989**, 30*, 3753.

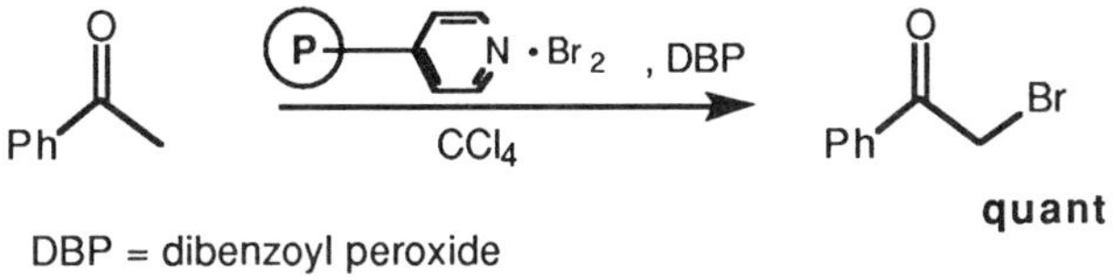

DBP = dibenzoyl peroxide

Sket, B.; Zupan, M. *Synth. Commun., **1989**, 19*, 2481.

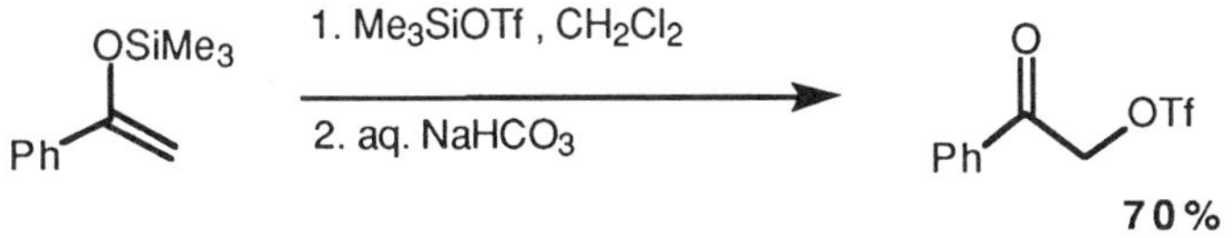

Moriarty, R.M.; Epa, W.R.; Penmasta, R.; Awasthi, A.K. *Tetrahedron Lett., **1989**, 30*, 667.

SECTION 370: HALIDE, SULFONATE - NITRILE

NO ADDITIONAL EXAMPLES

SECTION 371: HALIDE, SULFONATE - ALKENE

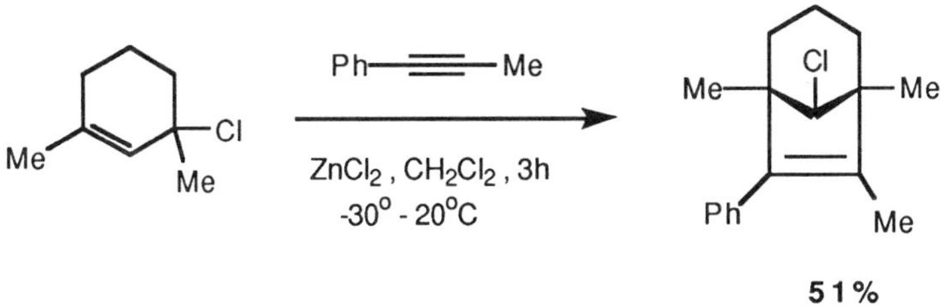

Gray, B.D.; McMillan, C.M.; Miller, A.; Moore, M. *Tetrahedron Lett., **1987**, 28*, 235.

C_6H_{13} / SiMe$_3$

1. DMF , ICl , 0°C , 2h

2. Al$_2$O$_3$

C_6H_{13} / I   68%

E:Z = <1:>99

<u>Tamao, K.</u>; Akita, M.; Maeda, K.; Kumada, M. *J. Org. Chem.*, *1987*, *52*, 1100.

Me / Ph / OH

1. PhCl , KHF$_2$, iPr$_2$NH
   (HF)$_n$-Py , 0°C

2. aq. KF

F / Ph   78%

E:Z = 100:0

Kanemoto, S.; Shimizu, M.; <u>Yoshioka, H.</u> *Tetrahedron Lett.*, *1987*, *28*, 663.

HgCl$_2$, I$_2$, 1.5h

0°C

Cl / I   62%

<u>Barluenga, J.</u>; Martínez-Gallo, J.M.; Nájera, C.; Yus, M. *J. Chem. Soc., Perkin Trans. I*, *1987*, 1017.

1. PhSeCl , CHCl$_3$, RT , 24h
2. SO$_2$Cl$_2$, 2h
3. PhH , 100°C , aq. NaHCO$_3$
4. 30% H$_2$O$_2$

Cl   68%

<u>Engman, L.</u> *Tetrahedron Lett.*, *1987*, *28*, 1463.

OMe / OSiMe$_3$

1. Cl,Me,Me,Cl,Cl , MeLi , heat

2. TBAF , THF

Cl / Me / Me / CO$_2$Me   50%

<u>Slougui, N.; Rousseau, G.</u> *Tetrahedron Lett.*, *1987*, *28*, 1651.

OBn

TBAF , MsF

Ph

t-BuMe₂SiO   OSiMe₂t-Bu

Ph   OBn

F   **65%**

Shimizu, M.; Nakahara, Y.; Kanemoto, S.; Yoshioka, H. *Tetrahedron Lett., 1987, 28,* 1677.

Ph   SnBu₃

$CF_3CF_2CF_2CF_2I$, 70°C

Pd(PPh₃)₄ , hexane

Ph n-C₄F₉

70%

Matsubara, S.; Mitani, M.; Utimoto, K. *Tetrahedron Lett., 1987, 28,* 5857.

Et

Ph   H

Me₃Si

1. Br₂ , CH₂Cl₂ , -78°C

2. NaOMe , MeOH

Ph   Et

Br

Oliva, A.; Molinari, A. *Synth. Commun., 1987, 17,* 837.

Ph ━━━ H

I₂/Al₂O₃ , Pet Ether , reflux

2h

Ph   I

I   H   **96%**

Larson, S.; Luidhardt, T.; Kabalka, G.W.; Pagni, R.M. *Tetrahedron Lett., 1988, 29,* 35. Hondrogiannis, G.; Lee, L.C.; Kabalka, G.W.; Pagni, R.M. *Tetrahedron Lett., 1989, 30,* 2069.

Me   Me

Me   CHO

1. $Ph-\overset{O}{\underset{N-Me}{S}}-CH_2F$, LDA

2. Al/Hg , AcOH , aq. THF , 4h

Me   Me

Me   F

76%

Boys, M.L.; Collington, E.W.; Finch, H.; Swanson, S.; Whitehead, J.F. *Tetrahedron Lett., 1988, 29,* 3365.

$$SnCl_4 , CH_2Cl_2 , -78°C \rightarrow RT$$

58%

Lolkema, L.D.M.; Hiemstra, H.; Mooiweer, H.H.; Speckamp, W.N. *Tetrahedron Lett.,* **1988**, *29*, 6365.

CF_3CH_2OTs

1. 2 eq. LDA , THF , -78°C
2. 1h , $\left(Ph \diagdown\diagup\diagdown\diagup \right)_3 B$ THF
3. AcOH , reflux

$F_2C=$ ~~~~~ Ph

90%

Ichikawa, J.; Sonoda, T.; Kobayashi, H. *Tetrahedron Lett.,* **1989**, *30*, 1641.

$[NaN(SiMe_3)_2 , Ph_3P-CH_2I^+ \, I^-]$
THF , -23°C

61%

Stork, G.; Zhao, K. *Tetrahedron Lett.,* **1989**, *30*, 2173.

Ph—≡≡

iPrI , BEt_3 , 25°C
hexane , 11h

(21   :   79)   81%

Ichnose, Y.; Matsunaga, S.; Fugami, K.; Oshima, K.; Utimoto, K. *Tetrahedron Lett.,* **1989**, *30*, 3155.
Takeyama, Y.; Ichinose, Y.; Oshima, K.; Utimoto, K. *Tetrahedron Lett.,* **1989**, *30*, 3159.

$Ph_3P^+CH_2I \, I^-$

1. NaN(SiMe_3)_2 , THF
   HMPT , DMF , -80°C
2. PhCHO

Ph ~~~ I   96%

Z:E = 98:2

Bestmann, H.J.; Rippel, H.C.; Dostalek, R. *Tetrahedron Lett.,* **1989**, *30*, 5261.

REVIEW:

"The Design and Applications of Free Radical Chain Reactions in Organic Synthesis"

Curran, D.P. Synthesis, **1988**, 417, 489.

SECTION 372: KETONE - KETONE

Corey, E.J.; Ghosh, A.K. Tetrahedron Lett., **1987**,.28, 175.

Chan, T.H.; Prasad, C.V.C. J. Org. Chem., **1987**, 52, 120.

Thomas, S.E. J. Chem. Soc., Chem. Commun., **1987**, 226.

Moriarty, R.; Prakash, O.; Duncan, M.P. *J. Chem. Soc., Perkin Trans. I, 1987*, 559.

Rigby, J.H.; Kotnis, A.S. *Tetrahedron Lett., 1987, 28*, 4943.

Yamamoto, K.; Kanoh, M.; Yamamoto, N.; Tsuji, T. *Tetrahedron Lett., 1987, 28*, 6347.

Degl'Innocenti, A.; Ricci, A.; Mordini, A.; Reginato, G.; Colotta, V. *Gazz. Chim. Ital., 1987*, *117*, 645.

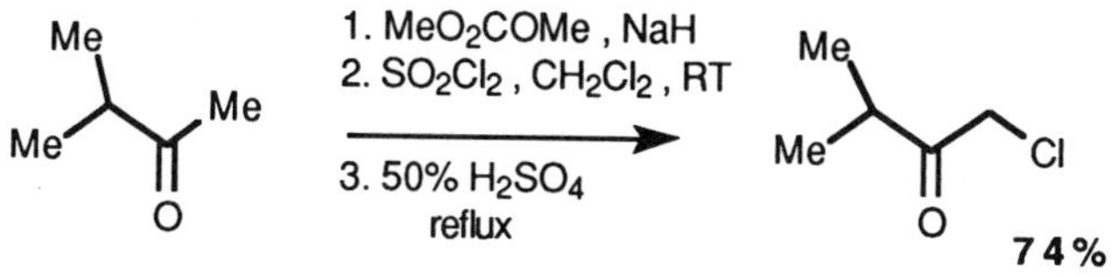

1. MeO$_2$COMe , NaH
2. SO$_2$Cl$_2$, CH$_2$Cl$_2$, RT

3. 50% H$_2$SO$_4$
   reflux

74%

DeKimpe, N.; De Cock, W.; Schamp, N. *Synthesis, **1987**,* 188.

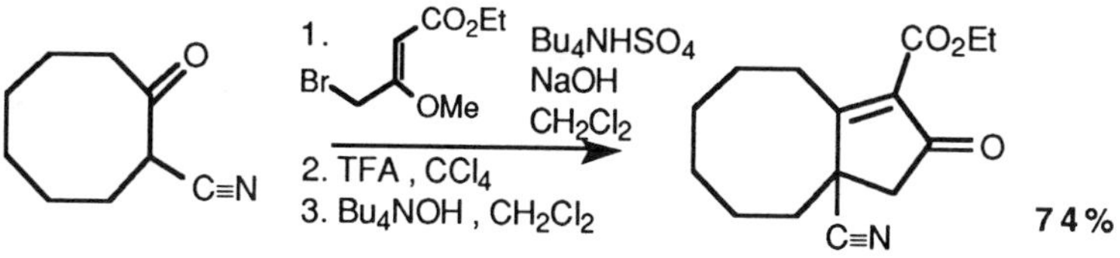

1. Br / CO$_2$Et / OMe   Bu$_4$NHSO$_4$
   NaOH
   CH$_2$Cl$_2$
2. TFA , CCl$_4$
3. Bu$_4$NOH , CH$_2$Cl$_2$

74%

Ognyanov, V.I.; Hesse, M. *Helv. Chim. Acta*, **1987**, *70*, 1393.

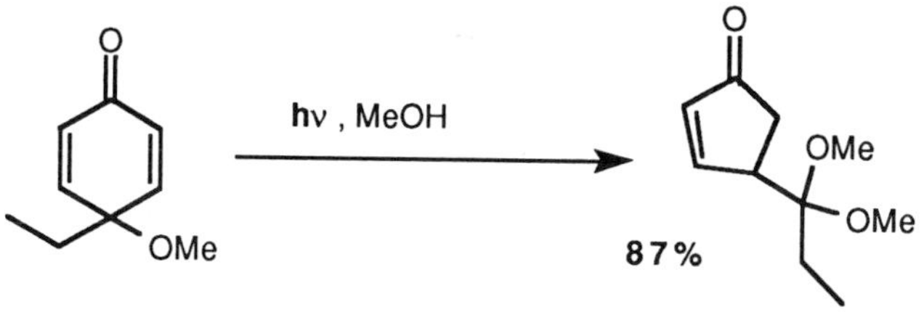

hν , MeOH

87%

Taveras Jr., A.G. *Tetrahedron Lett.*, **1988**, *29*, 1103.

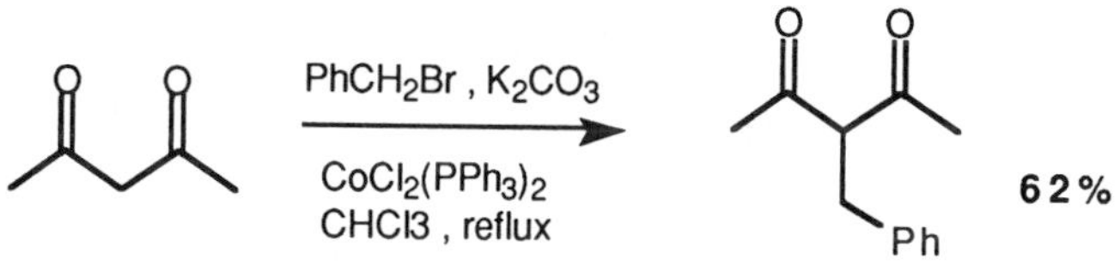

PhCH$_2$Br , K$_2$CO$_3$

CoCl$_2$(PPh$_3$)$_2$
CHCl3 , reflux

62%

González, A.; Marquet, J.; Moreno-Mañas, M. *Tetrahedron Lett.*, **1988**, *29*, 1469.

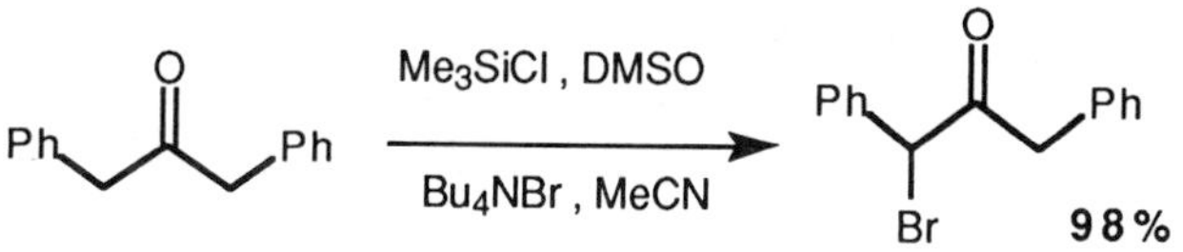

Me$_3$SiCl , DMSO

Bu$_4$NBr , MeCN

98%

Fraser, R.R.; Kong, F. *Synth. Commun.*, **1988**, *18*, 1071.

$$Ph\diagdown \overset{O}{\underset{Ph}{\bigsqcup}} \quad \xrightarrow[\text{reflux , 15h}]{\text{PCC , Py , CH}_2\text{Cl}_2} \quad Ph\overset{O}{\underset{O}{\bigsqcup}}Ph \quad \mathbf{85\%}$$

Bonadies, F.; Bonini, C. *Synth. Commun.*, **1988**, 18, 1573.

$$\diagup\diagdown\diagup\text{NO}_2 \quad \xrightarrow[\substack{\text{2. 30\% H}_2\text{O}_2\text{ , MeOH} \\ \text{K}_2\text{CO}_3\text{ , overnight} \\ 0^\circ\text{C} \rightarrow \text{RT}}]{\substack{\text{1. Al}_2\text{O}_3\text{ , RT ,} \\ \text{8h}}} \quad \text{60\%}$$

Ballini, R.; Petrini, M.; Marcantoni, E.; Rosini, G. *Synthesis*, **1988**, 231.

$$\xrightarrow[\substack{\text{PdCl}_2\text{/DBU/PPh}_3\text{ , reflux} \\ \text{2. 1\% aq. H}_2\text{SO}_4\text{ , 60}^\circ\text{C} \\ \text{2h}}]{\substack{1. \text{Cl}\diagup\diagup\diagdown_O\diagdown_O\text{Me , THF , 18h}}} \quad \mathbf{88\%}$$

Gu, X.-P.; Okuhara, T.; Ikeda, I.; Okahara,M. *Synthesis*, **1988**, 535.

$$\xrightarrow[\text{MeCN , RT}]{\substack{\text{OSiMe}_3 \\ \diagup\diagdown\text{Me} \quad \text{, CAN , NaHCO}_3}} \quad \mathbf{75\%}$$

Bachiocchi, E.; Casu, A.; Ruzziconi, R. *Tetrahedron Lett.*, **1989**, 30, 3707.

$$\xrightarrow[\text{H}_2\text{ , 24h}]{\text{Hg}^\circ\text{ , h}\nu\text{ (254 nm)}} \quad \mathbf{83\%}$$

Boujamra, C.G.; Crabtree, R.H.; Ferguson, R.R.; Muedas, C.A. *Tetrahedron Lett.*, **1989**, 30, 5583.

Aoki, S.; Fujimaura, T.; Nakamura, E.; Kuwahima, I. *Tetrahedron Lett., 1989, 30,* 6541.

Vankar, Y.D.; Saksena, R.K.; Bawa, A. *Chem. Lett., 1989,* 1241.

Otera, J.; Niibo, Y.; Nozaki, H. *J. Org. Chem., 1989, 54,* 5003.

Jun, J.-G.; Suh, S.; Shin, D.G. *J. Chem. Soc., Perkin Trans. I, 1989,* 1349.

SECTION 373: KETONE - NITRILE

Yeh, M.C.P.; Knochel, P. *Tetrahedron Lett., 1988, 29,* 2395.

PhCHO $\xrightarrow[\text{Ac}_2\text{O}]{\text{KCN , }p\text{-TsOH , PCC}}$

72%

Kang, S.-K.; Sohn, H.-K.; Kim, S.-G. *Org. Prep. Proceed. Int., 1989, 21, 383.*

SECTION 374: KETONE - ALKENE

For the oxidation of allylic alcohols to alkene ketones, see Section 168 (Ketones from Alcohols and Phenols)

For the oxidation of allylic methylene groups ($C=C-CH_2 \rightarrow C=C-C=O$), see Section 170 (Ketones from Alkyls and Methylenes).

For the alkylation of alkene ketones, also see Section 177 (Ketones from Ketones) and for conjugate alkylations see Section 74E (Alkyls form Alkenes).

1. LTMP , ether , 0°C , 30 min

2. 0°C , 2h ,

3. 4% aq. KOH , MeOH , 80°C

40%

Bonnert, R.V.; Jinkins, P.R. *J. Chem. Soc., Chem. Commun., 1987, 6.*

pyrolysis

78%

Fleming, A.; Sinari-Zingde, G.; Natchus, M.; Hudlicky, T. *Tetrahedron Lett., 1987,.28, 167.*

1. MeMgl , CuCl

2. FVP (500°C) , 10^{-2} torr

≈ 80% , S

Klunder, A.J.H.; Huizinga, W.B.; Sessink, P.J.M. Zwanenburg, B. *Tetrahedron Lett.*, *1987, 28*, 357.

$Co_2(CO)_8$, CO , PhH

RT (4h) , reflux (4d)

35%

Schore, N.E.; Knudsen, M.J. *J. Org. Chem., 1987, 52*, 569.

1. $Hg(OAc)_2$,

2. 150°C

83%

McKenzie, T.G. *Org. Prep. Proceed. Int., 1987, 19*, 435.

, $PdCl_2$, CuCl

1 atm O_2 , Na_2HPO_4
DMF , 50°C

75%

Hosokawa, T.; Ohta, T.; Kanayama, S.; Murahashi, S. *J. Org. Chem., 1987, 52*, 1758.

1. NaIO$_4$, aq. MeOH , 60°C , 5h

2. KOt-Bu , CCl$_4$/t-BuOH , 50°C

3. p-TsOH - Py , aq. acetone
   reflux

45%

unconjugated:conjugated = 91:9

Matsuyama, H.; Miyazawa, Y.; Takei, Y.; Kobayashi, M. *J. Org. Chem., 1987, 52,* 1703.

Co$_2$(CO)$_8$, CO
MeI , NaOH , PhH
H$_2$O

CTAB

54%   20%

Miura. M.; Akase, F.; Shinohara, M.; Nomura, M. *J. Chem. Soc., Perkin Trans. I, 1987,*
1021.

CH$_3$CHO , Cp$_2$ZrH2

NiCl$_2$, 130°C , 8h

71%

Nakano, T.; Irifune, S.; Umano, S.; Inada, A.; Ishii. Y.; Ogawa. M. *J. Org. Chem., 1987,*
52, 2239.

1. LDA , THF

2. 5% Pd(PPh$_3$)$_4$, THF

$$\overset{O}{\underset{}{P}}(OEt)_2$$

Br

3. 5% aq. KOH , reflux, 8.5h

76%

Welch. S.C.; Assercq, J.-M.; Loh, J.-P.; Glase, S.A. *J. Org. Chem., 1987, 52,* 1440.

Magnus, P.; Principe, L.M.; Slater, M.J. *J. Org. Chem., 1987, 52*, 1483.

Brunnelle, W.H.; Rafferty, M.A.; Hodges, S.L. *J. Org. Chem., 1987, 52*, 1603.

Z:E = 3:2

Mitsudo, T.; Kadokura, M.; Watanabe, Y. *J. Org. Chem., 1987, 52*, 3186.

Matsuoka, R.; Horiguchi, Y.; Kawajima, I. *Tetrahedron Lett., 1987, 28*, 1299.

DeShong, P.; Slough, G.A.; Rheingold, A.L. *Tetrahedron Lett., 1987, 28*, 2229.

H—≡—Ph
$\xrightarrow{\text{1. MeMn(CO)}_5\text{, 6KBar}}{\text{2. HCl, MeCN}}$

Me—C(O)—CH=CH—Ph

+

(cyclopentenone with Ph and Me substituents)

(3.2   :   1)   54%

DeShong, P.; Sidler, D.R.; Slough, G.A. *Tetrahedron Lett., 1987, 28,* 2233.

(cyclopropane with CO₂Me, Me₃SiO, t-Bu)
$\xrightarrow{\text{1. MeLi}}{\substack{\text{2. H}_2\text{O} \\ \text{3. 2N aq. HCl} \\ \text{pentane}}}$
t-Bu—C(O)—CH₂—CH=C(Me)(Me)

60%

Bretsch, W.; Reißig, H.-U. *Liebigs. Ann. Chem., 1987,* 175.

(bicyclic dioxolane structure with HO, C≡C-Bu)
$\xrightarrow{\substack{\text{Pd(O}_2\text{CCF}_3)_2 \\ \text{CH}_2\text{Cl}_2 \\ \text{RT, 12h}}}$
(indanone dioxolane with =CH-Bu)

91%

isomer ratio = 36:1

Liebeskind, L.S.; Mitchell, D.; Foster, B.S. *J. Am. Chem. Soc., 1987, 109,* 7908.

Ph—C(O)—CH₃
$\xrightarrow{\substack{\text{PhCHO, TiCl}_4\text{, NEt}_3 \\ \text{CH}_2\text{Cl}_2\text{,0°C, 30 min}}}$
Ph—C(O)—CH=CH—Ph

88%

Harrison, C.R. *Tetrahedron Lett., 1987, 28,* 4135.

Ph—S(O)—CH(Cl)—CH₂—CH₂—SPh
$\xrightarrow{\substack{\text{1. LDA} \\ \text{2. PhCHO} \\ \text{3. }t\text{-BuOK} \\ \text{4. PhSeNa, EtOH} \\ \text{5. mCPBA} \\ \text{6. aq. KOH}}}$
CH₂=CH—C(O)—CH₂—Ph

57%

Satoh, T.; Kumagawa, T.; Sagimoto, A.; Yamakawa, K. *Bull. Chem. Soc., Jpn., 1987, 60,* 34.

Fuentes, A.; Marinas, J.M.; Sinisterra, J.V. *Tetrahedron Lett., 1987, 28*, 4541.

the sand was dried over P_2O_5 (100°C , 1h , 0.08 Torr)

Boaventura, M.-A.; Drouin, J. *Synth. Commun., 1987, 17*, 975.

Tamura, Y.; Yakura, T.; Haruta, J.; Kita, Y. *J. Org. Chem., 1987, 52*, 3927.

Hayashi, M.; Mukaiyama, T. *Chem. Lett., 1987*, 1283.

Echavarren, A.M.; Stille, J.K. *J. Am. Chem. Soc., 1988, 110*, 1557.

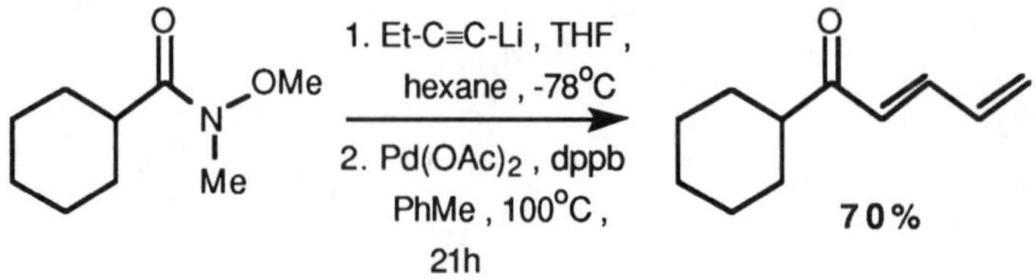

1. Et-C≡C-Li , THF ,
   hexane , -78°C
2. Pd(OAc)₂ , dppb
   PhMe , 100°C ,
   21h

70%

Trost, B.M.; Schmidt, T. *J. Am. Chem. Soc.*, *1988*, *110*, 2301.

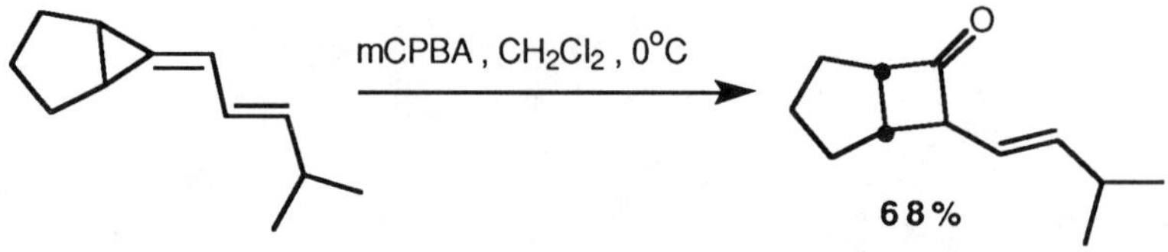

mCPBA , CH₂Cl₂ , 0°C

68%

75:25 exo:endo

McCullough, D.W.; Cohen, T. *Tetrahedron Lett.*, *1988*, *29*, 27.

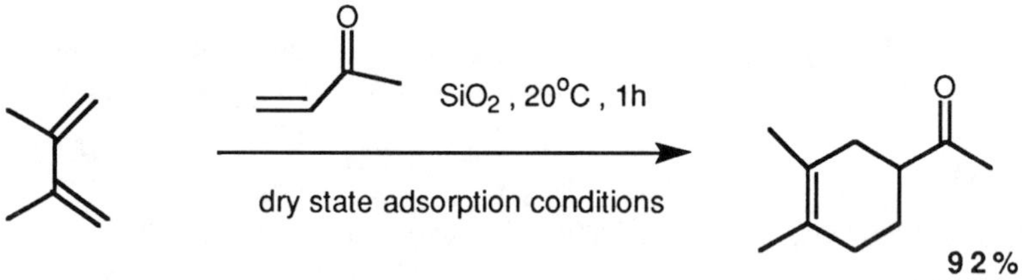

SiO₂ , 20°C , 1h

dry state adsorption conditions

92%

Veselovsky, V.V.; Gybin, A.S.; Lozanova, A.V.; Moiseenkov, A.M.; Smit, W.A.; Caple, R. *Tetrahedron Lett.*, *1988*, *29*, 175.

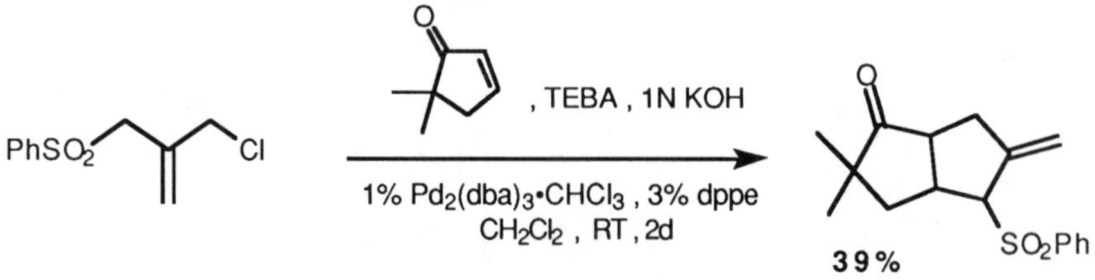

PhSO₂ — Cl

, TEBA , 1N KOH

1% Pd₂(dba)₃•CHCl₃ , 3% dppe
CH₂Cl₂ , RT , 2d

39%

SO₂Ph

Breuilles, P.; Uguen, D. *Tetrahedron Lett.*, *1988*, *29*, 201.

Me
$$\text{1. } \overset{\text{Me}}{\underset{\text{Me}}{||||}} \text{---Co}_2\text{(CO)}_6 \text{ , PhMe}$$
98°C , 35d

2. $H_2NCH_2CH_2NH_2$, EtOAc

23 %

Krafft, M.E. *Tetrahedron Lett.*, *1988*, 29, 999.

$$\text{RuH}_2\text{(PPh)}_3 \text{ , PhMe}$$
110°C , 24h

83 %

Ma, D.; Lin, Y.; Lu, X.; Yu, Y. *Tetrahedron Lett.*, *1988*, 29, 1045.

+   Me   Me

$$\text{HOCH}_2\text{CH}_2\text{OH}$$
100°C , 30 min

OCO₂Me

MeO₂CO   Me   Me

H   **quant**

Dunamsit, T.; Hoekstra, W.; Pentaleri, M.; Liotta, D. *Tetrahedron Lett.*, *1988*, 29, 3745.

Me

1. MeLi

2. EtOH , H⁺

OBn

Me

57 %

>98% ee

Meyers, A.I.; Sturgess, M.A. *Tetrahedron Lett.*, *1988*, 29, 5339.

$$\text{2 eq. Mn(OAc)}_3$$
$$\text{1 eq. Cu(OAc)}_2$$
AcOH . 25°C

CO₂Et

Cl   CO₂Et   Cl   CO₂Et

+

50 %   18%

Merritt, J.E.; Sasson, M.; Kates, S.A.; Snider, B.B. *Tetrahedron Lett.*, *1988*, 29, 5209.

1. Pd(OAc)₂, PPh₃, PhH
6h, reflux,
2. H₂O, 85°C, 1h

57%

Huang, Y.; Lu. X. *Tetrahedron Lett., **1988**, 29*, 5663.

CH₂Cl₂, 0°C
PhI⁺-CH₂ Ph

90%

Zhdankin, V.V.; Tykwinski, R.; Caple, R.; Berglund, B.; Koz'min, A.S.; Zefirov. N.S. *Tetrahedron Lett., **1988**, 29*, 3703.

C₅H₁₁, PPh₃
Ni(COD)₂, BEt₃

(85   :   15) 68%

Binger, P.; Schäfer, B. *Tetrahedron Lett., **1988**, 29*, 4539.

1. BzCl, CuI, PdCl₂(PPh₃)₂
2. 10 atm CO₂, NEt₃, 70°C
3. 50°C, autoclave

58%

Inoue, Y.; Ohuchi, K.; Imaizumi, S. *Tetrahedron Lett., **1988**, 29*, 5941.

1. SmI₂, THF
2. PhCHO
3. H₂O

75%

Zhang, Y.; Liu, T.; Lin, R. *Synth. Commun., **1988**, 18*, 2003.

Me₃SiO group reacting with
$$\text{Me}_3\text{SiO} \quad / \quad \text{Cl} \quad , \text{AgF} , 15°C$$
$$\xrightarrow{\text{MeCN/EtOH , 4h}}$$

53%

<u>Ryu, I.</u>; Suzuki, H.; Ogawa, A.; Kambe, N.; <u>Sonoda, N.</u> *Tetrahedron Lett., 1988, 29,* 6137.

$$\xrightarrow[\substack{\text{NEt}_3 , \text{MeOH} , \text{MeCN/PhH} \\ 100°C}]{600 \text{ psi CO} , 5\% \text{ Cl}_2\text{Pd(PPh}_3)_2}$$

C_5H_{11}

CO_2Me

90%

<u>Negishi, E.</u>; Wu, G.; Tour, J.M. *Tetrahedron Lett., 1988, 29,* 6745.

$$\xrightarrow[\substack{4\text{-}t\text{-butylcathecol} , 5\text{NaHCO}_3 \\ \text{Z:E} = 1:3}]{\text{MeCHClCHClMe} , 125°C}$$

97%

<u>Jacobi, P.A.</u>; Armacost, L.M.; Kravitz, J.I.; Martinelli, M.J. *Tetrahedron Lett., 1988, 29,* 6869.

Me₃SiO   OSiMe₃

$$\xrightarrow{\text{TBAF , THF}}$$

Ph

72%

Barinelli, L.S.; <u>Nicholas, K.M.</u> *J. Org. Chem., 1988, 53,* 2114.

$$\xrightarrow{\text{mCPBA , CH}_2\text{Cl}_2 , 25°C}$$

PhO₂S

98%

<u>Padwa, A.</u>; Chiacchio, U.; Kline, D.N.; Perumattam, J. *J. Org. Chem., 1988, 53,* 2238.
<u>Padwa, A.</u>; Kline, D.N.; Permattam, J. *Tetrahedron Lett., 1987, 28,* 913.

Andersson, C.-M.; Hallberg. A. *J. Org. Chem., 1988, 53*, 4149.

Paquette. L.A.; Reagan, J.; Schrieber. S.L.; Teleha, C.A. *J. Am. Chem. Soc., 1989, 111*, 2330.

Barluenga. J.; Aznar, F.; Cabal, M.P.; Valdes, C. *Tetrahedron Lett., 1989, 30*, 1413.

Haack. R.A.; Beck, K.R. *Tetrahedron Lett., 1989, 30*, 1605.

Ma, D.; Lu. X. *Tetrahedron Lett., 1989, 30*, 2109.

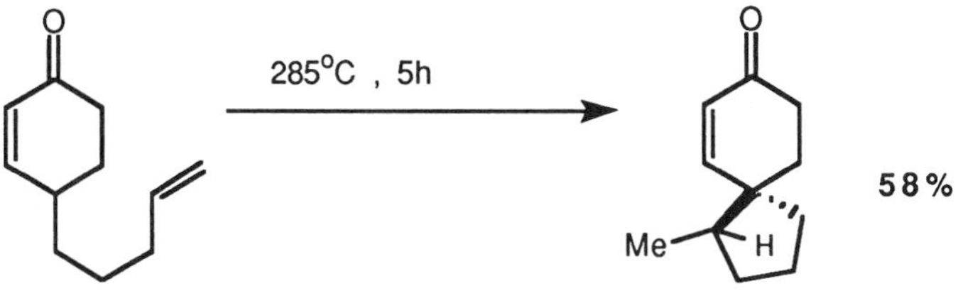

O
285°C , 5h
O
Me
H
58%

Kende, A.S. ; Newbold, R.C *Tetrahedron Lett.*, ***1989***, *30*, 4329.

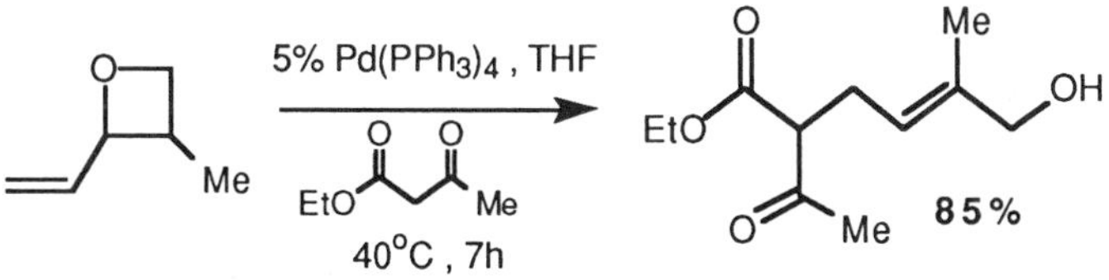

O
Me
5% Pd(PPh3)4 , THF
EtO
Me
40°C , 7h
O
Me
OH
EtO
O
Me
85%

Larock, R.C.; Stolz-Dunn, S.K. *Tetrahedron Lett.*, ***1989***, *30*, 3487.

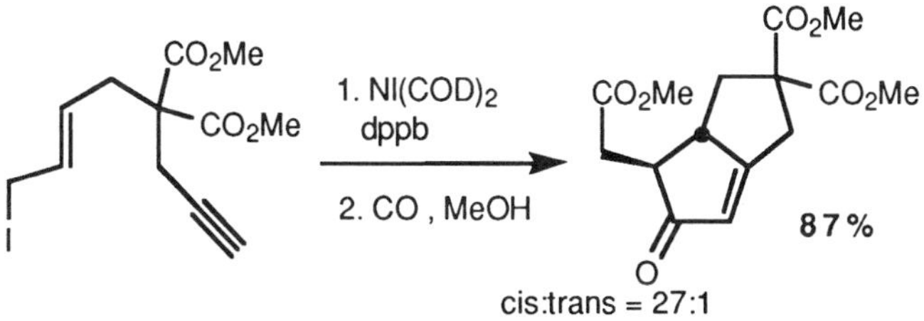

CO2Me
CO2Me
I
1. Ni(COD)2
dppb
2. CO , MeOH
CO2Me
CO2Me
CO2Me
O
87%
cis:trans = 27:1

Oppolzer, W.; Keller, T.H.; Bedoya-Zurita, M.; Stone, C. *Tetrahedron Lett.*, ***1989***, *30*, 5883.

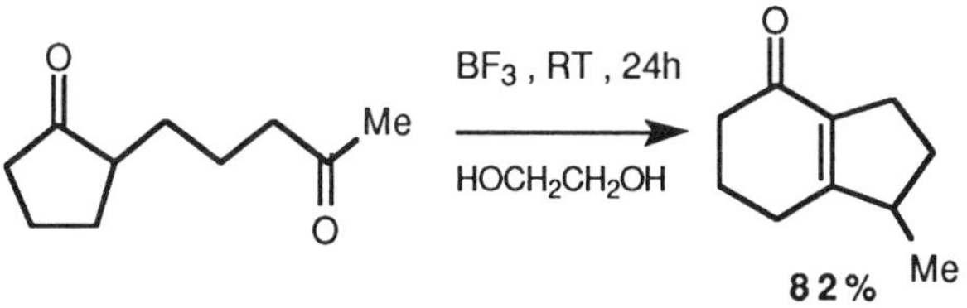

O
Me
O
BF3 , RT , 24h
HOCH2CH2OH
O
Me
82%

Miyao, Y.; Tanaka, M.; Suemune, H.; Sakai, K. *J. Chem. Soc., Chem. Commun.*, ***1989***, 1535.

OSiMe3

$$
\begin{bmatrix}
\text{i.} & \text{SiO}_2/\text{Pd(NH}_3)_4\text{Cl}_2 , \text{NH}_3 \\
\text{ii.} & 632^\circ\text{K , 1h} \\
\text{iii.} & 623^\circ\text{K , H}_2 , \text{1h}
\end{bmatrix}
$$

N-methyl-2-pyrrolidinone
O_2 , 333°K , 24h

90%

Baba, T.; Nakano, K.; Nishiyama, S.; Tsuruya, S.; Masai, M. *J. Chem. Soc., Chem. Commun.*, **1989**, 1697.

$C_{10}H_7$
O
EtO
$C\equiv C$-Bn
OH

138°C , *p*-xylene

$C_{10}H_7$
Bn
EtO
O

65%

Foland, L.D.; Karlsson, J.O.; Pelrri, S.T.; Schwabe, R.; Xy, S.L.; Patil, S.; Moore, H.W. *J. Am. Chem. Soc.*, **1989**, *111*, 975.

1. Me_3Si
S-Ph
O

2. H_3O^+

61%

Williams, R.V.; Lin, X. *J. Chem. Soc., Chem. Commun.*, **1989**, 1872.

O
Me
I
SnBu3

AIBN , Bu$_3$SnH

PhH , reflux , 72h

O
Me **85%**

Baldwin, J.E.; Adlington, R.M.; Robertson, J. *Tetrahedron*, **1989**, *45*, 909.
Baldwin, J.E.; Adlington, R.M.; Robertson, J. *J. Chem. Soc., Chem. Commun.*, **1988**, 1404.

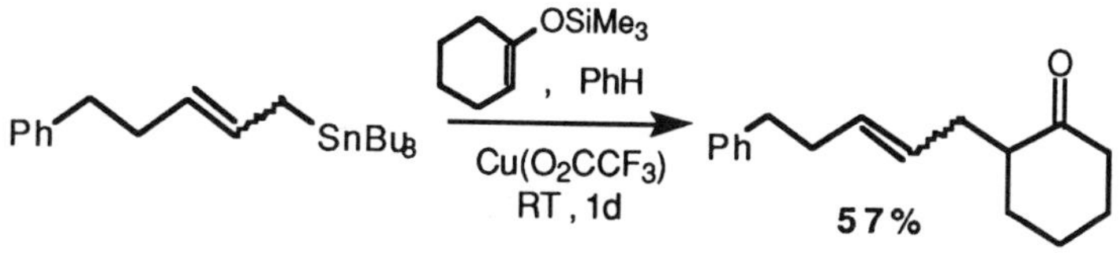

Takeda, T.; Ogawa, S.; Koyama, M.; Kato, T.; Fujiwara, T. *Chem. Lett.*, **1989**, 1257.

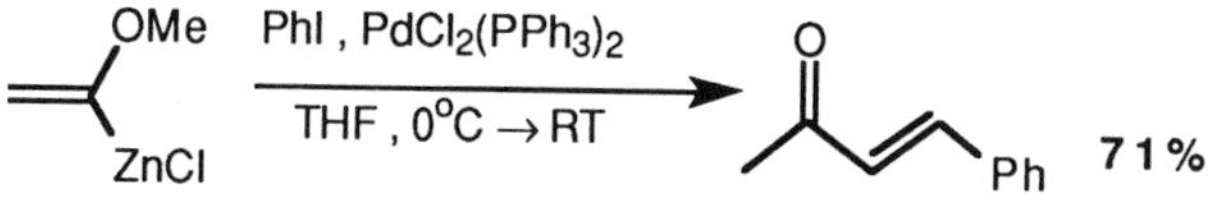

Ogima, M.; Hyuga, S.; Hara, S.; Suzuki, A. *Chem. Lett.*, **1989**, 1959.

REVIEW:

"New Synthetic Reactions of Allyl Alkyl Carbonates, Allyl β-Keto Carboxylates and Allyl Vinylic Carbonates Catalyzed by Palladium Complexes"

Tsuji, J.; Minami, I. *Acc. Chem. Res.*, **1987**, *20*, 140.

SECTION 375: NITRILE - NITRILE

NO ADDITIONAL EXAMPLES

SECTION 376: NITRILE - ALKENE

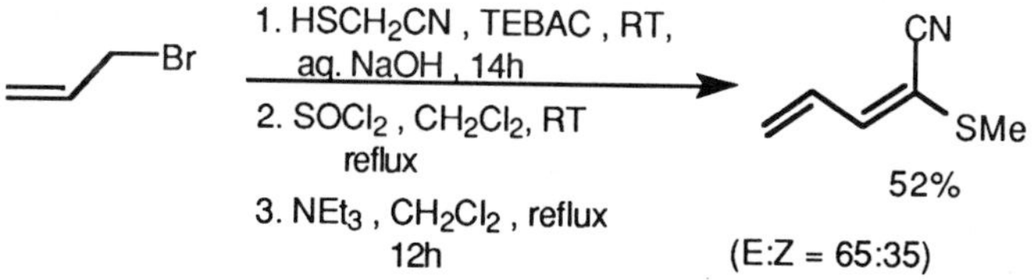

Stévenart-Demesmaeker, N.; Merényl, R.; Viehe, H.G. *Tetrahedron Lett.*, **1987**, *28*, 2591.

SECTION 377: ALKENE - ALKENE

$$\text{propyl-CHO} \xrightarrow[\substack{\text{2. } n\text{-Bu}_3\text{SnH , AIBN , heat} \\ \text{3. Ac}_2\text{O , NET}_3\text{ , DMAP} \\ \text{4. TBAF}}]{\substack{\text{1. } n\text{-BuLi ,} \quad \text{Me}_3\text{Si}\text{---}\equiv\text{---H}}} \text{product}$$

SnBu₃

30% overall
Z:E = 8:1

Nativi, C. Taddei, M. *Tetrahedron Lett.*, **1987**, *28*, 347.

$$\text{Ph-CHO} \xrightarrow[\substack{\text{3. CH}_2=\text{CHMgBr} \\ \text{4. AcOH , NaOAc}}]{\substack{\text{1. Me}_3\text{SiCH}_2\text{MgCl} \\ \text{2. CrO}_3 \cdot 2\text{ Py , CH}_2\text{Cl}_2}} \text{Ph product}$$

50%

Brown, P.A.; Bonnert, R.V.; Jenkins, P.R.; Selim, M.R. *Tetrahedron Lett.*, **1987**, *28*, 693.

$$\text{Ph}\diagdown\diagdown\text{I} \xrightarrow[\substack{\text{PdCl}_2(\text{MeCN})_2\text{ ,0.1h}}]{\text{Bu}_3\text{Sn}\diagup\diagdown\text{Ph , DMF , 25°C}} \text{Ph}\diagdown\diagup\diagdown\text{Ph}$$

71%

Stille, J.K.; Groh, B.L. *J. Am. Chem. Soc.*, **1987**, *109*, 813.

$$\xrightarrow[\text{THF , 24h , 25°C}]{\text{Cp}_2\text{ZrCl}_2 - \text{Mg}^0 - \text{HgCl}_2}$$

60%

Nugent, W.A.; Thorn, D.L.; Harlow, R.L. *J. Am. Chem. Soc.*, **1987**, *109*, 2788.

$$\xrightarrow[\substack{\text{2. TMSCI} \\ \text{3. 240°C}}]{\text{1. } n\text{-BuLi , -105°C}} \text{SiMe}_3$$

25%

Chou, T.; Tso, H.-H.; Tao, Y.-T.; Lin, L.C. *J. Org. Chem.*, **1987**, *52*, 244.

Trost, B.M.; Tour, J.M. *J. Am. Chem. Soc.*, *1987*, *109*, 5268.

Baldwin, J.E.; Bennett, P.A.R.; Forrest, A.K. *J. Chem. Soc., Chem. Commun.*, *1987*, 250.

Yamada, S.; Suzuki, H.; Naito, H.; Nomoto, T.; Takayama, H. *J. Chem. Soc., Chem. Commun.*, *1987*, 332.

. Rao, S.A.; Periasamy, M. *J. Chem. Soc., Chem. Commun.*, *1987*, 495.

Arase, A.; Hoshi, M. *J. Chem. Soc., Chem. Commun.*, *1987*, 531.

Du, C.-J.F.; Hart. H. *J. Org. Chem.*, *1987*, *52*, 4311.

Ishii, T.; Kawamura, N.; Matsubara. S.; Utimoto, K.; Kozima. S.; Hitomi, T. *J. Org. Chem.*, *1987*, *52*, 4418.

Chou. T.; You, M.-L. *J. Org. Chem.*, 1987, *52*, 2224.

Goldbach, M.; Jäkel, E.; Schneider, M.P. *J. Chem. Soc., Chem. Commun.*, *1987*, 1434.

Gilchrist. T.L.; Summersell, R.J. *Tetrahedron Lett.*, *1987*, *28*, 1469.

Ph₃As⁺-CH₂CH=CHCHO Br⁻

$$\xrightarrow[\substack{25°C}]{K_2CO_3 \text{ , ether-THF (tr. } H_2O)}$$

79%

Huang, Y.Z.; Shi, L.; Yang, J.; Zhang, J. *Tetrahedron Lett.*, *1987*, *28*, 2159.

$$\xrightarrow[\substack{60°C}]{10\% \text{ Ni(COD)}_2 \text{ , 2 eq. PPh}_3}$$

(21   :   1) 92%

Wender, P.A.; Ihle, N.C. *Tetrahedron Lett.*, *1987*, *28*, 2451,

1. LiCHBr₂ , THF , TMP
   BuLi , -78°C - RT
2. LiH , THF , reflux
3. Ac₂O

53%

Kowalski, C.J.; Lal, G.S. *Tetrahedron Lett.*, *1987*, *28*, 2463.

$$\xrightarrow[\substack{\text{pentane/hexane}\\ \text{reflux , 24h}}]{Re_2O_7 \text{ , } Al_2O_3 \text{ , } Me_4Sn}$$

68%

Warwel, S.; Kätker, H.; Rauenbusch, C. *Angew. Chem. Int. Ed., Engl*, *1987*, *26*, 702.

n-PrS⌒⌒SiPh₃

1. BuLi
2. Ti(OiPr)₄
3. ⬡-CHO
4. MeMgI ,
   NiCl₂(PPh₃)₂

62%

62 EE : 24 ZZ : 9 EE : 5 EZ

Ikeda, Y.; Ukai, J.; Ikeda, N.; Yamamoto, H. *Tetrahedron*, *1987*, *43*, 731.

Ishiyama, T.; Miyaura, N.; Suzuki. A. Chem. Lett., 1987, 25.

Trost. B.M.; Lee, D.C. J. Am. Chem. Soc., 1988, 110, 7255.

O'Connor, B.; Zhang, Y.; Negishi. E. Tetrahedron Lett., 1988, 29, 3903

Alexakis. A.; Jachiet, D Tetrahedron Lett., 1988, 29, 217.

Kasahara. A.; Izumi, T.; Kudou, N. Synthesis, 1988, 704.

E:Z = 60:40

Cazes, B.; Colovray, V.; Gore, J. *Tetrahedron Lett.*, *1988*, *29*, 627.

Bäckvall, J.-E.; Nájera, C.; Yus, M. *Tetrahedron Lett.*, *1988*, *29*, 1445.

Fan, C.; Cazes, B. *Tetrahedron Lett.*, *1988*, *29*, 1701.

Georg, G.I.; Kant, J.; He, P.; Ly, A.M.; Lampe, L. *Tetrahedron Lett.*, *1988*, *29*, 2409.

Friess, B.; Cazes, B.; Gore, J. *Tetrahedron Lett.*, *1988*, *29*, 4089.

Pornet, J.; Rayadh, A.; Miginiac. L. *Tetrahedron Lett., 1988, 29,* 3065.

Fiandanese, V.; Marchese, G.; Mascolo, G.; Naso. F.; Ronzini, L. *Tetrahedron Lett., 1988, 29,* 3705.

Oppolzer, W.; Bedoya-Zurita, M.; Switzer, C.Y. *Tetrahedron Lett., 1988, 29,* 6433.

Tolstikov, G.A.; Miftakhov, M.S.; Danilova, N.A.; Vel'der, Ya.L. *J. Org. Chem., U.S.S.R., 1988, 24,* 396.

Trost. B.M.; Tometzki, G.B. *J. Org. Chem., 1988, 53,* 915.

Tokuda, M.; Endate, K.; Sugionome, H. *Chem. Lett.*, *1988*, 945.

Takeyama, H.; Suzuki, T. *J. Chem. Soc., Chem. Commun.*, *1988*, 1044.

Prieto, J.A.; Larson, G.L.; Berrios, R.; Santiago, A. *Synth. Commun.*, *1988*, *18*, 1385.

Karabelas, K.; Hallberg, A. *J. Org. Chem.*, *1988*, *53*, 4909.

Hatanaka, Y.; Hiyama, T. *J. Org. Chem.*, *1989*, *54*, 268.

Tamao, K.; Kobayashi, K.; Ito, Y. *J. Am. Chem. Soc., 1989, 111,* 6478.

Tanaka, H.; Kosaka, A.; Yamashita, S.; Morisaki, K.; Torii, S. *Tetrahedron Lett., 1989, 30,* 1261.

Arcadi, A.; Bernocchi, E.; Burini, A.; Cacchi, S.; Marinelli, F.; Pietroni, B. *Tetrahedron Lett., 1989, 30,* 3465.

Stille, J.K.; Sweet, M.P. *Tetrahedron Lett., 1989, 30,* 3645.

Bailey, W.F.; Ovaska, T.V.; Leipert, T.K. *Tetrahedron Lett., 1989, 30,* 3901.

7% Pd(PPh₃)₄ , AcOH
80°C , 8h

49%

Oppolzer, W.; Swenson, R.E.; Pachinger, W. *Helv. Chim. Acta,* **1989,** *72,* 14.
Oppolzer, W.; Gaudin, J.-M. *Helv. Chim. Acta,* **1987,** *70,* 1477.

PdCl₂(MeCN)₂
HMPT
RT
1h

90%

Tolstikov, G.A.; Miftakhov, M.S.; Danilova, N.A.; Vel'der, Ya.L.; Spirikhin, L.V.
Synthesis, **1989,** 633.

, 2.5% Pd(OAc)₂
3 eq. KOAc , Bu₄NCl
25°C , DMF , 72h

66%

Larock, R.C.; Gong, W.H. *J. Org. Chem.,* **1989,** *54,* 2047.

Pd(OAc)₂ , NEt₃
, DMSO , 3h
60°C

87%

Andersson, C.-M.; Hallberg, A. *J. Org. Chem.,* **1989,** *54,* 1502.

IrH₅(P[iPr₃])₂ , PhH
60°C , 24h

92%

Ma, D.; Yu, Y.; Lu, X. *J. Org. Chem.,* **1989,** *54,* 1105.

Trost, B.M.; Edstrom, E.D.; Carter-Petillo, M.B. *J. Org. Chem.*, *1989*, *54*, 4489.
Trost, B.M.; Lee, D.C.; Rise, F. *Tetrahedron Lett.*, *1989*, *30*, 651.

Yanagi, T.; Oh-E., T.; Miyaura, N.; Suzuki, A. *Bull. Chem. Soc., Jpn.*, *1989*, *62*, 3892.

Zhang, Y.; Negishi, E. *J. Am. Chem. Soc.*, *1989*, *111*, 3454.

REVIEWS:

"Camphor Derivatives as Chiral Auxiliaries in Asymmetric Synthesis"

Oppolzer, W. *Tetrahedron*, *1987*, *43*, 1969.

"Orthoquinodimethanes"

Charlton, J.L.; Alauddin, M.M. *Tetrahedron*, *1987*, *43*, 2873.

"Retro-Diels-Alder Strategy in Natural Product Synthesis"

Ichihara, A. *Synthesis*, *1987*, 207.

SECTION 378: OXIDES - ALKYNES

Me$_3$Si━━━━SiMe$_3$ $\xrightarrow[\begin{array}{c}\text{MeCN/CH}_2\text{Cl}_2\\\text{RT , 1h}\end{array}]{\text{NTFB , NHFP}}$ Me$_3$Si━━━NO$_2$

70%

NTFB = nitronium tetrafluoroborate
NHFP = nitronium hexafluorophosphate

<u>Schmitt, R.J.</u>; Battaro, J.C.; Malhotra, R.; <u>Bedford, C.D.</u> *J. Org. Chem.*, *1987*, *52*, 2294.

SECTION 379: OXIDES - ACID DERIVATIVES

Me-C(=O)-S-(CH$_2$)$_{10}$-COOH $\xrightarrow[\text{2. aq. K}_2\text{CO}_3]{\text{1. oxone , MeOH , RT , 1h}}$ K$^+$ $^-$O$_3$S-(CH$_2$)$_{10}$-COOH

92%

<u>Reddie, R.N.</u> *Synth. Commun.*, *1987*, *17*, 1129.

SECTION 380: OXIDES - ALCOHOLS, THIOLS

PhSO$_2$-CH$_2$-CH(OMe)$_2$ $\xrightarrow[\text{2. PhCHO , -78}^\circ\text{C - RT}]{\begin{array}{c}\text{1. 2 eq BuLi , THF , -78}^\circ\text{C}\\\text{1h}\end{array}}$ PhSO$_2$-C(=CH-OMe)-CH(OH)-Ph

76%

<u>Simpkins, N.S.</u> *Tetrahedron Lett.*, *1987*, *28*, 989.

Ph$_2$P(=O)-Me $\xrightarrow[\text{2. }\triangle\text{-Me (epoxide)}]{\text{1. BuLi}}$ Ph$_2$P(=O)-CH$_2$CH$_2$-CH(OH)-Me

70%

Wallace, P.; <u>Warren, S.</u> *J. Chem. Soc., Perkin Trans. I*, *1988*, 2971.

SECTION 381: OXIDES - AMIDES

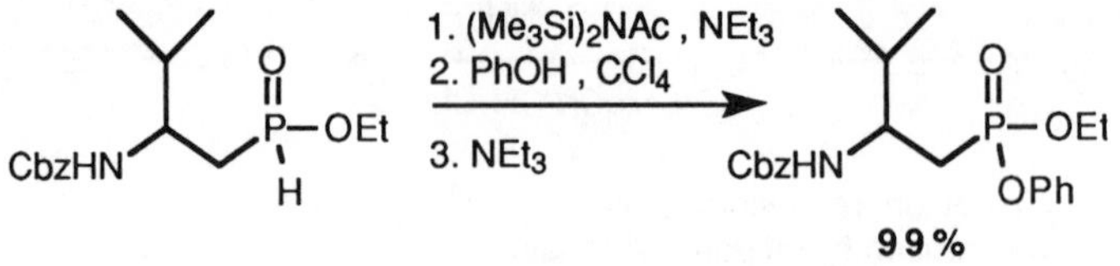

1. (Me₃Si)₂NAc , NEt₃
2. PhOH , CCl₄

3. NEt₃

CbzHN ... P—OEt / O / H

CbzHN ... P—OEt / O / OPh

99%

Sampson, N.S.; Bartlett, P.A. *J. Org. Chem.*, **1988**, *53*, 4500.

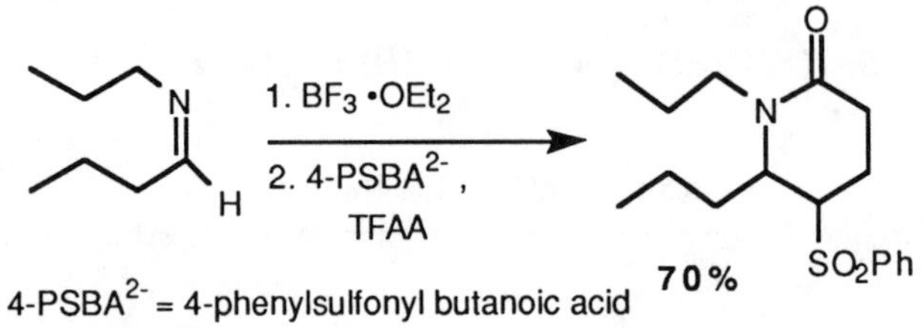

1. BF₃ •OEt₂

2. 4-PSBA²⁻ ,
   TFAA

4-PSBA²⁻ = 4-phenylsulfonyl butanoic acid   **70%**   SO₂Ph

Thompson, C.M.; Green, D.L.C.; Kubas, R. *J. Org. Chem.*, **1988**, *53*, 5389.

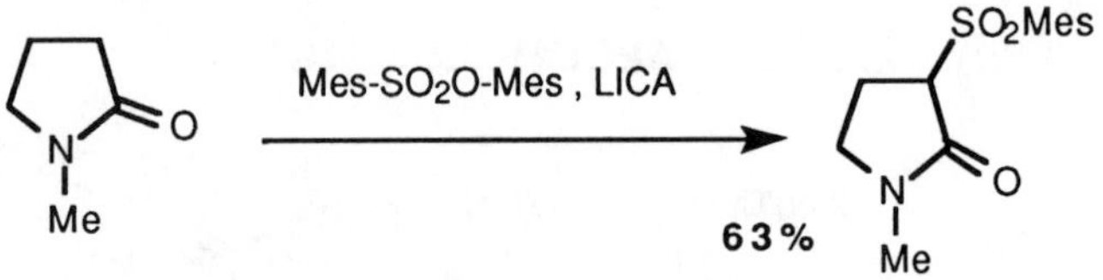

Mes-SO₂O-Mes , LICA

SO₂Mes

63%

Stewart, J.D.; Pinnick, H.W. *Heterocycles*, **1987**, *25*, 213.

SECTION 382: OXIDES - AMINES

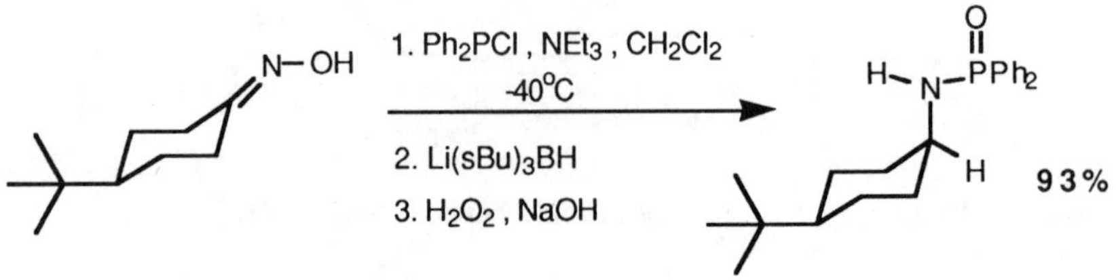

N—OH

1. Ph₂PCl , NEt₃ , CH₂Cl₂
   -40°C

2. Li(sBu)₃BH
3. H₂O₂ , NaOH

H—N—PPh₂ / O

H   **93%**

Hutchins, R.O.; Rutledge, M.C. *Tetrahedron Lett.*, **1987**, *28*, 5619.

Pyne, S.G.; Griffith, R.; Edwards, M. *Tetrahedron Lett.*, *1988*, *29*, 2089.

Ronan, B.; Marchalin, S.; Samuel, O.; Kagan, H.B. *Tetrahedron Lett.*, *1988*, *29*, 6101.

Kim, T.H.; Oh, D.Y. *Synth. Commun.*, *1988*, *18*, 1611.

Bäckvall, J.-E.; Rise, F. *Tetrahedron Lett.*, *1989*, *30*, 5347.

SECTION 383: OXIDES - ESTERS

3:2 mixture of isomers

Carretero, J.C.; DeLombaert, S.; Ghosez, L. *Tetrahedron Lett.*, **1987**, *28*, 2135.

Thompson, C.M. *Tetrahedron Lett.*, **1987**, *28*, 4243.

Koser, G.F.; Lodaya, J.S.; Ray III, D.G.; Kokil, P.B. *J. Am. Chem. Soc.*, **1988**, *110*, 2987.

Szymonifka, M.J.; Heck, J.V. *Tetrahedron Lett.*, **1989**, *30*, 2869; 2873.

Burgess, K.; Henderson, I. *Tetrahedron Lett.*, **1989**, *30*, 3633.

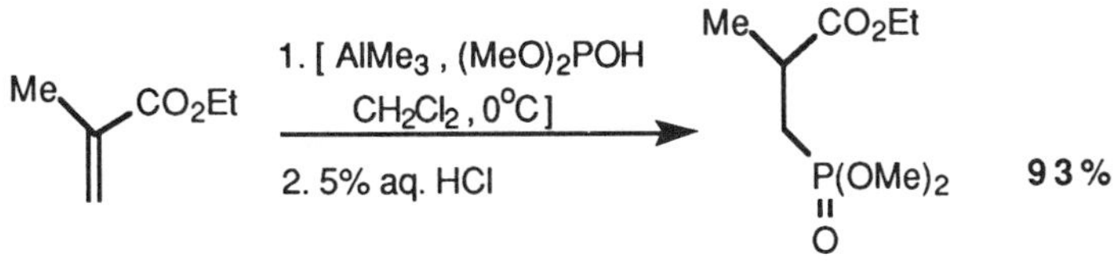

Me CO₂Et → Me CO₂Et

1. [AlMe₃ , (MeO)₂POH
   CH₂Cl₂ , 0°C]
2. 5% aq. HCl

$P(OMe)_2$
O **93%**

Green, K. *Tetrahedron Lett.*, *1989*, 30, 4807.

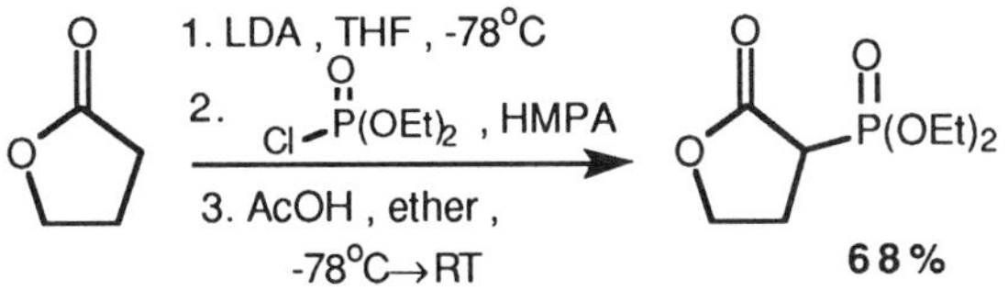

1. LDA , THF , -78°C
2. Cl–P(OEt)₂ , HMPA
3. AcOH , ether ,
   -78°C→RT

$P(OEt)_2$ **68%**

Jackson, J.A.; Hammond, G.B.; Wiemer, D.F. *J. Org. Chem.*, *1989*, 54, 4750.

SECTION 384: OXIDES - ETHERS, EPOXIDES, THIOETHERS

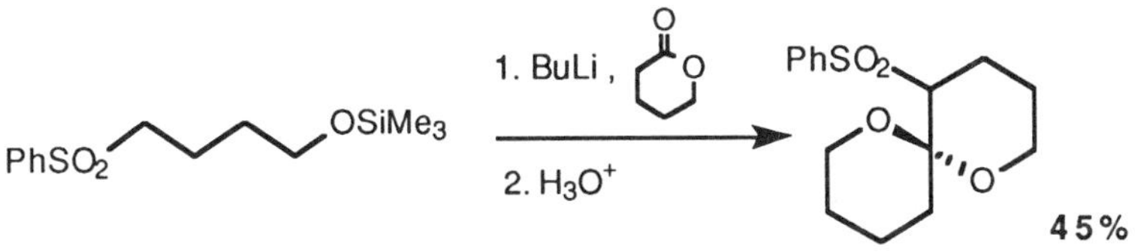

PhSO₂ OSiMe₃

1. BuLi ,
2. H₃O⁺

PhSO₂

45%

Brimble, M.A.; Officer, D.L.; Williams, G.M. *Tetrahedron Lett.*, *1988*, 29, 3609.

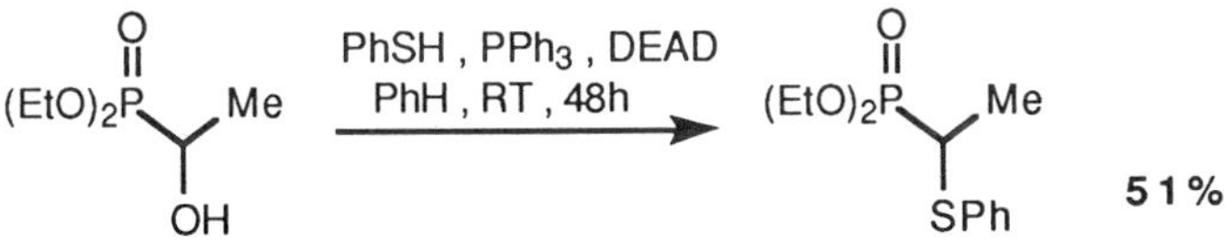

(EtO)₂P Me   PhSH , PPh₃ , DEAD   (EtO)₂P Me
OH   PhH , RT , 48h   SPh   **51%**

Gajada, T. *Synthesis*, *1988*, 327.

SECTION 385: OXIDES - HALIDES, SULFONATES

94% ee at sulfur

Satoh, T.; Oohara, T.; Ueda, Y.; Yamakawa. K. *Tetrahedron Lett.*, *1988*, *29*, 313.

Purrington. S.T.; Pittman, J.H. *Tetrahedron Lett.*, *1987*, *28*, 3901.

de C. Alpoim, M.C.M.; Morris, A.D.; Motherwell, W.B.; O'Shea, D.M. *Tetrahedron Lett.*, *1988*, *29*, 4173.

cis:trans = 6:1

Chuang. C.-P.; Ngoi, T.K.J. *Tetrahedron Lett.*, *1989*, *30*, 6369.

SECTION 386: OXIDES - KETONES

Monteiro, H.J. *Tetrahedron Lett., 1987, 28*, 3459.

Thomsen, M.W.; Handwerker, B.M.; Katz, S.A.; Belser, R.B. *J. Org. Chem., 1988, 53*, 906.

Im = imidazoyl

Ibarra, C.A.; Rodríguez, R.C.; Fernández Monreal, M.C.; García Navarro, F.J.; Martín Tesorero, J. *J. Org. Chem., 1989, 54*, 5620.

Hong, S.; Chang, K.; Ku, B.; Oh, D.Y. *Tetrahedron Lett., 1989, 30*, 3307.

Boeckman Jr., R.K.; Walters, M.A.; Koyano, H. *Tetrahedron Lett., 1989, 30*, 4787.

Pei, H.-J.; Farnsworth, D.W.; Saavedra, J.E. *Synth. Commun., 1989, 19,* 207.

Ali, S.M.; Tanimoto, S. *J. Org. Chem., 1989, 54,* 2247.

Roussis, V.; Wiemer, D.F. *J. Org. Chem., 1989, 54,* 627.

SECTION 387: OXIDES - NITRILES

Fang, J.-M.; Chen, M.-Y. *Tetrahedron Lett., 1987, 28,* 2853.

Kim, D.Y.; Oh, D.Y. *Synth. Commun., 1987, 17,* 953.

Et-C≡N → 1. 2 eq. LDA , THF , -78°C
2. (EtO)₂P(=O)Cl , THF
-78 → -30°C
3. H₂O

Me—CH(CN)—P(OEt)₂=O   73%

Kandil, A.A.; Porter, T.M.; Slessor, K.N. *Synthesis, 1987*, 411.

SECTION 388: OXIDES - ALKENES

[structure with Br on sulfolene ring] Bu₂CuLi , THF , -78°C
2h → Bu—[sulfolene ring]   70%

Chou, T.; Hung, S.C.; Tso, H.-H. *J. Org. Chem., 1987, 52*, 3394.

C₈H₁₇—C(=C(PhSe)—)SO₂Tol Bu₂CuLi , THF
0°C , 30 min → C₈H₁₇—C(=C(Bu)—)SO₂Tol   75%

Back, T.G.; Collins, S.; Krishna, M.V.; Law, K.-W. *J. Org. Chem., 1987, 52*, 4258.

Me—[C₆H₄]—I [SO₂ ring] , Pd(OAc)₂
NEt₃ , Bu₄NBr , PhH
95h → Me—[C₆H₄]—[SO₂ ring]   65%

Harrington, P.J.; DiFiore, K.A. *Tetrahedron Lett., 1987, 28*, 495.

[allyl chain] 1. TolSO₂⁻Na⁺ •4 H₂O , I₂
MeOH , CH₂Cl₂ ,RT
15 min
2. NEt₃ , MeOH → [chain]—SO₂Tol   94%
E:Z = 78:22

Inomata, K.; Sasaoka, S.; Kobayashi, T.; Tanaka, Y.; Igarashi, S.; Ohtani, T.; Kinoshita, H.; Kotake, H. *Bull. Chem. Soc., Jpn., 1987, 60*, 1767.

Fillion, H.; Hseine, A.; Pera, M.-H.; Dufaud, V.; Refouvelet, B. *Synthesis, 1987*, 708.

E:Z = 74:26

Carretero, J.C.; Demillequand, M., M.; Ghosez, L. *Tetrahedron , 1987, 43,* 5125.

Kameyama, M.; Shimezawa, H.; Satoh, T.; Kamigata, N. *Bull. Chem. Soc., Jpn., 1988, 61,* 1231.

E:Z = 10:0

Baudin, J.-B.; Julia, S.A. *Tetrahedron Lett., 1988, 29,* 3251, 3255.

Santelli-Rouvier, C. *Synthesis, 1988*, 64.

Calogeropoulou, T.; Wiemer, D.F. *J. Org. Chem.*, *1988*, *53*, 2295.

anti:syn = 90:10

Chaigne, F.; Gotteland, J.-P.; Malacria, M. *Tetrahedron Lett.*, *1989*, *30*, 1803.

1. LiN(SiMe₃)₂ , THF , -78°C
2. Me₃SiCl , -78°C → RT
3. BuLi , -78°C

4. (PhO)₂P(=O)Cl
-78°C → RT

64%

Koerwitz , F.L.; Hammond, G.B.; Wiemer, D.F. *J. Org. Chem.*, *1989*, *54*, 738 , 743.

0°C , 5 min

ether-PhH

65%

Bäckvall, J.-E.; Plobeck, N.A.; Juntunen, S.K. *Tetrahedron Lett.*, *1989*, *30*, 2589.

Holt, D.A.; Erb, J.M. *Tetrahedron Lett.*, *1989*, 30, 5393.

Hiroi, K.; Kurihara, Y. *J. Chem. Soc., Chem. Commun.*, *1989*, 1778.

Chang, K.; Ku, B.; Oh, D.Y. *Synth. Commun.*, *1989*, 19, 1891.

Lee, J.W.; Oh, D.Y. *Synth. Commun.*, *1989*, 19, 2209.

Labadie, S.S. *J. Org. Chem.*, *1989*, 54, 2496.

SECTION 389: OXIDES - OXIDES

73%

Pietrusiewicz, K.M.; Zabłocka, M. *Tetrahedron Lett.*, *1988*, *29*, 1987, 1991.

>97%

Golding, P.; Millar, R.W.; Paul, N.C.; Richards, D.H. *Tetrahedron Lett.*, *1988*, *29*, 2731.

1. NaH , THF , RT , 30 min

2. (*t*-Bu)$_2$P(=O)Cl , 0°C , 72h

3. TFA , PhH , RT , 18h

69%

Dhawan, B.; Redmore, D. *Synth. Commun.*, *1988*, *18*, 327.

Anderson, M.B.	267	Arcelli, A.	043
Anderson, P.C.	335	Archelas, A.	037
Andersson, C.-M.	281	Aritomi, K.	213
	419	Arliński, R.	108
	432	Armacost, L.M.	457
	458	Armesto, D.	234
	471		251
Andersson, P.G.	390		372
Ando, A.	007	Armistead, D.M.A.	032
	340	Armstrong, A.	190
Ando, H.	044	Arnaud, A.	012
Ando, K.	020	Arrivo, S.M.	396
	106	Arseniyadis, S.	149
Ando, M.	232	Arya, P.	417
	308	Asano, F.	240
Andreini, B.P.	305	Asao, T.	355
Andreoli, P.	137	Asaoka, M.	109
Andrews, S.W.	425	Asensio, G.	218
Anelli, P.L.	061		302
	240	Ashby, E.C.	210
Angell, E.C.	133	Ashida, T.	383
Angle, S.R.	098	Ashiya, H.	236
	315	Aso, Y.	118
Angoh, A.G.	265		143
Antel, J.	110		189
Antonioletti, R.	424		200
Anwar, S.	318		414
Aoe, K.	115	Assercq, J.-M.	450
	362	Astrab, D.P.	321
Aoki, O.	318	Atkinson, R.F.	281
	319	Atkinson, R.S.	158
Aoki, S.	083	Attanasi, O.A.	064
	398	Aubert, C.	351
	447	Aubé, J.	138
Aoyama, T.	004		139
	241	Augelli-Szafran, C.E.	191
	280	Aurrecoechea, J.M.	148
Arai, H.	335	Avignon-Tropis, M.	352
Arai, K.	134	Awasthi, A.K.	439
Araki, K.	195		006
Araki, S.	024	Ay, M.	155
	025	Azerad, R.	039
	027	Aznar, F.	433
	046		458
	118	Azzouzi, A.	324
	409		
Araldi, G.	028	**B**	
Arase, A.	263		
	463	Baba, A.	017
Arcadi, A.	388		043
	398		179
	470		327

Baba, S. 092
Baba, T. 460
Babiak, K.A. 107
Babler, J.H. 175
419
Baboulene, M. 326
Babu, J.R. 205
Babu, S. 147
Baccolini, G. 156
Bach, R.D. 355
Bachi, M.D. 407
Bachiocchi, E. 446
Baciocchi, E. 294
420
Back, T.G. 481
Bacos, D. 158
Baghavan, M. 397
Bagheri, V. 082
177
392
Bagler, V. 176
Bahsas, A. 145
386
Bailey, P.D. 064
Bailey, W.F. 093
097
470
Bairgrie, L.M. 187
Baizer, M.M. 110
Baker, B.E. 077
095
433
Baker, G.R. 391
Bakos, T. 058
Bakouetila, M. 428
Bakshi, R.K. 254
Bakthavatchalam, R. 044
Bal Reddy, K. 430
Balasubramanian, K.K. 225
Balavoine, G. 041
Baldoli, C. 192
Baldridge, R. 169
183
Baldwin, J.E. 298
309
359
360
367
374
460
463

Baldwin, S.W. 138
Balenkova, E.S. 364
Balicki, R. 160
161
244
Balko, T.W. 394
Ballesteros, A. 384
Ballini, R. 107
161
255
295
446
Ballistreri, F.P. 006
Balme, G. 114
390
Bambal, R. 109
Ban, Y. 115
164
Bando, T. 404
Bandodakar, B.S. 235
Banfi, L. 375
Banfi, S. 240
Banno, H. 065
352
Barak, G. 008
238
Baraldi, P.G. 342
Barbaro, G. 378
Barbeaux, P. 048
Barbieri, G. 336
Barcelo, G. 014
Barco, A. 342
Bargar, T.M. 385
Barinelli, L.S. 457
Barltrop, J.A. 191
Barluenga, J. 027
115
132
171
218
244
268
273
284
301
302
321
328
351
374
384
386

Barluenga, J.	388	Baudin, J.-B.	098
	433		482
	440	Bauld, N.L.	286
	458		288
Barner, B.A.	129	Baum, J.S.	381
Barney, C.L.	385	Bawa, A.	447
Barre, M.	100	Bay, E.	218
Barrett, A.G.M.	014	Bayod, M.	433
	323	Bazureau, J.P.	390
Barrios, H.	241	Bäckvall, J.-E.	079
Barros, M.T.	236		114
Barros, S.M.	267		252
Barry, J.	176		390
Barta, M.	123		404
Bartels, B.	430		433
Bartlett, P.A.	474		467
Bartoli, D.	185		475
	413		483
Bartoli, G.	144	Bäuml, E.	063
	161		186
	162	Beak, P.	129
Bartoli, J.F.	037		361
Bartoli,.D.	293	Beard, M.	169
Barton, D.H.R.	080		183
	120	Beck Jr., K.R.	292
	145	Beck, G.R.	008
	149	Beck, K.R.	458
	152	Beckett, R.P.	323
	161	Becking, L.	126
	290	Beckwith, A.L.J.	414
Barua, N.C.	101	Bedford, C.D.	473
	257	Bednarshi, M.D.	321
	278	Bedoya-Zurita, M.	368
Basavaiah, D.	104		389
	206		459
	331		468
	342	Bedworth, P.V.	284
		Beebe, T.R.	169
			183
Basha, A.	121	Behling, J.R.	025
	128		107
	150		109
Baskaran, S.	050	Beifuss, U.	110
	347	Belan, A.	038
	397	Belaud, C.	369
Bastos, C.	091	Belicchi, M.F.	019
Basu, M.K.	010	Bell, T.W.	082
Bates, T.F.	435	Bellamy, F.	273
Battaglia, A.	378	Bellassoued, M.	312
Battaro, J.C.	473		314
Battioni, P.	037	Belletire, J.L.	144
Baudin, J.-B.	001		177

Belletire, J.L.	306	Biggs, K.R.	435
Belleville, D.J.	288	Bigi, F.	028
Bellis, G.P.	262	Bih, Q.-R.	063
Belotti, D.	367	Binger, P.	456
Belser, R.B.	479	Bird, C.W.	412
Ben-David, Y.	180	Björkling, F.	171
Benati, L.	256	Black, D.J.	183
Benderly, A.	127	Black, T.H.	166
Benedetti, F.	033		167
Benetti, S.	342		172
Bennett, P.A.R.	463		314
Bennett, S.M.	265		396
Bennett, W.D.	310	Blair, I.A.	057
Bergbreiter, D.E.	223	Blanton, J.R.	223
Berglund, B.	456	Blechert, S.	162
Berk, S.C.	109	Bloch, R.	043
	234		112
Bernardi, A.	329	Bloodworth, A.J.	215
Bernatchez, M.	036	Bloom, S.H.	276
	315	Blough, B.E.	032
Bernocchi, E.	470		350
Bernstein, M.A.	335	Boaventura, M.-A.	453
Berrada, S.	395	Bocelli, G.	028
Berriet, J.	267	Boeckman Jr., R.K.	029
Berrios, R.	469		168
Bertenshaw, S.	360		479
Bertolini, G.	341	Boger, D.L.	016
Bertounesque, E.	312		121
Bertrand, M.P.	391		250
Beslin, P.	180		084
Bestmann, H.J.	155	Bois-Choussy, M.	051
	442	Boivin, J.	290
Betschart, C.	373		205
Beugelmans, R.	084	Boivin, T.L.	020
Bégué, J.-P.	351	Boldrini, G.P.	284
Bharathi, P.	267	Boleslawski, M.P.	056
Bharathi, T.K.	104	Bolitt, V.	079
	331		038
	342	Bolte, J.	066
Bhat, K.S.	336	Bolton, G.L.	446
	348	Bonadies, F.	034
Bhat, N.G.	266	Bonini, C.	446
Bhatt, M.	205		448
Bhatt, M.V.	052	Bonnert, R.V.	462
Bhupathy, M.	411		055
	437	Borchardt, R..	248
Bianco, A.	032	Borkowsky, S.L.	192
	222	Borredon, M.E.	196
Bickelhaupt, F.	434		199
Bicknell, A.J.	284	Borselli, G.	392
Biehl, E.R.	185	Bortolini, O.	237
Biffi, C.	061		

Bosch, E.	407
Bosch, P.	215
	372
Bosco, M.	144
	162
Bou, V.	123
Boub, A.B.	273
Boubia, B.	273
Boucher, R.J.	422
Bouda, H.	196
	199
Bougauchi, M.	066
Boujamra, C.G.	446
Boukouvalas, J.	178
Boulanger, R.	315
Boumendjel, A.	067
Bowyer, K.J.	215
Boyer, J.H.	036
Boykin, D.W.	101
Boys, M.L.	441
Brackenridge, I.	190
Bradsher, C.K.	087
Brady, W.T.	418
Bram, G.	176
	239
Branchaud, B.P.	280
Brand, M.	212
Brandt, C.A.	223
Braslau, R.	294
Braun, M.	347
Braunstein, P.	187
Bray, B.L.	299
Breitgoff, D.	332
Brestensky, D.M.	101
Bretsch, W.	452
Breuilles, P.	454
Brégeault, J.-M.	313
Brieva, R.	134
	301
Brimble, M.A.	477
Brinkman, H.R.	173
Brinkmeyer, R.S.	394
Britton, T.C.	009
Broida, N.	399
Broka, C.A.	096
	430
Brook, M.A.	031
Brookhart, M.	118
Brooks, D.W.	045
	121
	150

Brooks, D.W.	207
	329
Brower, K.R.	211
Brown, D.	303
Brown, D.S.	190
	197
	426
Brown, H.C.	022
	039
	040
	049
	059
	254
	266
	325
	336
	348
Brown, J.D.	076
	327
Brown, J.M.	088
Brown, M.J.	420
Brown, P.A.	462
Brown, S.H.	067
Brown, S.L.	033
Brownbridge, P.	178
	201
Bruneau, C.	409
Brunnelle, W.H.	451
Brunner, H.	100
Bruno, M.	426
Brückner, R.	352
Bryson, T.A.	252
Bræden, J.E.	236
Buchanan, R.A.	197
Buchbauer, G.	124
Buchert, M.	115
Buchman, O.	010
Buchwald, S.L.	165
	263
Buckman, B.O.	287
Buisson, D.	039
Bulman Page, P.C.	045
Bunadies, F.	424
Bunce, R.A.	103
Bunce, R.A.	242
Bunnelle, W.H.	277
Buntain, G.A.	152
Buono, G.	020
Buono, G.	040
Burbaum, B.W.	232
Bures, E.J.	196

Burgess, K.   050
                               319
                               371
                               403
                               476
Burgett, P.M.   139
Burini, A.   470
Burke, S.D.   032
Burmistrov, K.S.   144
Burns, S.J.   432
Burton, D.J.   096
                               394
Burton, G.   284
Busacca, C.A.   371
Buss, D.   270
Butler, I.R.   230
Butler, W.M.   109
Butsugan, Y.   024
                               025
                               027
                               046
                               118
                               409
Buxton, S.R.   199
Byers, J.H.   096

                C

Cabal, M.P.   458
Cacchi, S.   093
                               388
                               398
                               470
Cahiez, G.   047
                               111
                               234
                               137
Calet, S.   133
                               179
                               307
Caley, B.A.   394
Calogeropoulou, T.   483
Cambanis, A.   063
Cambie, R.C.   394
Campbell, A.L.   107
                               109
                               147
Campbell, J.A.   369
Campestrini, S.   237
Campi, E.   356
Campos, P.J.   218
                               301

Campos, P.J.   302
Camps, F.   100
                               208
                               215
                               258
                               399
Canal, G.   301
Cancellon, J.M.   115
Cano, A.C.   064
                               244
Canonne, P.   036
                               315
Capdevielle, P.   259
                               261
Caple, R.   231
                               454
                               456
Caporusso, A.M.   004
                               106
Capperucci, A.   251
Caputo, R.   255
Carboni, B.   350
Cardani, S.   329
Cardellicchio, C.   232
Carey, J.T.   344
Caringi, J.J.   306
Carlsen, P.H.J.   236
Carlton, L.   427
Carpenter, J.F.   348
Carpita, A.   305
                               433
Carretero, J.C.   178
                               476
                               482
Carrie, R.   157
Carta, M.P.   131
Carter-Petillo, M.B.   472
Casey, M.   112
Casiraghi, G.   019
Casnati, G.   019
                               028
Castañeda, A.   205
Castro, B.   014
Casu, A.   446
Casucci, D.   308
Caubere, P.   016
                               041
Cauliez, P.   308
Cava, M.P.   134
Cazes, B.   467
Cederbaum, F.E.   265
Celerier, J.-P.   158

Celerier, J.P.	375	Chatgilialoglu, C.	225
Cesarotti. E.	089		226
Cha. J.K.	416	Chattopadhyay. T.K.	166
Cha. J.S.	059	Chaussard, J.	275
	060	Chavant, P.-Y.	047
	065	Chen, C.	026
	068		274
Chabardes. P.	357		344
Chabert, P.	273	Chen, C.-S.	169
Chaigne, F.	483	Chen, H.C.	243
Chamberlin, A.R.	185	Chen, J.	013
	276	Chen, K.-M.	315
Chambers, M.R.I.	258	Chen, M.-H.	097
Chambert, P.	322	Chen, M.-Y.	480
Chamchaang, W.	132	Chen, O.-Y.	085
Chamorro, E.	215	Chen, Q.-Y.	004
Chan, C.	305	Chen, S.-F.	387
Chan. D.M.T.	401	Chen, Y.	138
Chan, M.-C.	223	Chen, Y.-C.J.	185
Chan, T.C.	087	Chen, Y.-S.	362
Chan. T.H.	021	Chen. Z.-C.	166
	036	Chenault. J.	002
	417	Cheng. C.H.	214
	418	Cheng, J.-W.	083
	443	Cheng, K.-M.	223
Chan, T.H.P.	415	Cheng, M.-C.	255
Chander, M.C.	300	Cheng, W.	076
Chandrasekaran. S.	050	Cherepy, N.	073
	244	Cheshire, D.R.	340
	247	Cheskis, B.A.	237
	254	Chi, D.Y.	436
	347	Chiacchio, U.	457
	397	Chianelli, D.	413
	409	Chiba, M.	090
Chandrasekhar, S.	160		331
Chandrasekharan, J.	039	Chiba. T.	172
Chanet-Ray, J.	140		259
Chang, C.-T.	097	Chidambaram, N.	050
	393		244
Chang, H.-T.	110		247
Chang, H.S.	255		409
Chang, K.	479	Chiericato Jr., G.	083
	484	Chiesa, A.	089
Chang, V.H.-T.	094	Chikada, S.	056
Chang, Y.-K.	008	Chikamatsu, H.	044
Chanson, E.	159	Chikashita. H.	069
Charbonnier, F.	302	Chion, H.S.	288
	429	Chisholm. M.H.	284
Charles. G.	214	Chizhov, O.S.	237
Charlton, J.L.	472	Cho, B.T.	039
Chatani. N.	266		040
Chatgilialoglu, C.	111	Cho, I.H.	239

Choi, J.-K.	008
Choi, S.-C.	382
	395
Choi, Y.	280
Chong, J.M.	034
Chou, S.-S.P.	267
Chou, T.	462
	464
	481
Choudary, B.M.	162
	181
	267
Christensen, D.	293
Chuang, C.-P.	478
Chuche, J.	431
Chung, C.K.	225
Chung, K.N.	190
Chung, S.	239
Church, G.	057
Church, L.A.	117
Ciattini, P.G.	093
	360
Cirillo, P.F.	427
Citterio, A.	080
Civitarese, G.	420
Claremon, D.A.	260
Clark, T.J.	117
Classon, B.	208
Claxton, E.E.	176
Cleary, D.G.	192
	288
Clerici, F.	233
Clinet, J.C.	041
Clive, D.L.J.	265
	340
Clos, N.	091
Coffen, D.L.	131
Coghlan, M.J.	394
Cohen, T.	023
	277
	411
	437
	454
Colclough, E.	270
Cole, T.E.	049
	254
Coleman, R.S.	016
Coll, J.	100
	399
Collignon, N.	371
Collin, J.	046

Collin, J.	115
	235
Collingnon, N.	062
Collington, E.W.	441
Collingwood, S.P.	207
Collins, S.	022
	243
	481
Colombo, L.	329
	341
Colonna, S.	295
Colotta, V.	444
Colovray, V.	467
Comasseto, J.V.	223
	267
Combellas, G.	275
Combrink, K.D.	004
Comins, D.L.	075
	082
	127
	327
	385
	387
Comisso, G.	091
Concellón, J.M.	268
	273
	284
	351
Condon, M.E.	381
Conn, R.E.	337
Conrad, P.C.	197
Conte, V.	294
Contento, M.	137
Convert, O.	313
Cook, J.D.	377
Cooke Jr., M.P.	104
	108
Cooke, D.	183
Cooley, J.H.	133
Cooper, C.S.	329
Coppi, L.	023
	414
	417
Corey, E.J.	020
	281
	369
	397
	443
Corley, E.G.	126
Cormier, R.A.	227
Cormons, A.	387
Cornia, M.	019

Corriu, R.J.P.	059	Dai-Ho, G.	145
	060	Daignault, S.	057
	105	Dailey, W.P.	312
Cort, A.D.	038	Dakka, G.	207
Cortez, C.	404	Dakka, J.	008
Cossío, F.P.	122	Dakka, J.	238
Cossy, J.	029	Dallemer, F.	235
	053	Dalpozzo, R.	144
	123		162
	256	Damour, D.	385
	367	Danheiser, R.L.	045
	370		203
	411		224
Costa, A.	189	Daniels, R.G.	308
Costanza, C.	128	Danilova, N.A.	468
Court, J.J.	243		471
Courtois, G.	163	Danks, T.N.	149
Couturier, D.	073	Dannecker, R.	110
	308	Darcy, M.G.	293
Couty, F.	070	Das, J.	347
Cox, P.	196	Datel, T.	214
Córdoba, A.A.	244	Datta, A.	331
Crabtree, R.H.	067	Dauben, W.G.	283
	446		438
Creswell, M.W.	431	Daude, N.	422
Crich, D.	208	Daumas, M.	062
Crimins, M.T.	253		273
Crisp, G.T.	084	Daves Jr., G.D.	098
Cristobal, B.I.	102	Davies, A.P.	207
Crooks III, W.J.	219	Davies, H.M.L.	117
Crouse, D.J.	222	Davies, S.G.	116
Crout, D.H.G.	346		323
Crute, T.D.	350	Davis, A.P.	318
Cruz, W.O.	313	Davis, C.S.	074
Cuadrado, F.	102	Davis, F.A.	294
Cullen, W.R.	151		297
	230		340
Cunkle, G.T.	308		381
Curran, D.P.	097		432
	393	Davisson, M.E.	033
	409	Davoust, D.	314
	443	De Cock, W.	445
Cyr, D.R.	034	de C. Alpoim, M.C.M.	478
Ceković, Z.	016	de Groot, A.	420
		de Groot, Ae.	354
D		de Kanter, F.J.J.	434
		de la Cal, M.T.	102
d'Incan, E.	067	de Lange, B.	326
	044	Deardorff, D.R.	333
da Silva, E.	290	Deaton, D.N.	032
Dabard, R.	090	DeFays, I.	183
Dad, M.M.	418	DeGiovani, W.F.	083

Degl'Innocenti, A.   251
                               392
                               444
Degueil-Castaing, M.   175
DeJeso, B.   175
DeKimpe, N.   142
                               372
                               383
                               445
Del Buttero, P.   192
del Valle, L.   106
Delahunty, C.M.   028
Delair, P.   320
Delgado, F.   244
Dellaria, J.F.   378
Delmas, M.   192
                               196
                               199
DeLombaert, S.   476
Dembechm, P.   303
Demerseman, P.   037
DeMesmaeker, A.   095
                               233
Demillequand, M., M.   482
Demir, A.S.   182
Demnitz, F.W.J.   168
Denis, A.   114
Denis, J.-M.   149
Denmark, S.E.   052
Denne, I.   350
Dent, W.   374
Depres, J.-P.   306
DeShong, P.   345
                               398
                               451
                               452
Deshpande, V.H.   166
Desmond Longford, C.P.   417
Desmond, R.   356
Despres, L.   123
Devanne, D.   399
Devasagayaraj, A.   153
DeVos, A.   209
DeWinter, A.J.   249
Dhanak, D.   323
Dhanoa, D.S.   103
Dhar, D.N.   417
Dharma Rao, P.   206
Dhawan, B.   485
Dhimane, H.   373
                               388
Dhumrongvaraporn, S.   265

Dicko, A.   326
Dieter, J.W.   104
Dieter, R.K.   107
                               184
DiFabio, R.   034
Differding, E.   380
DiFiore, K.A.   481
DiFuria, F.   237
                               294
Dimas, L.   157
Dimock, S.H.   035
Ding, S.   370
DiRaddo, P.   075
Dittami, J.P.   126
Dixneuf, P.H.   358
                               363
                               366
                               399
                               409
Dixon, N.J.   351
Dlubala, A.   180
Dmitrikova, L.V.   144
Doad, G.J.S.   191
Doken, K.   260
Dolle, R.E.   181
                               371
Dombroski, M.A.   286
Donnelly, D.M.X.   152
Dormand, A.   270
Dorohkov, V.I.   121
Dorow, R.L.   309
Dorsey, B.D.   345
Dorta, R.L.   130
                               190
Dostalek, R.   442
Doubleday, W.   023
Doussot, J.   156
Dowd, P.   382
                               395
Doyle, M.M.   069
Doyle, M.P.   139
                               176
                               177
                               392
Dragovich, P.S.   154
Drewniak, M.   148
Drewniaki, M.   128
Drouin, J.   453
Drumwright, R.E.   103
Du, C.-J.F.   464
DuBay III, W.J.   166
                               172

Dubina, V.L.	144	Egashira, H.	236
Dubois, J.-E.	312	Egert, E.	425
Duburs, G.	389	Eggert, U.	422
Dufaud, V.	482	Eguchi, S.	084
Dufour, M.	324	Ehlis, T.	289
Dufresne, C.	228	Ehrenkaufer, R.E.	091
Duggan, M.E.	123	Eichinger, P.C.H.	283
	413	Einhorn, C.	200
	430		242
Duhamel, L.	274	Einhorn, J.	037
	314		242
Duhamel, P.	309	Eisch, J.J.	284
Dujardin, G.	411	El Alami, N.	369
Dulcere, J.P.	406	El Ali, B.	313
	430	El Boudadili, A.	270
Dumont, W.	204	El Kaim, L.	051
	249	Elder, J.S.	284
Dunamsit, T.	455	Elgendy, S.	242
Duncan, M.P.	006	Elinson, M.N.	184
	090		252
	132	Ellenberger, S.R.	137
	340	Elliott, J.D.	339
	418	Ellis, K.L.	413
	420	Ellman, J.A.	009
	444		309
Dunlap, N.K.	108	Ellsworth, E.L.	025
Dunn, M.M.	333		035
Duñach, E.	320	Elmore, P.R.	235
Dupin, J.-F.E.	002	Elsevier, C.J.	267
Durandetti, S.	022	Elsheimer, S.	011
Dureanceau, S.J.	332	Endate, K.	469
Durst, H.P.	294	Enders, D.	031
		Endu, T.	379
E		Engel, P.	308
		Engler, T.A.	004
Eaton, J.T	009		250
Eaton, J.T.	012	Engman, L.	393
Eaton, P.E.	207		440
	308	Enholm, E.J.	027
Eberlin, M.N.	382		333
Ebihara, K.	111	Enomoto, S.	253
Ebner, M.	134	Epa, W.R.	439
Echavarren, A.M.	034	Erb, J.M.	484
	081	Erdbrügger, C.F.	425
	353	Erdik, E.	048
	453		155
Eddine, J.J.	309	Ernst, B.	095
Edington, C.	339		209
Edstrom, E.D.	472		233
Edwards, J.D.	177	Erzhanov, K.B.	423
Edwards, M.	475	Escaffre, P.	068
Effenberger, F.	026		

Esteban, S.	251
Estermann, H.	359
Etter, J.B.	114
	329
	348
Evain, E.J.	133
Evans, D.A.	009
	049
	309
	337
Evans, R.D.	436

F

Fabiano, E.	141
Fabrissin, S.	033
Fadel, A.	255
Failla, S.	006
Falck, J.R.	056
	079
	273
Fallis, A.G.	032
	202
	412
Fan, C.	467
Fan, R.	047
Fan, W.-Q.	070
Fang, J.-M.	110
	480
Fang, Q.	165
Fañanás, F.J.	132
Fañanás, J.	027
Farina, S.M.F.	077
Farnsworth, D.W.	480
Farooq, O.	074
	077
	184
	235
Fasseur, D.	073
Fasseur, D.	308
Faulkner, D.J.	332
Fauve, A.	038
Fava, G.G.	019
Feghouli, A.	041
Feghouli, G.	016
Feigenbaum, A.	092
Ferguson, R.R.	446
Feringa, B.L.	109
	326
Ferland, J.-M.	057
Fernandez, I.	161
Fernandez-Simon, J.L.	115

Fernández Monreal, M.C.	479
Fernández, J.R.	171
	321
Fernández-Simon, F.	268
Fernández-Simón, J.	351
Fernández-Simón, J.L.	273
	284
Ferraboschi, P.	033
Ferraz, H.M.C.	223
Ferre, E.	100
Ferreira, J.T.B.	313
Ferreri, C.	255
Ferrero, M.	386
Ferringa, B.L.	144
Fiandanese, V.	093
	232
	468
Fiandanesse, V.	234
Fiecchi, A.	033
Fields, J.D.	167
Fields, S.C.	125
Filimonov, V.D.	253
Fillion, H.	368
	482
Finato, B.	080
Finch, H.	441
Finet, J.-P.	086
	120
	145
	152
Fiorani, T.	080
Firouzabadi, H.	239
	244
Fisher, S.A.	116
Fishpaugh, J.R.	184
Fitjer, L.	269
Fitzmaurice, N.J.	356
Fleming, A.	448
Fleming, F.F.	261
Fleming, I.	030
	339
Fleming, S.A.	115
Flinn, A.	374
Flitsch, W.	364
Fobare, W.F.	145
	384
Fok, C.C.M.	279
Foland, L.D.	460
Folest, J.-C.	215
Foley, M.A.	387
Forrest, A.K.	463
Fort, Y.	016

Fort, Y.	041	Fugami, K.	434
Fortin, R.	057		442
Fortt, S.M.	208	Fuji, K.	177
Foster, B.S.	452		231
Fotsch, C.H.	276		341
Foubelo, F.	027	Fujihara, H.	373
	132	Fujii, M.	254
Foucaud, A.	428	Fujimaura, T.	447
Fourneron, J.-D.	037	Fujimoto, M.	122
Fournet, G.	114	Fujimura, O.	388
	390	Fujimura, T.	083
Fournier, J.	409		182
Fowler, F.W.	370	Fujinami, T.	183
Fox, D.P.	110		199
Fraenkel, G.	374		340
Francisco, C.G.	130		427
	190	Fujisaki, S.	171
Fraser, R.R.	445		216
Fray, M.J.	344		438
Freeman, F.	389	Fujisawa, T.	010
Freire, R.	190		041
Frejd, T.	261		257
	054		278
Frenette, R.	182	Fujita, A.	013
	425	Fujita, H.	373
Freudenberger, J.H.	319	Fujita, M.	027
Freyer, A.J.	124		324
Fréchet, J.M.J.	147	Fujita, S.	355
Friedrich, E.C.	118	Fujita, T.	184
Friese, C.	118	Fujiwara, M.	179
Friess, B.	467		327
Fringuelli, F.	203	Fujiwara, S.	135
Frisque-Hesbain, A.-M.	209	Fujiwara, T.	461
Froen, D.	231	Fujiwara, Y.	318
Fruedenberger, J.H.	319		319
Fry, A.J.	341	Fujiwsawa, T.	314
Fry, D.F.	144	Fukuota, S.	166
	306	Fukushima, M.	427
Fry, J.L.	234	Fukushima, S.	085
Fryzuk, M.D.	151	Fukuyama, J.M.	377
Fu, G.C.	049	Fukuzawa, S.	182
Fu, J.-M.	031		199
Fuchikami, T.	013		340
	180		427
Fuchs, P.L.	267	Funakoshi, K.	242
Fuentes, A.	271	Fundy, M.A.M.	070
	453	Funk, R.L.	066
Fugami, K.	082	Furstoss, R.	037
	176	Furuhashi, K.	203
	226	Furukawa, N.	373
	406	Furuta, K.	196
	429		358

Furutani, H.	204
	330
Furuya, M.	434
Fustero, S.	244
	328
	374
Fülling, G.	167
Fürstner, A.	333

G

Gadwood, R.C.	249
Gaeta, K.K.	223
Gaggero, N.	295
Gahman, T.C.	170
Gairns, R.S.	112
Gajada, T.	477
Gajda, T.	375
Gal, C.	211
Gallagher, T.	196
	384
Gallego, M.G.	372
Gallos, J.K.	301
Gallucci, J.	374
Gallulo, V.	157
Galoši, A.	226
Galzy, P.	012
Gambaro, M.	320
Ganbua, I.	122
Ganem, B.	174
	191
Gannon, S.M.	247
Gao, Y.	083
	160
	197
Gaoni, Y.	143
Garcia, J.	123
García Navarro, F.J.	479
García, J.M.	122
Garlaschelli, L.	269
Garner, P.	283
Garratt, P.J.	390
Garyali, K.	260
Gasdaska, J.R.	353
Gaset, A.	192
	196
	199
Gasol, V.	208
	215
	258
Gassman, P.G.	117
	338

Gassman, P.G.	432
Gaudemar, M.	314
Gaudin, J.-M.	368
	471
Gaul, M.D.	395
Gaus, P.L.	162
Gauthier, J.	057
Gauthier, J.Y.	350
Gauthier, S.	147
Gelman, D.P.	008
Gelmi, M.L.	233
Genet, J.P.	240
	377
	114
Gennari, C.	138
	341
Georg, G.I.	136
	467
Germani, R.	203
Gerritz, S.W.	162
Gheerbrant, E.	211
Ghosal, S.	298
Ghosez, L.	178
	209
	380
	476
	482
Ghosh, A.K.	443
Ghosh, S.	178
Ghosh, S.K.	268
Ghosh, T.	280
Ghribi, A.	337
Giacomini, D.	137
Gianferrara, T.	033
Giang, Y.-S.F.	418
Giannis, A.	328
Giblin, G.M.P.	282
Giese, B.	111
Gil, G.	100
Gilabert, D.M.	151
Gilbert, L.	043
Gilchrist, T.L.	464
Gilday, J.P.	073
	228
Gillard, J.W.	057
Giller, D.	226
Gilles, B.	035
Gingras, M.	036
Giorgianni, P.	378
Giovannini, A.	366
Girard, C.	112

Girard, Y.	337
Gislon, G.	138
Glase, S.A.	450
Gnonlonfoun, N.	189
Godfrey, A.	105
	421
Godtfredsen, S.E.	171
Goe, G.L.	056
Gogoll, A.	404
Goldbach, M.	464
Goldberg, Y.Sh.	264
Golding, B.T.	141
	207
Golding, P.	485
Goldstein, S.	276
Gonda, E.	417
Gondgaon, N.M.	162
Gong, W.H.	095
	433
	471
González, A.	445
González, A.M.	102
González, F.J	328
González, J.M.	218
	302
Goralski, C.T.	325
Gore, J.	390
	467
Goto, K.	163
Gotor, V.	134
	301
	384
Gotteland, J.-P.	483
Gottlieb, H.E.	399
Goudgaon, N.M.	050
	102
	157
Goulet, M.T.	033
Gourcy, J.G.	038
Gove, J.	114
Gowriswari, V.V.L.	104
	206
	331
	342
Graillot, Y.	326
Gramain, J.-C.	324
Gras, J.-L.	053
Grass, F.	217
Graves, D.M.	237
Gray, B.D.	439
Gray, D.	217
Gray, T.	380

Graziani, M.	043
Green, D.L.C.	474
Green, K.	477
Greene, A.E.	302
	306
	429
Greenspoon, N.	224
	221
Greenwood, T.D.	369
Grenier, L.	425
Gretz, E.	181
Greuter, H.	312
Gribble, G.W.	009
	012
	148
Gridnev, I.D.	364
Grieco, P.A.	145
	384
	386
Griffith, R.	475
Grigg, R.	073
Griller, D.	225
Grimaldi, J.	387
Grimshaw, J.	428
Grisenti, P.	033
Groh, B.L.	462
Grosjean, F.	373
Gross, R.S.	240
Grothaus, P.G.	045
Gu, X.	321
Gu, X.-P.	381
	419
	446
Guan, X.	136
Guanti, G.	375
Guerrero, A.	208
	215
	258
Guerrero, A.F.	438
Guerrini, A.	303
Guette, J.P.	326
Guggisberg, A.	182
Guibé-Jampel, E.	310
Guillemin, J.-C.	149
Guindon, Y.	057
	335
	337
	350
Guiry, P.J.	152
Guitart, J.	100
Gung, W.Y.	028
Guo, B.-S.	023

Guo, H.	297	Hammer, R.P.	052
Guo, Z.-W.	167	Hammond, G.B.	477
Gupta, A.K.	387		483
Gupta, K.B.	413	Hammoud, A.	304
Gupta, V.	050	Hampel, K.	364
Gupton, J.T.	203	Han, A.-L.	078
	332	Hanafusa, T.	266
	399		383
Gutierrez, C.G.	087	Hanamoto, T.	053
Gutierrez, E.	239	Hanaoka, M.	304
Guy, A.	156	Handa, Y.	316
	326	Handwerker, B.M.	116
Guziec Jr., F.S.	213		479
	229	Hanessian, S.	148
	241		210
Gułębiowski, A.	354		307
Gybin, A.S.	231	Hannon, F.J.	020
	454		106
		Haq, W.	014
H		Hara, D.	225
		Hara, H.	034
Ha, D.-C.	138		239
Ha, D.S.	215	Hara, R.	333
	438	Hara, S.	266
Haack, R.A.	458		273
Haaima, G.	406		410
Habaue, S.	096		461
Habermas, K.L.	052	Harada, A.	202
Habuš, I.	091	Harada, K.	200
Haddad, M.	375	Harada, T.	225
Hadjiarapoglou, L.	203		320
	393	Haraguchi, K.	336
Hagen, J.P.	356	Harding, K.E.	074
Hahn, C.S.	239		362
	342		376
Hahn, G.R.	015	Hardtmann, G.E.	315
Halazy, S.	204	Haridasan, V.K.	131
Hale, K.J.	226	Harirchian, B.	286
Hallberg, A.	098		288
	281	Harlow, R.L.	462
	419	Harn, N.K.	392
	432	Harring, L.S.	117
	458	Harrington, P.J.	481
	469	Harrington, R.E.	247
	471	Harris Jr., G.D.	123
Halloran, D.J.	281	Harris, A.R.	017
Halshimoto, S.	155		224
Halterman, R.L.	020		295
Hamanaka, S.	103	Harris, J.J.	356
	227	Harris, S.M.	298
Hamatani, T.	024	Harrison, C.R.	452
Hamel, N.	366	Harrison, L.W.	403

Harrison, L.W.	407	Hayashi, R.	026
Harrison, M.J.	064	Hayashi, T.	030
Hart, D.J.	138		048
Hart, H.	280		050
	464		091
Harter, W.G.	149		108
Hartung Jr., J.B.	154		175
Harusawa, S.	262		248
Haruta, J.	304		326
Haruta, J.	453		367
Harvey, R.G.	075		376
	404	Hayashida, H.	045
Harvey, S.M.	033		379
Hasebe, M.	206	Hayashizaki, K.	376
Hashimoto, H.	097	Hayata, T.	112
	214		250
Hashimoto, M.	055	Hazarika, M.J.	101
Hashimoto, S.	135	He, P.	467
	277	He, Y.-B.	004
	324	Heathcock, C.H.	125
Hashimoto, T.	229		425
	436	Hebeisen, P.	266
Hashmi, S.A.N.	014	Hebel, D.	415
Hata, E.	301	Heck, J.V.	476
Hatajima, T.	046	Hegedus, L.S.	154
	117	Heintz, M.	011
Hatanaka, Y.	083	Heintz, R.M.	008
	093	Heinze, P.L.	096
	246	Heitz, M.-P.	058
	281		209
	306		417
	469	Helle, M.A.	328
Hatayama, Y.	197	Helmchen, G.	410
Hatsuya, S.	119	Helquist, P.	122
	349		344
Hattori, A.	011		346
Hattori, K.	225	Henderson, I.	476
Hattori, T.	406	Hendi, M.S.	364
Hatyanaka, Y.	085		369
Haufe, G.	416	Hendi, S.B.	364
Haufe, G.	436		369
Hauser, F.M.	137	Hendrickson, J.B.	260
Hayakawa, H.	336	Henegar, K.E.	396
Hayakawa, K.	334	Hengge, A.	275
Hayama, T.	053	Henin, F.	099
Hayami, J.	324		411
Hayasaka, T.	101	Henningsen, M.C.	281
	111	Henry, J.R.	123
Hayashi, M.	107	Hensley, V.	169
	302	Hensley, V.	183
	412	Hernández, R.	378
	453	Herndon, J.W.	051

Herndon, J.W.	231	Ho, E.	387
Herrick, J.J.	082	Ho, K.M.	220
Herrinton, P.M.	420		223
Herrmann, R.	007	Ho, M.S.	225
Hershberger, S.	110	Ho, S.	119
Hershberger, J.	110	Hoashi, K.	078
Heslin, J.C.	423	Hodges, S.L.	451
Hesse, M.	136	Hoekstra, W.	303
	182		455
	185	Hoffer, R.K.	110
	366	Hoffman, C.	378
	445	Hoffman, R.V.	136
Heumann, A.	408		152
Hey, J.P.	106	Hoffmann, H.M.R.	346
Hidai, M.	076		422
	174	Hoffmann, P.	095
	393	Hoffmann, R.W.	321
Hiemstra, H.	442		350
	361	Hogan, K.T.	040
Higashiura, S.	407	Hohenhorst, M.	364
Higashiyama, K.	142	Hojo, M.	389
Higuchi, N.	335		404
Hiiro, T.	024		415
Hillery, P.	243	Holcomb, H.	172
Hilty, T.K.	392	Holland, H.L.	295
Hino, T.	028	Hollingsworth, D.R.	362
Hirama, M.	126	Holm, K.H.	199
	320	Holmquist, C.R.	400
Hirano, M.	236	Holt, D.A.	484
	238	Holtan, R.C.	248
Hirao, T.	400	Holton, R.A.	222
Hirashima, T.	170	Homma, K.	195
Hirata, Y.	230	Hon, M.-Y.	289
Hiratake, J.	311	Honda, T.	257
Hiroe, Y.	076	Hondrogiannis, G.	441
	174	Hong, S.	479
Hiroi, K.	065	Hong, S.Y.	008
	484	Hong, Y.	022
Hirotaq, K.	074		243
Hirst, G.C.	424	Hootelé, C.	382
Hissom Jr., B.R.	033	Hopf, H.	266
Hitomi, T.	464	Hopkins, M.H.	420
Hiyama, T.	027	Hopkins, R.B.	252
	083	Hopp, P.	111
	085	Hoppe, D.	020
	093		425
	246	Hori, K.	232
	281	Hori, Y.	287
	306		368
	324		402
	469	Horiguchi, Y.	345
Hlasta, D.J.	243		451

Horita, K.	055	Huang, X.	062
Horito, S.	214		102
Horiuchi, C.A.	318		272
	437		281
Horluchi, N.	266	Huang, Y.	431
Hormozi, N.	412		188
Hornback, J.M.	153		456
Horne, K.	105	Huang, Y.-Z.	026
Horspool, W.M.	234		124
	251		193
Hoshi, M.	263		272
	463		274
Hoshino, M.	035		344
	316		465
	343	Huang, Z.-Z.	144
Hoshino, Y.	021	Huber, D.	162
	097	Hudlicky, T.	047
Hosoi, A.	323		099
Hosoi, S.	177		120
Hosokawa, T.	068		289
	426		336
	449		448
Hosomi, A.	023	Hueso-Rodriguez, J.	368
	045	Huh, K.-T.	146
	078		151
	143	Huizinga, W.B.	449
	243	Hulce, M.	042
	249	Hull, C.	024
	379	Hull, P.	111
Hosomi, T.	377	Hung, C.W.	099
	382	Hung, S.C.	481
Hou, W.	202	Hunt, D.A.	113
Hou, Y.Q.	281	Hunter, D.H.	151
Hou, Z.	318	Hunter, R.	427
	319		430
Hoveyda, A.H.	049	Hunter, R.	433
Howard, P.N.	424	Hur, C.-U.	280
Högberg, T.	134	Hussoin, Md.S.	260
Hseine, A.	368	Husson, H-P.	149
Hseine, A.	482	Hutchins, R.O.	226
Hu, C.-J.	137		474
Hu, J.	202	Hürzeler, M.	061
Hu, L.-Y.	082	Hwang, C.-K.	336
Hu, N.X.	118		413
	143		424
	189		430
	414	Hyuga, S.	410
Hu, Y.	202		461
Hua, D.	183		
Huang, G.	297	**I**	
Huang, H.-W.	137		
Huang, P.Q.	149	Ibarra, C.A.	479

Ibata, T.	155	Imamoto, T.	046
Ibragimov, I.T.	231		114
Ichihara, A.	472		117
Ichihara, J.	383	Imanaka, T.	007
	246		260
	442		122
Ichikawa, Y.	393	Imi, K.	230
	226		428
	287	Imwinkelried, R.	369
	442	Inaba, M.	160
Igarashi, S.	481	Inada, A.	450
Igarashi, T.	245	Inagaki, T.	142
Ihle, N.C.	465	Inanaga, J.	220
Iida, M.	183		225
Iio, H.	271		299
	349		316
Iitaka, Y.	317		322
Ikariya, T.	089		330
Ike, K.	359		338
Ikeda, H.	335		047
Ikeda, I.	381	Inazu, T.	078
	419	Inenaga, M.	132
	446	Inokuchi, T.	193
Ikeda, K.	212		230
Ikeda, M.	287	Inomata, K.	061
	414		357
	429		481
Ikeda, N.	465	Inoue, H.	018
Ikeda, Y.	088	Inoue, M.	253
	465		369
Ikegami, S.	277	Inoue, S.	088
Ikemoto, Y.	024	.	335
	148	Inoue, Y.	456
	373	Inumushi, A.	302
	431	Iovel, I.G.	264
Ikenaga, K.	286	Iqbal, J.	028
Ikura, K.	434		339
Ila, H.	331		394
	387		395
	413		413
Ilijiv, D.	016	Irfune, S.	217
Ilzumi, Y.	302	.	450
Ilzumi, Y.	343	Iritani, K.	127
Imada, M.	179	Isaka, M.	047
	327	Isayama, S.	050
Imada, Y.	132		324
	154		414
	407		429
Imagire, J.S.	123	Ishida, M.	013
Imai, K.	230	Ishihara, K.	425
Imaizumi, S.	456	Ishihara, T.	338
Imaj, M.	222	Ishii, H.	213

Ishii, T.	464	Itoh, K.	097	
Ishii, Y.	013		169	
	075	Itoh, M.	248	
	076	Itoh, S.	200	
	089	Itoh, T.	041	
	170		421	
	180	Itô, S.	126	
	201	Iwabuchi, Y.	357	
	217	Iwaki, S.	187	
	235	Iwakuma, T.	126	
	317	Iwamoto, K.	419	
	393		422	
	450	Iwamoto, Y.	004	
Ishikawa, M.	086	Iwamura, H.	005	
	225		427	
Ishikawa, T.	005	Iwamura, M.	005	
	086	Iwasa, S.	177	
	097	Iwasaki, M.	076	
	466		393	
Ishizaki, K.	084	Iwasawa, N.	342	
Ishizone, H.	257		369	
Ishizuka, T.	323		371	
Islam, I.	397	Iwata, C.	122	
Isobe, Y.	074	Iwata, M.	173	
Isoe, S.	184	Iwata, S.	078	
	356	Iyer, P.S.	075	
	361	Iznaden, M.	379	
	426	Izukawa, H.	179	
Isogami, Y.	155	Izumi, T.	466	
Itabashi, K.	168	Izumi, Y.	403	
	194			
Ito, H.	024			**J**
	027			
	046	Jachiet, D	466	
	409	Jachiet, D.	353	
Ito, K.	179	Jacks, W.S.	400	
Ito, Y.	018	Jackson, J.A.	477	
	030	Jackson, R.F.W.	190	
	050	Jackson, W.R.	069	
	091		356	
	098	Jacobi, P.A.	119	
	108		457	
	175	Jacobsen, W.N.	310	
	316	Jacobson, A.E.	243	
	326		256	
	335	Jacoby, D.	375	
	367	Jadhav, P.K.	336	
	376	Jahangir	031	
	386	James, A.P.	088	
	389	James, B.R.	151	
	470	Jankowski, K.	429	
	069	Jansen, B.J.M.	354	

Jansen, J.F.G.A.	109	Julia, S.A.	482
	144	Jun, J.-G.	447
Jansson, K.	054	Jun, T.X.	229
Jaouadi, M.	014	Jung, I.B.	204
Jarglis, P.	170	Jung, K.-C.	236
Jastrebski, J.T.B.H.	358	Jung, M.E.	040
Jawalkar, D.G.	400	Jung, S.-H.	278
Jäger, V.	393	Jung, Y.-W.	160
Jäkel, E.	464	Junjappa, H.	331
Jean, M.	093		387
Jean, T.-S.	437		413
Jeffery, T.	300	Juntunen, S.K.	483
	303	Jurczak, J.	354
	305	Juršić, B.	226
Jefford, C.W.	007	Just, G.	171
	178	Just, G.	206
Jeffries, P.M.	162	Jørgensen, K.A.	186
Jeganathan, A.	182		293
	310		
Jenkins, P.R.	462	**K**	
Jenner, G.	032		
Jenny, C.-J.	078	Kaba, T.	018
Jeropoulos, S.	281	Kaba, T.	018
Jhingan, A.K.	072	Kabalka, G.W.	050
Jiang, J.	138		102
	155		157
Jin, Y.-Y.	166		162
Jinkins, P.R.	448		217
Jirovetz, L.	124		297
Joglar, J.	328		441
Joglar, J.	374	Kaczmarek, L.	159
Johnson, C.R.	046		161
	103		244
	113	Kadokura, M.	402
	282		451
Johnson, P.L.	097	Kadota, I.	084
Johnson, W.S.	339	Kadow, J.F.	113
Johnston, R.D.	056	Kagabu, S.	262
Johnstone, R.A.W.	080	Kagan, H.B.	024
Jolly , R.S.	127		046
Jones, R.J.	134		115
Jonradi, A.W.	319		235
Joseph, S.P.	028		291
	417		335
Joshi, B.V.	300		475
Joshi, N.N.	022	Kagechika, K.	360
Jousseaume, B.	159	Kagotani, M.	148
Jowett, I.C.	178	Kagotani, M.	210
Juge, S.	377	Kahn, B.E.	321
	240	Kahn, M.	360
Julia, S.A.	001	Kaji, A.	229
	098		254

Kajigaeshi, S.	171	Kant, J.	136
	216		467
	438	Kantam, M.L.	181
Kakinami, T.	216	Karabelas, K.	469
Kakinami, T.	438	Karady, S.	126
Kakushima, M.	182	Karaman, R.	234
Kalck, P.	068	Karim, M.R.	407
Kalyan, Y.B.	245	Karlsson, J.O.	460
Kamada, T.	184	Karrick, G.L.	089
	212	Kasahara, A.	466
Kamata, M.	278	Kasahara, I.	088
Kambe, N.	024	Kascheres, C.	382
	135	Kashimura, S.	048
	142		397
	197	Kashiwagi, H.	253
	227	Kashiwagi, K.	377
	419	Kaspar, J.	043
	434	Kasuda, K.	220
	457	Kataoka, Y.	021
Kameda, K.	007		245
Kametani, T.	257		271
Kameyama, M.	482		388
Kamigata, N.	169	Katayama, H.	424
	482	Kates, S.A.	455
Kamimura, A.	229	Kato, S.	013
	289	Kato, T.	461
Kamimura, J.	224	Katoh, S.	184
	227	.	212
Kamio, C.	270	Katoh, T.	241
Kamiya, Y.	046	Katritzky, A.R.	070
	117		128
Kan, T.	055		137
Kanayama, S.	068		148
	449		155
Kanazawa, T.	322		301
Kanda, N.	115		341
Kandil, A.A.	481	Katsuki, H.	196
Kaneda, K.	250	Katsuki, M.	085
Kanemasa, S.	146	Katsuki, T.	053
Kanematsu, K.	334		166
Kanemoto, S.	207	Katsumura, N.	409
	238	Katsurada, M.	122
	346	Katti, S.B.	014
	440	Katz, S.A.	116
	441		479
Kang, G.-J.	151	Katz, T.J.	313
Kang, J.	146	Katzenellenbogen, J.A.	436
Kang, K.K.	208	Kawa, M.	214
Kang, S.-K.	448	Kawada, K.	240
Kang, T.W.	359	Kawai, K.	134
Kanoh, M.	444	Kawai, M.	343
Kanschik, A.	269		403

Kawai, Y.   333
Kawajima, I.   398
                             451
Kawakita, T.   021
Kawamukai, H.   171
Kawamura, N.   091
                             464
Kawamura, S.   415
Kawano, H.   089
                             317
Kawasaki, H.   343
Kawasaki, N.   136
Kawase, M.   055
Kawase, Y.   371
Kawashima, M.   010
                             314
Kawata, M.   222
Kawazoe, Y.   052
Kätker, H.   465
Keay, B.A.   057
                             196
Keck, G.E.   096
                             334
Keefer, L.K.   181
Keinan, E.   104
                             214
                             221
                             224
Keller, T.H.   389
                             459
Kellogg, R.M.   376
Kellogg, R.P.   329
Kelly, B.J.   158
Kelner, R.   399
Kemmitt, R.D.W.   109
Kempf, D.J.   388
Kende, A.S.   238
                             266
                             298
                             459
Kennedy, K.   172
Kennedy, R.M.   222
Kenny, C.   047
                             318
Kernan, M.R.   332
Kerr, J.M.   090
Kervagoret, J.   051
Keshavamurthy, K.S.   417
Ketcha, D.M.   127
Keusenkothen, P.F.   129
                             324
Khai, B.T.   043

Khamsi, J.   145
                             152
Khan, K.M.   056
Khan, M.A.   339
Khan, S.A.   212
Khanapure, S.P.   185
Khanna, I.K.   147
Khurana, J.M.   210
Kibayashi, C.   256
Kibayashi, T.   217
Kibourn, M.R.   436
Kiely, D.E.   072
                             236
Kihlberg, J.   054
Kiji, J.   011
                             180
                             213
Kiji, S.   437
Kikukawa, K.   286
Kim, D.S.H.L.   389
Kim, D.Y.   276
                             480
Kim, H.   336
                             423
                             438
Kim, J.D.   060
Kim, J.E.   060
Kim, K.   243
                             382
Kim, K.E.   100
Kim, K.S.   146
                             204
                             239
                             248
                             342
Kim, S.   051
                             059
                             122
                             190
                             194
                             198
                             208
                             426
                             448
Kim, S.-G.   342
Kim, S.J.   194
Kim, S.S.   475
Kim, T.H.   416
Kim, Y.G.   204
Kim, Y.H.   243
                             248
                             255

Kim, Y.H.	294	Knochel, P.	001
	382		025
Kim, Y.S.	060		109
Kimobayashi, H.	088	Knochel, P.	234
Kimura, M.	427		243
Kimura, Y.	216		357
King, J.L.	268		447
King, S.A.	433	Knouzi, N.	157
King, S.M.	165	Knudsen, M.J.	449
Kinoshita, H.	481	Ko, S.Y.	316
Kinsman, R.	384	Kobayashi, H.	246
Kira, M.	028		442
	030	Kobayashi, K.	098
	042		157
	159		386
Kirk, J.M.	190		389
Kirki, O.	171		470
Kirkpatrick, E.	172	Kobayashi, M.	030
Kirschning, A.	035		135
Kirya, Y.	341		151
Kise, N.	359		450
	365	Kobayashi, N.	102
Kisfaludy, L.	131	Kobayashi, S.	021
Kishan Reddy, Ch.	210		303
Kita, Y.	304		342
	453		346
Kitahara, H.	355		412
Kitamura, M.	088		434
	329	Kobayashi, T.	307
Kitamura, T.	305		325
Kitaoka, M.	290		335
Kitazume, T.	338		338
Kitteringham, J.	045		481
Kiyoi, T.	188	Kobayashi, Y.	393
Kiyoshige, K.	043		405
	184	Koch, K.	238
Kjonaas, R.A.	110	Koch, S.S.C.	185
Klade, C.A.	203	Kochhar, K.S.	125
Klan, M.A.	413	Kochi, J.K.	216
Klang, J.A	284	Kocieński, P.	351
Kleijn, H.	358		430
Klein, P.	332		349
Kline, D.N.	457	Kodera, Y.	377
Klingler, F.D.	172		382
Klingstedt, T.	261	Kodomari, M.	212
Klix, R.C.	355		216
Klumpp, G.W.	434	Koerner, M.	107
Klunder, A.J.H.	449	Koerwitz , F.L.	483
Klunder, J.M.	291	Koga, K.	019
Knapp, S.	363		106
	373		317
Knobeloch, J.M.	396		343

Kogen, H.	361	Koyasu, Y.	076
Kohda, K	052		174
Kohmoto, S.	177	Koz'min, A.S.	456
Kohra, S.	023	Koziara, A.	141
	045		155
	143	Kozikowski, A.P.	053
	243		327
	249		344
	379	Kozima, S.	464
Koie, K.	427	Koziski, K.A.	322
Kokil, P.B.	476		427
Kolasa, T.	309	König, J.	026
Kolb, M.	002	Krafft, M.A.	219
Komaglou, K.	210	Krafft, M.E.	133
Komiya, S.	042		139
Konar, M.M.	043		421
Kondo, T.	077		455
	180	Kraus, G.A.	084
	241		415
Kondo, Y.	052	Krause, J.G.	247
Kondoh, H.	257	Kravitz, J.I.	119
Kondorsky, A.E.	423	Kravitz, J.I.	457
Kong, F.	445	Krämer, T.	020
Konishi, H.	011		425
	180	Kreager, A.	231
	213	Kreh, R.P.	063
Konradi, A.W.	319	Kreif, A.	411
Konstantinović, S.	392		048
	412		204
Kopping, B.	111		249
Korda, A.	130	Krieger, P.E.	056
Koreeda, M.	334	Krimer, M.Z.	245
Kosaka, A.	470	Krishna, A.	292
Koser, G.F.	476	Krishna, M.V.	481
Kosiba, M.L.	332	Kristen, G.	277
Kosugi, H.	290	Kropp, M.	224
Kosugi, M.	246	Krow, G.R.	251
	435	Kruse, L.I.	181
Kosukegawa, O.	180	Ku, B.	479
Kotake, H.	481		484
Kotian, K.D.	199	Kubas, R.	474
Kotlsuki, H.	145	Kucera, D.J.	205
Kotnis, A.S.	444	Kudav, N.A.	296
Kotsuki, H.	084	Kudou, N.	466
	393	Kudzma, L.V.	087
Koumoto, N.	257	Kulinkovich, O.G.	437
Kovacs, T.	003	Kulkarni, Y.S.	232
Kowalski, C.J.	465	Kumada, M.	440
Koyama, H.	203	Kumagawa, T.	452
Koyama, M.	461	Kumajima, I.	410
Koyama, Y.	005	Kumar, A.K.	166
Koyano, H.	479	Kumar, P.	260

Kumaraswamy, G.	292	LaMaire, S.J.	263
	327	Lampe, L.	467
Kumobayashi, H.	089	LaMunyon, D.H.	327
	329	Landge, A.B.	166
Kunieda, T.	323	Landolt, R.G.	235
Kunishima, M.	231	Lane, N.T.	417
Kunita, K.	387	Lang, R.W.	312
Kuntz, B.A.	022	Lanneau, G.F.	059
Kunz, T.	176		060
Kuo, H.-L.	267	Lao, J.-C.	183
Kurata, K.	229	Larcheveque, M.	039
Kurihara, T.	262		391
Kurihara, Y.	484	Lardicci, L.	080
Kuroboshi, M.	338		106
Kurth, M.J.	385	Lardicci, R.	004
Kurusu, Y.	025	Larock, R.C.	077
	026		095
	029		097
	401		147
Kutney, J.P.	151		192
Kuwahima, I.	447		234
Kuwajima, I.	047		351
	083		370
	105		403
	179		407
	241		433
	345		459
Kuwata, F.	397		471
Kuzuhara, H.	173	Larpent, C.	090
	308	Larson, G.L.	106
Kwart, L.D.	336		351
Kwast, E.	129		469
Kwiatkowski, S.	310	Larson, S.	441
Kwon, S.S.	065	Laswell, W.L.	151
Kyler, K.S.	055	Laszlo, P.	292
	298	Lathbury, D.	384
Kyung, S.-H.	074	Lathbury, D.C.	150
		Lau, C.K.	228
L		Laucher, D.	060
			076
Labadie, S.S.	484	Laumen, K.E.	332
Labar, D.	204	Laurent, A.	436
Labiad, B.	435	Lautens, M.	079
Laborde, E.	094	LaVaute, T.	042
	402	Lavigne, A.	259
Laboue, B.	234		261
Laboureur , J.L.	249	Law, K.-W.	481
	204	Leathers, T.	172
Lakin, D.	356	LeBlanc, C.	370
Lal, G.S.	465		411
Lal, S.G.	294	Lebold, S.A.	014
Lam, C.H.	220	Lecea, B.	122

Leclerc, G.	162	Leone-Bay, A.	218
Lecolier, S.	358	León, E.I.	378
LeCorre, M.	211	LePetit, J.	100
	390	Leport, L.	044
Lectka, T.	285	LeRoux, J.	390
LeDeit, H.	211	Lesage, M.	225
Leduc, P.	037		226
Lee, C.	117	Leuck, D.J.	403
Lee, D.C.	466		407
	472	Leung-Toung, R.	187
Lee, E.	280	Leutenegger, U.	102
	359	Levesque, G.	180
Lee, G.M.	176	Levorse, A.T.	363
Lee, H.K.	255		373
Lee, H.Y.	248	Lewis, E.J.	118
Lee, J.C.	060	Ley, S.V.	190
Lee, J.G.	208		197
	215		363
	438		401
Lee, J.W.	484		426
Lee, K.	254	Léonard, E.	320
	297	Lê, T	204
Lee, K.W.	060	Lhommet, G.	158
Lee, L.C.	441		375
Lee, M.C.	119	Li, C.-S.	367
Lee, P.H.	122		371
Lee, S.W.	040		214
Lee, T.A.	122	Li, J.	393
Lee, T.V.	277	Li, J.S.	415
	413		418
	422	Li, Q.-L.	070
Lee, W.-K.	129	Li, S.-W.	272
Lee, W.J.	096		274
Lee, Y.-S.	106	Li, W.-S.	063
Lee. K.	276	Li, Z.	104
Leenay, T.L.	438	Liang, Y.	157
LeGoffic, F.	012	Lichtenthaler, F.W.	170
	062		172
	273	Licini, G.	294
Lehnert, E.K.	086	Lidor, R.	212
Leipert, T.K.	470	Liebeskind, L.S.	353
Leismann, H.	032		452
Leitner, W.	100	Liguori, A.	261
Lellouche, I.	041	Lin, C.-C.	110
Lemaire, M.	156	Lin, L.C.	462
	326	Lin, R.	158
Lemor, A.	156		202
Lenarda, M.	043		456
Lenior, D.	275	Lin, S.	188
	285	Lin, X.	460
	321		455
Lenz, G.R.	128	Lincandro, E.	192

Lind, J.	131	Lou, W.-X.	069
Linde II, R.G.	333	Louie, M.S.	098
Linderman, R.J.	105	Loupy, A.	215
	172		239
	237	Lozanova, A.V.	454
	421	Lu, X.	195
Lindsell, W.E.	230		217
Lindstrom Jr., P.A.	070		237
Linstrumelle, G.	304		357
Linz, G.	410		408
Liotta, D.	303		431
	455		455
Liotta, L.J.	191		456
Lipshutz, B.	109		458
Lipshutz, B.H.	025		471
	035	Lu, Y.	234
	057	Luche, J.-L.	037
	086		099
	107		108
	112		200
	113		242
Lister, S.	196		320
Little, R.D.	110	Lue, P.	128
Liu, H.-J.	032	Luh, T.-Y.	087
	064		220
	214		223
	238		256
	256		272
	416		282
Liu, K.-T.	137	Luidhardt, T.	441
Liu, L.K.	063	Luke, G.P.	298
Liu, T.	318	Lukevics, E.	264
	456	Luly, J.R.	378
Liu, Z.	208	Lun, K.S.	225
Livinghouse, T.	127	Luna, H.	336
	380	Lundquist, J.T.	063
	383	Lunetta, S.E.	118
Llebaria, A.	399	Lunn, G.	181
Lloyd-Williams, P.	402		372
Lodaya, J.S.	476	Luo, F.-T.	083
Lodi, L.	020	Luo, W.	032
Loft, M.S.	035	Lupo Jr., A.J.	345
Loh, J.-P.	450	Lutomski, K.A.	060
Lohray, B.B.	031	Luz, M.C.	312
	160	Luzzio, F.A.	241
Lolkema, L.D.M.	442	Ly, A.M.	467
Lonikar, M.S.	105	Lyle, R.E.	415
Lorenz, K.	170	Lynch, J.K.	315
Lorenz, K.T.	288		
Lossener, K.	087	**M**	
Lou, J.-D.	061		
	069	Ma, D.	237

Ma, D.	357	Malinowski, M.	159
	408		161
	455		244
	458	Mallart, S.	377
	471	Mallick, I.M.	249
Ma, S.	202	Malpass, D.B.	284
	217	Maluleka, S.L.	314
Ma, Y.	310	Manage, A.C.	112
Mabone, M.	376	Mancuso, V.	382
Macdonald, S.J.F.	416	Mandai, T.	170
Macdonald, T.L.	028	Mandai, T.	222
	086	Mandal, A.K.	400
Machida, N.	269	Mandolini, L.	038
Machinaga, N.	256	Manfredini, S.	342
Mader, M.	122	Mangeney, P.	105
Madesclaire, M.	205		337
Madin, A.	102	.	352
Maeda, K.	440		373
	170	Mani, R.S.	129
Maeno, K.	438	Mann, A.	164
Maercker, A.	279		432
Maestracci, M.	012	Manna, S.	273
Maeta, H.	198	Mano, T.	201
Magnin, D.R.	094	Mansuy, D.	037
Magnus, P.	451	Manta, E.	416
Magnusson, G.	054	Mantlo, N.B.	075
Mah, R.	130	Manzocchi, A.	033
	186	Mar, E.K.	034
Mahapatro, S.N.	139	Marais, D.	012
Mahmood, K.	056	Marcantoni, E.	161
Mahmoodi, N.O.	177		295
Mahon, M.F.	196		446
Mahoney, W.S.	101	Marchalin, S.	475
Maier, W.F.	072	Marchese, G.	093
Maione, A.M.	054		232
Maiorana, S.	192		234
Maiti, S.N.	051		468
	159	Marcin, L.R.	123
Majerski, Z.	251	Marcinow, Z.	221
Majewski, M.	269	Marcuccilli, C.J.	419
Mak, K.T.	268	Marder, T.B.	401
Makabe, Y.	387	Marek, I.	352
Makamura, E.	047	Marfat, A.	131
Makhova, I.V.	184	Mariano, P.S.	145
	252		387
Maki, Y.	074	Marinas, J.M.	271
Makino, Y.	103	.	453
Makita, A.	175	Marinelli, F.	388
Malacria, M.	483		398
Malamas, M.S.	345		470
Malanga, C.	080	Marino, J.P.	094
Malhotra, R.	473		402

Markiewicz, W.T.	054
Markov, V.I.	144
Marman, T.H.	376
Marquardt, D.J.	153
Marquet, J.	445
Marren, T.J.	103
Marrs, P.S.	093
Marsella, J.A.	325
Marsh, B.K.	288
Marshall, J.A.	028
	036
	299
	350
Marston, C.R.	056
Martelli, G.	137
Marth, C.	271
Martin, C.	313
Martin, J.	313
Martin, J.-A.F.	251
Martin, J.D.	416
Martin, J.F.	372
Martin, S.F.	151
Martinelli, M.J.	457
Martinez, A.G.	208
Martinez, J.	014
Martín Tesorero, J.	479
Martín, V.S.	311
Martínez-Alcazar, P.	251
Martínez-Gallo, J.M.	440
Maruoka, K.	021
	030
	049
	065
	106
	352
	392
	421
Maruyama, K.	414
Marwah, A.K.	167
Marwah, P.	167
Marx, J.N.	063
Maryanoff, B.E.	275
	285
Marzabadi, M.R.	362
Masai, M.	460
Masamune, H.	316
Masamune, S.	397
Mascolo, G.	468
Masnovi, J.	231
Mason, T.J.	017
Massebiau, M.-C.	044
Massiot, G.	067

Massy, D.J.R.	205
Masuda, Y.	263
Masui, Y.	131
Masuyama, Y.	025
	026
	029
	401
Mathvink, R.J.	250
Matsubara, S.	127
	238
	306
	427
	441
	464
Matsuda, H.	017
	043
	179
	327
Matsuda, K.	146
Matsuda, S.	304
Matsuda, T.	286
Matsuda, Y.	320
Matsui, K.	306
Matsui, S.	303
	377
Matsukawa, M.	322
Matsuki, T.	118
Matsumoto, K.	135
	155
	324
	424
Matsumoto, T.	085
	198
Matsumoto, Y.	050
Matsumura, Y.	108
	175
	184
	198
	212
	322
Matsunaga, S.	184
	442
Matsuoka, H.	245
Matsuoka, M.	175
Matsuoka, R.	451
Matsuoka, Y.	379
Matsuura, T.	201
Matsuyama, H.	169
	450
Matsuzaka, H.	076
	174
	393

Matsuzawa, K.   194
Matsuzawa, S.   047
Matt, D.   187
Matteson, D.S.   051
Matthews, D.P.   385
Maumy, M.   259
Maumy, M.   261
Maycock, C.D.   236
Mayr, H.   063
                         186
Mazdiyasni, H.   045
Mazzieri, M.R.   279
Márquez, C.   064
                         244
McAdam, D.P.   209
McCallum, J.S.   353
McCarthy, J.R.   385
McCauley Jr., J.P.   226
McCauley, M.D.   227
McCombie, S.W.   334
McCullough, D.W.   278
                         454
McDonald, C.   172
McDonald, C.E.   169
McFarlane, K.L   399
McGarvey, G.J.   351
McGill, J.M.   379
Mcharek, S.   029
Mchich, M.   053
McIntosh, M.C.   123
McKean, D.R.   279
McKenna, E.G.   272
McKenzie, J.R.   172
McKenzie, T.C.   416
McKenzie, T.G.   449
McKillop, A.   061
McLean, W.N.   080
McMahon Jr., W.A.   415
McMillan, C.M.   439
McMillen, D.   169
                         183
McMurry, J.E.   098
                         270
                         283
                         285
                         285
                         319
McNab, H.   380
Mead, K.   028
Mead, K.T.   299
Mechoulam, R.   211

Mederer, K.   155
Medich, J.R.   046
Medina, J.C.   055
Mehta, G.   418
Mehta, S.   210
Meier, G.P.   197
Meier, I.K.   268
Meier, M.S.   280
Meijs, G.F.   224
                         283
                         414
Mellor, J.M.   332
Melnick, M.J.   124
Menendez, M.   137
Menezes, R.   377
Menezes, R.F.   370
Menicagl, R.   113
Menicagli, R.   080
                         096
                         254
Mercier, J.   313
Mereyala, H.B.   168
Merényl, R.   461
Merino, I.   388
Merkushev, E.B.   218
Merritt, J.E.   455
Metz, W.A.   334
Metzner, P.   395
Mewshaw, R.E.   439
Meyers, A.I.   060
                         076
                         115
                         327
                         371
                         455
Micetich, R.G.   051
Michael, J.P.   427
Michael, M.   011
Michaelis, R.   404
Michell, M.B.   046
Michelot, D.   300
Middleton, D.S.   365
                         422
                         435
Midland, M.M.   023
                         041
Miftakhov, M.S.   468
                         471
Miginiac, L.   163
                         357
                         385
                         468

Miginiac, P.	330	Misra, R.A.	240
Miginiac, Ph.	045	Mitani, M.	441
Migita, T.	246	Mitchell, D.	452
	435	Mitchell, J.C.	215
Mignani, G.	217	Mitsuda, N.	228
Mihailović, M.L.	392	Mitsuda, Y.	163
	412	Mitsudo, T.	287
Mihoubi, M.N.	406		368
Mikaelian, G.S.	231		402
Mikhail, G.K.	418		451
Milenkov, B.	182	Mitsue, Y.	359
Miles, W.H.	173		381
Millar, R.W.	485	Miura, K.	176
Miller, A.	439		226
Miller, J.A.	339	Miura, M.	011
Miller, J.F.	332		097
Miller, M.J.	309		161
	378		169
Miller, R.D.	396		450
Miller, S.R.	059	Miwa, Y.	358
	288	Miyai, T.	331
Miller, T.A.	057	Miyajima, Y.	435
Mills, L.S.	061	Miyakoshi, T.	400
Mills, S.G.	356	Miyama, N.	097
Milowsky, A.S.	276	Miyamoto, O.	335
Milstein, D.	180	Miyamoto, S.	170
	401	Miyane, T.	188
Mimura, S.	405	Miyao, Y.	459
Mimura, T.	302	Miyasaka, T.	336
Minami, I.	245	Miyashi, T.	278
	272	Miyashita, M.	035
	405		343
	461		420
Minami, K.	414	Miyata, K.	219
Minato, A.	082	Miyaura, N.	085
Minato, M.	025		266
	252		466
Minato, M.	405		472
Minomura, M.	245	Miyazaki, Y.	277
Minowa, N.	019	Miyazawa, M.	341
Mioskowski, C.	056	Miyazawa, Y.	450
	058	Miyoshi, N.	024
	079		142
	209		197
	273	Miyura, N.	086
	322	Mizobuchi, T.	271
	417	Mizumo, K.	287
Mishra, P.	420	Mizuno, M.	272
Misiti, D.	398	Mizuno, T.	170
Misner, J.W.	147	Moberg, C.	408
Misra, P.K.	014	Mochizuki, K.	406
		Mochizuki, M.	066

Modena, G.   237
                            294
Moeller, K.D.   362
Moerlein, S.M.   218
Moëns, L.   110
Mohan, L.   277
Mohr, P.   347
Moimas, F.   091
Moiriarty, R.M.   006
Moise, C.   270
Moiseenkov, A.M.   237
                            454
Molander, G.A.   047
                            114
                            117
                            318
                            329
                            348
                            425
                            428
Molinari, A.   441
Molloy, K.C.   196
Moloney, M.G.   095
Monahan III, R.   303
Monahan, L.C.   380
Montanari, F.   061
                            240
Monteiro, H.J.   479
Montevecchi, P.C.   256
Montury, M.   326
Moody, C.J.   396
                            423
Mooiweer, H.H.   361
                            442
Mook Jr., R.   264
Moore, H.W.   460
Moore, M.   439
Moorman, A.E.   292
Mordini, A.   023
                            444
Moreau, J.J.E.   105
Morel, D.   217
Morena, E.   360
Moreno-Mañas, M.   445
Morera, E.   093
Moretó, J.M.   399
Moretti, R.   107
                            250
Mori, E.   136
Mori, M.   115
                            136
                            360

Mori, M.   400
Mori, W.   257
Mori, Y.   048
Moriarty, R.M.   006
                            090
                            132
                            340
                            418
                            420
                            439
                            444
Morikawa, S.   165
Morimoto, K.   007
Morimoto, T.   039
                            090
                            236
                            238
                            331
Morisaki, K.   470
Morita, N.   355
Morita, Y.   069
Moriwake, T.   160
Moriwaki, M.   216
Moriwaki, M.   438
Morris Jr., P.E.   072
                            236
Morris, A.D.   478
Morris, D.   169
                            183
Morris, T.H.   199
Mortlock, S.V.   024
Morton, H.E.   057
Moskal, J.   067
Motherwell, W.B.   478
Mottaghineiad, E.   239
                            244
Mouloud, H.A.H.   215
Moussa, A.   269
Moustrou, C.   391
Moyano, A.   302
                            429
Mozumi, M.   362
Mucha, B.   346
Muchow, G.   020
Mudurro, J.M.   083
Muedas, C.A.   446
Muira, K.   406
Mukai, C.   304
Mukaiyama, T.   019
                            050
                            107
                            302

Mukaiyama, T.	303	Murray, R.W.	277
	324	Murugaverl, B.	153
	334	Muruyama, K.	278
	342	Musgrage, B.B.	327
	346	Musllam, H.A.	418
	347	Muzart, J.	099
	377	Muzart, J.	247
	412		303
	434	Müller, P.	151
	453	Müller, U.	404
	195	Myers, A.G.	070
Mukkanti, K.	162	Mysorekar, S.V.	269
Mullins, M.J.	197	Młochowski, J.	259
Munakata, M.	135		
Munawar, M.A.	212	N	
Munyemana, F.	209		
Murahashi, S.	068	Nag, A.	100
	132	Naga, S.	159
	154	Nagahama, N.	408
	154	Nagahara, S.	065
	170	Nagai, K.	088
	272	Nagakawa, K.	200
	292	Nagami, K.	304
	359	Nagao, Y.	132
	377		187
	381		232
	382	Nagata, R.	201
	383	Nagata, T.	196
	387	Nagendrappa, G.	235
	407	Nagumo, S.	183
	426	Nahmed, E.M.	032
	449	Naito, H.	463
Murahashi, S.I.	131	Najafi, M.R.	432
Murai, A.	195	Nakagawa, Y.	335
Murai, S.	024	Nakahara, T.	388
	197	Nakahara, Y.	436
	419		441
Murai, T.	013	Nakai, E.	034
	197	Nakai, T.	034
Murakami, M.	347	Nakajan, S.	426
Murata, S.	161	Nakajima, H.	160
	412	Nakajima, M.	019
Murata, T.	356		317
Murayama, E.	114	Nakajima, N.	055
	245	Nakajima, S.	134
	249		164
Murayama, T.	224	Nakajo, E.	316
Mure, M.	200	Nakamura, E.	083
Murphy, M.	381		094
Murphy, P.J.	316		179
Murray, P.W.	293		345
Murray, R.W.	153		398

Nakamura, E.   447
Nakamura, K.   046
                      331
                      333
Nakamura, T.   027
                      169
                      233
Nakanishi, A.   182
                      183
Nakanishi, H.   435
Nakanishi, T.   066
Nakano, K.   460
Nakano, M.   146
Nakano, T.   075
                      450
Nakashima, M.   369
Nakata, T.   405
Nakatani, H.   414
                      429
Nakatani, K.   398
Nakayama, J.   316
Nam, D.   362
                      376
Namy, J.L.   235
Naota, T.   170
                      272
Narasaka, K.   369
                      371
Narayana, C.   153
Narayanan, B.A.   277
Narisano, E.   375
Naruse, Y.   055
Naruto, S.   133
Nash, J.J.   281
Naso, F.   093
                      234
                      468
Natalie Jr., K.J.   369
Natchus, M.   448
Natchus, M.G.   099
Natesh, A.   036
Nativi, C.   462
Naujoks, E.   266
Nazer, B.   059
Nájera, C.   440
                      467
Neeson, S.J.   072
Negishi, E.   003
                      059
                      082
                      092
                      094

Negishi, E.   265
                      288
                      289
                      457
                      466
                      472
Negoro, T.   088
Negre, M.   326
Nelson, D.J.   299
Neumann, W.P.   227
Newbold, R.C   459
                      266
Newcomb, M.   153
Newkome, G.R.   391
Nezhat, L.   112
Nezu, J.   228
Nédélec, J.-Y.   215
Ng, C.T.   223
Ng, F.W.   169
Ng, J..   109
Ng, J.S.   107
Ngoi, T.K.J.   478
Ngooi, T.K.   167
Nguyen, D.   109
Ni, Z.   195
Ni, Z.-J.   087
                      256
                      272
Ni, Z.-T.   282
Nicholas, K.M.   104
                      457
Nicolaou, K.C.   336
                      413
                      424
                      430
Nigam, S.C.   044
                      164
Nihira, T.   175
Niibo, Y.   056
Niibo, Y.   447
Nikifurov, A.   124
Nikishin, E.I.   184
Nikishin, G.I.   252
                      423
Nikolaides, N.   174
Nikolic, N.A.   417
Nil, K.   200
Nisar, M.   245
                      405
Nishi, K.   304
Nishi, M.   018
Nishi, T.   361

Nishida, A.	045	Novak, L.	003
Nishida, N.	381	Novi, M.	077
	419	Nowakowska, B.	054
Nishiguchi, I.	170	Noyori, R.	088
Nishiguchi, T.	066		329
	240	Nozaki, H.	056
	269		238
	270		346
Nishimura, A.	342		431
Nishimura, K.	132		432
Nishimura, M.	159		447
Nishio, T.	016	Nozaki, K.	221
Nishioka, T.	311		226
Nishioka, Y.	290		287
Nishiumi, W.	011		343
	180		407
Nishiyama, E.	330	Nozawa, K.	134
Nishiyama, S.	460	Nozoe, S.	323
Nishiyama, Y.	103	Nudelman, A.	399
	227	Nugent, R.A.	381
Nishiyama, Y.	419	Nugent, W.A.	332
Nishizawa, H.	393		462
Nitta, K.	026	Nugiel, D.A.	424
Niwa, M.	232	Nunn, D.S.	398
Niwa, S.	018	Nuñez, M.TY.	311
	325	Nurmi, T.T.	093
Nobel, D.	187	Nutaitis, C.F.	148
Node, M.	177	Nuzillard, J.-M.	067
	341	Nyangullu, J.M.	064
Noe, R.	169		238
	183		416
Nokami, J.	170	Nye, S.A.	074
	222		
	342	**O**	
Nomoto, T.	463		
Nomura, M.	011	O'Bryan, E.	183
	052	O'Connor, B.	094
	097		171
	161		466
	169	O'Connor, K.J.	029
	450	O'Connor, S.	127
Nonishita, K.	106		385
Nonoshita, K.	065	O'Dell, D.E.	028
	352	O'Donnell, M.J. (Ed.),	310
Norman, M.H.	125	O'Donnell, M.J.	310
Normant, J.F.	105	O'Neal, S.	087
	268	O'Neil, I.A.	360
	337	O'Reilly, N.J.	338
	352	O'Shea, D.M.	478
	373	Oae, S.	204
North, M.	374	Obara, Y.	246
Nouguier, R.	053	Obayashi, M.	027

Obrecht, D.	189	Oh, D.Y.	297
Ochi, M.	084		475
	145		479
	196		480
Ochiai, H.	027		484
	175		484
	233	Oh, S.Y.	060
Ochiai, M.	132		060
	187	Oh-E., T.	472
	232	Ohashi, Y.	408
Oda, J.	311	Ohba, M.	345
	311	Ohe, K.	249
Oda, K.	404		405
Oda, T.	131	Ohfune, Y.	367
Oell;Erba,C.	077	Ohgo, Y.	239
Officer, D.L.	477	Ohi, S.	359
Ofosu-Asante, K.	089		365
Ogaki, M.	322	Ohki, H.	026
Ogasawara, K.	357		349
Ogawa, A.	024	Ohkuma, T.	329
	103		384
	135	Ohlmeyer, M.J.	050
	142		319
	227		371
	419	Ohno, A.	331
	434		333
	457	Ohno, M.	021
Ogawa, H.	170		084
Ogawa, K.	018	Ohno, R.	302
Ogawa, M.	013	Ohsawa, T.	228
	170	Ohshiro, Y.	200
	201	Ohta, A.	066
	217	Ohta, H.	102
	235	Ohta, T.	068
	256		088
	450		323
Ogawa, S.	461		449
Ogawa, T.	215	Ohtani, T.	481
	290	Ohuchi, K.	456
Ogawa,M.	075	Oiarbide, M.	099
Ogima, M.	461		253
Ognyanov, V.I.	366	Oikoua, K.	192
	445	Oishi, T.	228
Oguchi, T.	013		320
Ogura, F.	118		405
	143	Ojima, I.	091
	189		130
	200		354
	414	Oka, S.	157
Ogura, H.	167		331
Oh, D.Y.	254		333
	276	Okahara, M.	381

Okahara, M.	419	Ono, N.	254
	446		289
Okamoto, S.	122	Onoda, Y.	249
Okamoto, T.	135	Onomura, O.	198
	155		322
	157	Oohara, T.	150
	324		200
	438		478
Okamoto, Y.	202	Ooi, T.	065
Okano, T.	011	Ooka, Y.	426
	180	Ookawa, A.	018
.	213	Oon, S.-M.	139
Okawara, M.	379	Opitz, G.	289
Okazaki, H.	432	Oppolzer, W.	250
Okazaki, R.	138		288
Okazoe, T.	245		368
	270		389
	271		459
	429		468
Okimoto, M.	172		471
	259		472
Okinaga, T.	317	Orfanopoulos, M.	220
Oku, A.	225	Orsini, F.	179
	320	Ortar, G.	093
Okuda, F.	135		360
	151	Ortiz, M.J.	234
Okuhara, T.	446		372
Okukado, N.	092	Ortíz, B.	241
Okumoto, H.	193	Osakada, K.	194
Okura, K.	011	Oshima, K.	082
Okura, S.	431		176
Olah, G.A.	074		221
	075		226
	077		238
	184		287
	199		343
	235		346
	291		406
	436		407
Olah, J.A.	077		429
Olano, B.	244		434
	374		442
Oliva, A.	441	Oshiro, Y.	400
Oliver, T.F.	181	Osowska-Pacewicka, K	155
Olsen, R.J.	032		306
Onaka, M.	302	Osterhout, M.H.	379
	343	Osuka, M.	200
	403	Osumi, K.	193
Onan, K.D.	139	Ota, T.	335
Ono, A.	224	Otaka, K.	078
	227	Otake, K.	026
Ono, N.	229		029

Otani, S. 324
Otera, J. 056
431
447
432
Otimoto, K. 176
Otsubo, T. 143
189
200
414
118
Otsuji, Y. 287
Ott, J. 071
Ötvös Jr., L. 131
Ouchabane, R. 152
Ouimet, N. 425
Oumar-Mahamat, H. 391
Ousset, J.B. 322
Out, G.J.J. 434
Ovaska, T.V. 470
Overman, L.E. 205
287
386
420
424
Owen, T.C. 191
Ozaki, K. 102
Ozawa, K. 335
Ozbalik, N. 080
149

P

Pabon Jr., R.A. 288
Pac, C. 143
Pachinger, W. 471
Padmaja, A. 116
Padmanabhan, S.I. 290
Padwa, A. 374
457
Page, P.C.B. 257
300
Pagni, R.M. 217
441
Pak, C.S. 040
Palacios, F. 386
388
Pale, P. 32
Pale-Grosdemange, C. 123
Paley, R.S. 094
Palit, S.K. 100
Palkowitz, A.D. 020

Palkowitz, A.D. 348
Palmer, M.A.J. 348
Palmieri, G. 162
Palomo, C. 099
122
253
Palumbo, G. 255
Panaioli, S. 038
Pandey, G. 163
292
327
Pando, C. 215
Panek, J.S. 104
Pansegrau, P.D. 076
Panunzio, M. 137
Papadopoulos, S. 084
Papagni, A. 192
Paquette, L.A. 073
228
288
458
Parab, V.L. 130
Paradisi, M.P. 294
Park, G. 391
Park, J. 382
Park, J.H. 051
059
198
208
426
Park, P. 344
Park, S.B. 100
Park, W.S. 039
040
Parker, D.T. 145
Parker, K.A. 322
427
Parrinello, G. 068
279
Parsons, P.J. 150
435
Parvez, M. 176
Passacantilli, P. 032
222
Pat-saev, A.K. 423
Patel, M. 121
Patel, M.N. 130
Patel, S.V. 082
Paterson, I. 345
Patil, S. 460
Patin, H. 090
Patravale, B.S. 130

Patricia, J.J.	093	Pete, J.-P.	092
	119		099
Paul, N.C.	485		367
Pearson, M.M.	177	Petit, A.	176
Pearson, W.H.	149	Petit, Y.	391
	255	Petrier, C.	099
Pecunioso, A.	096	Petrillo, G.	077
	113	Petrini, M.	107
	254		161
Pedersen, S.F.	154		255
	319		295
	325		446
Pei, H.-J.	480	Petty, C.M.	191
Pelizzoni, F.	179	Péra, M.-H.	368
Pellon, P.	036	Périchon, J.	011
Pelrri, S.T.	460		022
Pelter, A.	178		067
	242		215
	270		320
Penmasta, R.	006	Pfaltz, A.	102
	439	Pham, T.N.	210
Pennanen, S.I.	222	Phillips, B.T.	260
Pennetreau, P.	292	Pho, H.Q.	139
Pentaleri, M.	455	Piccolo, O.	012
Peperzak, R.M.	354	Picotin, G.	045
	420		330
Pera, M.-H.	482	Pienemann, T.	156
Perez, D.	104	Pierce, J.D.	103
	214	Piermatteri, A.	294
	224	Piers, E.	093
Perez-Ossorio, R.	234		261
	251	Pietroni, B.	470
	372	Pietrusiewicz, K.M.	297
Periasamy, M.	007		485
	153	Pike, P.	110
	317	Pikul, S.	369
	463	Pilarski, B.	137
Perichon, J.	029	Pilipauskas, D.R.	147
	044	Pillai, V.N.R.	131
Perisamy, M.	008	Pilli, R.A.	380
	210	Pinhas, A.R.	132
Perlmutter, P.	069	Pinhey, J.T.	095
	356	Pinnick, H.W.	125
	376		474
Perni, R.B.	014	Pinto, I.	150
Perron, F.	170	Piotrowski, A.M.	284
	414	Pirrung, M.C.	398
	426	Pittman, J.H.	210
Perrot, M.	059	Pittman, J.H.	478
Perrot, M.	060	Piva, O.	099
Perumattam, J.	457		303
Pete, J.-P.	029		

Pizzo, F.   203
Planinšek, Z.   008
Plaquevent, J.,-C.   314
Plattner, J.J.   378
Ple, G.   274
Plobeck, N.A.   483
Plummer, B.F.   137
Pock, R.   186
Podder, R.K.   116
Pogrebnoi, S.I.   245
Pohmakotr, M.   307
Poindexter, M.K.   313
Poirier, J.-M.   296
                               411
Pole, P.   053
Poli, G.   329
Polizzi, C.   004
                               106
Ponce, Y.Z.   151
Pons, D.   240
Pons, J.-M.   275
                               285
Poquet-Dhimane, A.   092
Pornet, J.   357
                               385
                               468
Portella, C.   029
                               379
Porter, J.R.   277
                               390
Porter, N.A.   094
Porter, T.M.   481
Portnoy, M.   180
Posner, B.A.   268
Posner, G.A.   113
Posner, G.H.   281
Poss, A.J.   058
Potts, K.T.   074
Pougny, J.R.   352
Pouilhès, A.   365
Poupart, M.-A.   425
Powell, G.   169
Powers,.D.B.   020
Prakash Rao, H.S.   418
Prakash, C.   057
Prakash, G.K.S.   199
Prakash, O.   340
                               418
                               444
Prasad Rao, K.R.K.   044
Prasad, C.S.N.   180
Prasad, C.V.C.   336

Prasad, C.V.C.   413
                               443
Prasad, G.   333
Prasad, K.   315
Prati, L.   089
Preston, P.N.   230
Pri-Bar, I.   010
Price, J.D.   120
                               336
Priepke, H.   352
Prieto, J.A.   469
Principe, L.M.   451
Pristach, H.A.   033
Procter, G.   316
                               330
Prodger, J.C.   257
Pruitt, J.R.   168
Pulido, F.J.   102
Pullaiah, K.C.   335
Purrington, S.T.   210
                               478
Pyne, S.G.   475

                      Q

Qian, C.   284
Qiu, A.   284
Qiu, W.   029
                               271
Quesnelle, C.A.   057
Quici, S.   061
                               240
Quimpère, M.   429

                      R

Rabideau, P.W.   089
                               221
                               276
Racok, J.S.   151
Radhakrishna, A.S.   044
Radzik, D.M.   281
Rafferty, M.A.   451
Raggon, J.W.   325
Ragoussis, N.   314
Rajadhyaksha, S.N.   277
Rajagopalan, S.   432
RajanBabu, T.V.   332
Rajeeswari, S.   134
Ram, S.   091
                               163
                               227

Ramachandran, P.V.	039	Reddy, D.B.	116
Ramaiah, P.	399	Reddy, K.R.	418
Ramakanth, S.	283	Reddy, M.V.R.	116
Raman, K.	395	Reddy, N.L.	339
Ramana, M.M.V.	296	Reddy, N.P.	181
Ramanathan, H.	126	Reddy, P.P.	259
Rambabu, M.	260	Reddy, P.S.	116
Ramesh, M.	080	Reddy, P.S.N.	259
	149	Redersen, S.F.	372
Ramondenc, Y.	274	Redmore, D.	485
Ramos Tombo, G.M.	071	Reed, J.N.	150
Ramos, A.	372	Reed, K.L.	203
Randad, R.S.	348		399
Ranu, B.C.	010	Reed, R.T.	235
	244	Reetz, M.T.	074
Rao, A.S.	399		201
Rao, B.R.	043	Reeves, H.D.	242
Rao, G.S.	167	Refouvelet, B.	482
Rao, G.S.K.	073	Reger, D.L.	410
Rao, J.M.	304	Reginato, G.	251
Rao, K.K.	267		392
Rao, M.N.	260		444
Rao, M.S.C.	073	Reißig, H.-U.	452
Rao, S.A.	317	Reich, H.J.	248
Rao, S.A.	463	Reichert, D.E.C.	430
Rapoport, H.	164	Reichlin, D.	069
	168	Reissig, H.-U.	115
Rathbone, D.L.	346		176
Rathke, M.W.	081	Reitz, A.B.	275
Rathore, R.	254		285
Ratovelomanana, V.	304	Remuson, R.	324
Raucher, S.	299	Renaldo, A.F.	233
Rauenbusch, C.	465		279
Rautenstrauch, V.	169	Renaud, P.	061
Ravard, A.	314	Repič, O.	315
Ravi, D.	168	Rettig, S.J.	230
Ray III, D.G.	476	Reuter, D.C.	035
Ray, J.E.	004		112
Ray, S.C.	116	Revis, A.	392
Rayadh, A.	357	Rey, A.W.	151
	468	Reydellet, V.	346
Raychaudmuri, S.R.	178	Reynolds, D.W.	288
Raynham, T.M.	368	Rhee, C.K.	042
Rämsby, S.	134	Rheingold, A.L.	398
Reagan, J.	458		451
Rebelledo, F.	134	Rhodes, R.A.	101
Rebiere, F.	291		147
Rebolledo, F.	300	Ricci, A.	251
Reddie, R.N.	473		303
Reddy, A.V.N.	051		392
	159		414
Reddy, C.P.	354		

Ricci, A.	417	Romanovich, A.,Ya.	237
	432	Romberger, M.L.	278
	444	Romeo, A.	054
Rice, K.C.	243	Romeo, G.	261
	256	Romero, J.R.	083
Richard, C.S.	161	Romine, J.L.	115
Richards, D.H.	485	Romines, K.R.	117
Rico, J.G.	285	Romney-Alexander, T.M.	262
	319	Ron, E.	232
Riecke, R.D.	408	Ronan, B.	475
Riego, J.M.	189	Ronzini, L.	093
Rieke, R.D.	035		232
	288		468
	321	Rosenthal, S.	045
Rieker, W.F.	076		300
Rieth, K.	289	Rosenzweig, H.S.	374
Rigaudy, J.	152	Rosini, G.	107
Rigby, J.H.	408		161
	444		446
Righi, G.	032	Roskamp, E.J.	154
Righi, G.	222		325
Rigo, B.	073		372
	308		400
Rihter, B.	231	Rossi, K.	097
Riley, D.P.	008	Rossi, L.M.	233
Rippel, H.C.	442	Rossi, R.	305
Risaliti, A.	033		433
Rise, F.	264	Rotchford, J.	150
	472	Rotello, V.M.	377
	475	Rounds, W.D.	009
Rishton, G.M.	128		012
Rivas-Enterrios, J.	312	Roush, W.R.	020
Rivera, V.	064		348
Rizzacasa, M.A.	403	Rousseau, G.	310
Roberts, B.P.	226		440
Roberts, D.W.	297	Roussis, V.	480
Robertson, J.	460	Rowlands, M.	178
Robinson, N.G.	309		242
Robyr, C.	368	Roy, S.	186
Rochin, C.	291	Rozen, S.	211
Rockell, C.J.M.	422		212
Roden, F.S.	277		415
Rodrigues, K.E.	150	Rubiera, C.	321
Rodriguez, J.	406	Ruckle, R.	145
	430	Ruholl, H.	007
Rodriguez, M.L.	416	.	312
Rodríguez, R.C.	479	Ruhter, G.	305
Rodríquez, M.A.	302	Ruiz, M.O.	208
Roekens, B.	380	Ruiz-Hitzky, E.	239
Roemmele, R.C.	164	Ruiz-Perez, C.	416
Romain, I.	112	Rulin, F.	336
Romann, A.J.	312	Ruppin, C.	358

Ruppin, C.	399
Russell, A.T.	316
	330
Russowsky, D.	380
Ruther, M.	110
Rutledge, M.C.	474
Rutledge, P.S.	394
Ruzziconi, R.	294
	420
	446
Rüeger, H.	151
Rühlmann, A.	110
Rybczynski, P.J.	398
Ryu, I.	135
	197
	434
	457

S

Saavedra, J.E.	480
Saba, A.	364
Sabol, M.R.	037
Saboureau, C.	011
	067
Saburi, M.	089
	317
Sadeghi, M.M.	141
Saednya, A.	280
Saegusa, T.	165
	188
	248
	408
Sagimoto, A.	452
Sago, N.	088
Saha, A.K.	168
Said, S.B.	259
Saiganesh, R.	225
Saiki, M.	179
Saimoto, H.	256
	346
Saito, I.	201
Saito, K.	200
	262
Saito, S.	160
Saito, Y.	214
	412
Sakae, M.	122
Sakaguchi, M.	027
Sakai, K.	183
	242
	404

Sakai, K.	459
Sakai, S.	182
	183
	199
	340
	427
Sakai, T.	219
Sakaitani, M.	367
Sakakibara, Y.	262
Sakakura, T.	063
	307
Sakashita, H.	429
Sakata, H.	105
Sakito, Y.	156
Saksena, R.K.	447
Sakuma, K.	005
Sakurai, H.	028
	030
	042
	078
	159
Salaün, J.	255
Salaün, J.	310
Salazar, J.A.	378
Saleh, S.	057
Salomon, M.	055
Salomon, M.F.	186
Salomon, R.G.	178
	186
Salvador, J.M.	136
Samarai, L.I.	121
Samizu,K.	357
Sammes, P.G.	191
Sampson, N.S.	474
Sampson, P.	407
Samuel, O.	475
San Filippo, L.J.	213
	229
Sanchez, R.	042
	123
Sanda, F.	359
	365
Sandali, C.	156
Sanderson, P.E.J.	030
Sandhoff, K.	328
Sandoval, C.	241
Sane, P.V.	295
Sanghavi, N.M.	130
Sankar Lal, G.	432
Sanner, C.	039
Sanner, M.A.	328
Sano, H.	049

Sano, H.	246
	435
Sano, T.	346
Sant, J.Y.	386
Santa, L.E.	025
Santamaria, J.	152
Santaniello, E.	033
Santelli, M.	233
	275
	285
	430
Santelli-Rouvier, C.	482
Santi, R.	080
Santiago, A.	469
Saraf, S.D.	437
Sargent, M.V.	403
Sarkar, A.	043
	100
Sarkar, D.C.	010
	244
Sarkar, R.K.	116
Sarkar, S.K.	100
Sarkar, T.K.	116
	268
	337
Sarma, J.	290
Sarma, M.R.	267
Sarmah, P.	278
	257
Sartori, G.	028
Sarussi, S.J.	175
Sasaki, K.	200
Sasaki, M.	046
	115
Sasaki, T.	084
Sasaki, Y.	363
	366
Sasaki,H.	086
Sasaoka, S.	481
Sashiwa, H.	256
Sassaman, M.B.	199
Sasson, M.	455
Sasson, Y.	008
	207
	238
Sastry, C.V.R.	167
Satapathi, T.K.	337
Satayanarayana, N.	008
Sato, F.	405
Sato, K.	030
	042
	335

Sato, K.	410
Sato, M.	085
Sato, S.	065
Sato, T	010
	112
	114
	245
	249
	250
	314
	431
	432
Sato, Y.	146
Satoh, H.	216
Satoh, J.Y.	318
Satoh, M.	086
Satoh, T.	150
	200
	248
	419
	422
	452
	478
	482
Satoh, Y.	266
	273
Satomi, H.	248
Satô, H.	080
Satyanarayana, K.	244
	409
Satyanarayana, N.	153
Saunders, D.G.	056
Sausinš, A.	389
Sauvêtre, R.	268
Savariar, S.	045
	224
Savelli, G.	203
Savignac, P.	297
	371
Savignac, Ph.	062
Savoia, D.	366
Sawada, S.	415
Sawaki, Y.	427
Sawamura, M.	175
	376
Sawyer, J.S.	086
Saxena, R.K.	204
Sayo, N.	329
Sánchez-Obregón, R.	241
Scamuzzi, B.	433
Scettri, A.	424
Schaer, B.H.	131

Schaffhausen, J.G.	197	Schwartz, J.	268
Schaller, C.	425	Schwartz, M.A.	128
Schamp, N.	142	Scilimati, A.	167
	372	Scola, P.M.	123
	383		140
	445	Scolastico, C.	329
Schauble, J.H.	436	Scott, R.	073
Schaumann, E.	035	Scott, W.	042
	118	Scott, W.J.	081
Schäfer, B.	456		098
Schäfer, H.-J.	007	Screttas, C.G.	135
	156	Scriven , E.F.V.	163
	312	Seconi, G.	303
	126	Seddighi, M.	239
	404		244
Scheigetz, J.	228	Seebach, D.	061
Schimperna, G.	138		231
	329		359
	341		373
Schinzer, D.	353		393
Schlecht, M.F.	336	Seidel, B.	393
	423	Seitz, T.	201
	438	Sekiya, K.	094
Schlessinger, R.	174	Sekiya, K.	179
Schmid, B.	071	Selim, M.R.	462
Schmidt, B.	373	Sellén, M.	079
Schmidt, K.	087	Selski, D.	169
Schmidt, S.J.	181	Selva, A.	364
Schmidt, S.P.	207	Semmelhack, M.F.	119
Schmidt, T.	433	Sen, A.	181
	454	Senanayaker, C.	408
Schmitt, R.J.	473	Senboku, H.	092
Schmolka, S.	404	Senet, J.-P.	014
Schnatter, W.F.K.	231	Sengupta, S.	341
Schneider, M.P.	332	Sennyey, G.	014
	464	Seo, W.	106
Schobert, R.	237	Seong, C.M.	097
	278	Sera, A.	424
Schofield, C.	360	Serizawa, H.	266
Schofield, C.J.	359	Serra-Zanetti, F.	064
Schore, N.E.	449	Seshamma, T.	116
Schramm, S.B.	033	Sessink, P.J.M.	449
Schreiber, S.L.	033	Set, L.	340
	458	Severnak, S.A.	087
Schultz, A.G.	247	Sha, C.-K.	437
Schultz, P.	168	Shambayati, S.	333
Schultz-van Itter, N.	223	Shanklin, M.S.	139
Schumm, J.S.	396	Shapiro, M.J.	315
Schuster, G.B.	224	Sharma, G.V.M.	160
Schwabe, R.	460	Sharma, G.V.M.	267
Schwartz, C.E.	197	Sharma, M.	267
		Sharma, M.M.	295

Sharma, S.	309	Shimizu, K.	084
Sharp, M.J.	031	Shimizu, M.	207
	076		338
	386		362
Sharpless, K.B.	083		363
	160		436
	197		440
	291		441
	316	Shimizu, S.	358
Shashidhar, M.S.	052	Shin, D.-S.	056
Shaw, A.N.	371	Shin, D.G.	447
Shaw, C.	134	Shing, T.K.M.	402
Shekhani, M.S.	056	Shinkai, I.	356
Sheldrick, G.M.	110	Shinmyozu, T.	078
Shellhammer Jr., A.J.	103	Shinoda, K.	049
Shen, T.	096		392
Shen, Y.	026	Shinoda, T.	277
	193	Shinohara, M.	450
	271	Shinohara, T.	426
	274	Shioiri, T.	004
	274		007
	344	Shioiri, T.	241
Sheng, H.	188		280
Sheppard, A.C.	297		340
	340	Shiokawa, M.	405
	381	Shiota, T.	292
Sheu, J.	370		383
Shi, L.	124	Shirahama, H.	055
	193	Shirai, H.	075
	274	Shiraishi, H.	318
	465		319
Shi, L.-L.	272	Shirakawa, E.	376
Shi, L.-L.	274	Shirasaka, T.	021
Shibasaki, M.	136		421
	164	Shono, T.	048
	360		184
Shibata, I.	017		198
	043		212
Shibutani, T.	373		322
Shigemasa, Y.	256		359
Shih, J.G.	436		365
Shih, M.-H.	137		397
Shim, S.B.	243	Shook, D.A.	381
Shima, K.	109	Short, R.P.	397
	143	Shoup, T.M.	015
Shimagaki, M.	405		050
	034		231
	066	Shridhar, D.R.	167
	239	Shubert, D.C.	428
Shimezawa, H.	482	Shymanska, M.V.	264
Shimizu, I.	213	Sibille, S.	022
	245		029

Sibille, S. 044
067
Siddiq, M. 212
Siddiqui, M.A. 142
Sidler, D.R. 345
398
452
Siegel, C. 345
Siegel, M.G. 085
Sih, C.J. 167
169
Silverman, I.R. 339
Silverstein, R.M. 312
Simmons, D.P. 069
Simon, C.D. 427
430
Simon, E.S. 321
Simoni, D. 342
Simonyan, S.O. 231
Simpkins, N.S. 197
282
365
422
435
473
Simpson, J.H. 305
Sinai-Zingde, G. 099
Sinari-Zingde, G. 448
Sindona, G. 261
Singaram, B. 325
Singh, B.B. 044
Singh, G. 413
Singh, J. 379
Singh, M. 153
240
293
Singh, M.P. 051
Singh, R. 206
Singh, S.P. 204
Sinhababu, A.K. 055
Sinisterra, J.V. 271
453
Sisko, J. 124
Sjogren, E.B. 337
Skarzewski, J. 259
Skattebøl, L. 199
Sket, B. 439
Skuy, D. 069
Slagt, M. 434
Slater, M.J. 451
Slattery, D.K. 011
Sledeski, A.W. 178

Slessor, K.N. 481
Slough, G.A. 398
451
452
Slougui, N. 440
Smaardijk, Ab.A. 023
Smallridge, A.J. 356
Smiley, P.M. 173
Smit, W.A. 231
245
454
Smith III, A.B. 108
226
345
438
Smith, C.A. 238
298
Smith, E.H. 199
282
Smith, K. 242
Smith, M.B. 129
157
324
360
368
370
377
Smoniu, I. 220
Smyth, M.S. 058
Snider, B.B. 119
232
253
286
287
455
Snieckus, V. 031
076
142
Soai, K. 101
111
325
018
Soares, C.J. 385
Sock, O. 011
Soderquist, J.A. 283
Soderquist, J.L. 378
Soejima, T. 048
Sohn, H.-K. 448
Solarz, T.L. 203
Somers, P.K. 336
Somerville, R.F. 394
Someswara Rao, C. 260

Somfai, P.	122
Song, H.	095
	192
Song, Y.-H.	342
Songster, M.	137
Sonoda, N.	024
	103
	135
	142
	197
	227
	419
	434
	457
Sonoda, T.	246
	442
South, M.S.	386
Sowell, G.	110
Spagnoli, P.	355
Spagnolo, P.	159
Sparks, M.A.	104
Speckamp, W.N.	361
	442
Spencer, H.K.	087
Spero, D.M.	427
Spevak, P.	051
	159
Spicer, L.D.	163
	227
Spielberger, C.	169
	183
Spinazzé, P.G.	057
Spirikhin, L.V.	471
Spivey, A.C.	360
Spletzer, E.	097
Spotnitz, R.M.	063
Spreafico, F.	012
Springer, J.P.	174
Srebnik, M.	022
	049
	211
	254
Sridar, V.	085
	095
Srikrishna, A.	188
	335
	403
Srinivas Rao, C.	300
Srinivasan, P.S.	260
Srivastava, P.R.	394
Srivastava, R.G.	008
Srivastava, R.R.	413

Stabchansky, S.	127
Stafford, J.J.	283
Stang, P.J.	166
	305
Stankiewicz, J.	108
Stanoeva, E.	142
Stauch, H.	136
Stauch, H.	185
Stavber, S.	008
	217
Steckhan, E.	223
Steele, B.R.	135
Steigerwald, M.	119
Stevenson, P.	073
Stevenson, P.J.	072
Stewart, J.D.	125
	474
Stewart, L.J.	217
Stévenart-Demesmaeker, N.	461
Stichter,H.	434
Stick, R.V.	209
Still, W.C.	202
Stille, J.K.	034
	068
	081
	279
	305
	353
	401
	453
	462
	470
Stinn, D.E.	095
Stock, L.M.	089
Stolte, M.	169
Stolz-Dunn, S.K.	351
	459
Stone, C.	389
	459
Stoner, E.J.	203
Stork, G.	070
	130
	186
	264
	398
	442
Stotle, M.	183
Strickland, D.	150
Strologo, S.	080
Ström, P.	134
Stryker, J.M.	101
	101

Studabaker, W.B.	118	Suzukamo, G.	156
Stults, J.S.	365	Suzuki, A.	085
Sturgess, M.A.	455		086
Su, W.	397		097
Suárez, A.R.	279		266
Suárez, E.	130		273
	190		410
	378		461
Subba Rao, Y.V.	162		466
Subramanian, P.K.	308		472
Suchismita	226	Suzuki, H.	022
Suckling, C.J.	090		119
Sudha Rani, K.	163		215
Suemitsu, R.	246		216
Suemune, H.	183		290
	404		406
	459		457
Sugahara, T.	131		463
Sugai, K.	405	Suzuki, I.	129
Sugano, Y.	133	Suzuki, K.	034
Sugihara, T.	357		082
Sugimori, J.	369		085
	371		194
Sugimoto, A.	422		198
Sugimura, H.	193		239
⌐	092		317
	152		341
Sugionome, H.	469		384
Sugita, N.	249	Suzuki, M.	246
	405		357
Sugiura, T.	272	Suzuki, N.	224
Suh, S.	447		227
Sulikowski, G.A.	108	Suzuki, T.	035
Sulmon, P.	372		043
	383		198
Sumiya, R.	165		343
	408		412
Sumiya, T.	246		420
Summers, J.B.	150		469
Summersell, R.J.	464	Suzuki, Y.	257
Sun, C.-M.	267	Sviridov, S.V.	437
Sunami, M.	342	Swanson, D.R.	059
Sunderbabu, G.	188		265
Sunjić, V.	091	Swanson, S.	441
Sura, T.P.	296	Sweeney, J.B.	360
Surya Prakash, G.K.	075	Sweet, M.P.	470
	184	Swenson, R.E.	471
	436	Switzer, C.Y.	468
Surzur, J.-M.	.391	Syper, L.	017
Susla, M.	341	Szantay, C.	003
Sustmann, R.	111	Szura, D.P.	149
Suya, K.	065	Szymonifka, M.J.	476

T

Taber, D.F.	236
	395
Tabuchi, T.	299
Taddei, M.	023
	164
	303
	414
	417
	432
	462
Tafesh, A.M.	096
Taffer, I.M.	226
Tagami, K.	290
Tagliavini, E.	020
Tait, B.D.	282
Tajima, K.	245
Tajima, O.	406
Tajira, A.	355
Takacs, B.E.	431
Takacs, J.M.	286
	328
	431
Takada, H.	222
Takagi, K.	190
	262
	424
Takagi, M.	203
Takagi, Y.	041
Takagishi, S.	193
Takahara, J.P.	025
	029
Takahara, P.M.	319
Takahashi, A.	290
Takahashi, H.	021
	039
	117
	142
	331
	405
Takahashi, N.	406
Takahashi, O.	203
Takahashi, S.	092
	202
	212
Takahashi, T.	003
	092
Takahashi, T.	265
	289
Takahata, H.	301

Takahata, H.	362
Takai, K.	021
	026
	238
	245
	270
	271
	388
	429
Takaishi, N.	232
Takaki, K.	339
Takamatsu, T.	362
Takamine, K.	318
	319
Takano, S.	061
	357
Takashika, H.	246
Takasu, M.	055
Takasugi, J.J.	283
Takaya, H.	088
	088
	329
Takayama, H.	463
Takayanagi, H.	167
Takeda, A.	204
	219
	330
	405
	167
Takeda, T.	461
Takei, H.	109
Takei, Y.	450
Takemoto, Y.	215
Takemura, H.	078
Takenoshita, H.	195
Takeuchi, S.	239
Takeyama, H.	469
Takeyama, Y.	442
Takeyasu, T.	266
Taki, H.	170
	272
Takido, T.	168
	194
Takiguchi, T.	406
Takimoto, S.	166
Takiyama, N.	046
	114
Talbert, J.	234
Tam, S.W.	225
Tamagawa, H.	075
Tamai, S.	119
Tamao, K.	030

Tamao, K.	082	Tanimoto, S.	480
	098	Tanino, K.	241
	316	Tanino, K.	410
	335	Tanis, S.P.	325
	386	Tanke, R.S.	378
	389	Tankguchi, H.	318
	440	Tankguchi, Y.	154
	470	Tanko, J.	289
Tamaru, Y.	027	Tanner, D.	122
	112	Tantayanon, S.	077
	175	Tao, F.	202
	233	Tao, Y.-T.	462
	389	Tapolzcay, D.J.	435
	404	Tarasov, V.A.	231
	415	Tarazi, S.	362
Tamm, C.	347	Tardif, S.	228
Tamura, R.	119	Tarnchampoo, B.	375
Tamura, Y.	304	Taschner, M.J.	183
	453	Tashiro, M.	075
Tanaka, C.	379	Tata, J.R.	174
Tanaka, E.	362	Taunton, J.	176
Tanaka, F.	177	Taura, Y.	242
Tanaka, H.	024	Tavasco, C.	020
	148	Taveras Jr., A.G.	445
	336	Taverns, A.G.	247
	373	Tay, M.K.	062
	388		371
	431	Tayano, T.	273
	470	Taylor, N.J.	401
Tanaka, K.	022	Taylor, R.J.	396
	044		423
	078	Taylor, S.K.	033
Tanaka, M.	063	Tazaki, M.	335
	183	Tejero, T.	373
	242	Teleha, C.A.	458
	307	Tellier, F.	268
	401	Tenaglia, A.	038
	459		191
Tanaka, T.	216	Teng, M.	370
Tanaka, Y.	334	Teranaka, T.	405
	481	Terando, N.H.	394
Tang, Q.	084	Terao, K.	125
Tange, H.	044	Terkle-Huslig, T.	235
Tanigawa, H.	112	Terranova, E.	038
Tanigawa, Y.	154	Terrett, N.K.	365
Taniguchi, H.	319		422
Taniguchi, Y.	154	Testaferri, L.	185
	407		293
Tanikaga, R.	135		413
Tanimoto, S.	013	Tetzlaff, C.	277
	157	Teulade, M.-P.	297
	354	Teuting, D.R.	353

Thebtaranonth, C.	375	Tokoroyama, T.	271
Thebtaranonth, Y.	375		349
Theis, W.	396	Tokuda, M.	152
Thenappan, A.	394		469
Therien, M.	337	Tolstikov, G.A.	468
Thetford, D.	191		471
Thiebault, A.	275	Tomaselli, G.A.	006
Thiéry, A.	012	Tomás, M.	384
Thomas, A.	413	Tometzki, G.B.	468
Thomas, E.J.	024	Tominaga, Y.	023
Thomas, R.D.	435		045
Thomas, S.E.	149		078
	365		143
	443		243
Thomi, S.	250		249
Thompson, C.M.	474		379
	476	Tomioka, K.	019
Thompson, D.W.	417		106
Thompson, N.	428		317
Thomsen, M.W.	116	Tomita, S.	061
	479		357
Thorez, A.	068	Tomlinson, G.D.	430
Thorn, D.L.	462		433
Thornton, E.R.	345	Tomooka, K.	194
Thurkauf, A.	243		317
	256	Tonn, C.E.	311
Thurston, J.	415	Toone, E.J.	321
Tian, W.-S.	383	Topoleski, K.	011
Tidwell, T.T.	187	Torii, S.	024
Tiecco, M.	185		148
	293		192
	337		230
	413		373
Tietze, L.F.	110		388
Tilley, J.W.	131		431
Timony, P.E.	218		470
Tingoli, M.	185	Toriyama, M.	168
	293	Torra, N.	173
	413	Torres, E.	351
Tishcenko, I.G.	437	Torrini, I.	294
Tius, M.A.	321	Toshimitsu, A.	125
Tobin, T.	310	Tosi, G.	064
Toda, F.	043	Toupet, L.	157
	044	Tour, J.M.	284
	078		457
	184		463
Toda, S.	287	Toy, A.	195
Todesco, P.E.	144	Toyoda, J.	155
Tohjima, K.	360	Tozawa, Y.	355
Tokitoh, N.	138	Trabelsi, M.	411
Tokles, M.	107	Trivellas, A.	027
		Trombini, C.	020

Trometer, J.D.	350	Tsuji, J.	245
Trost, B.M.	079		252
	191		272
	264		405
	279		405
	288		461
	294	Tsuji, T.	025
	305		347
	306		444
	315	Tsuji, Y.	077
	344		135
	418		146
	433		151
	454		180
	463		241
	466	Tsukamoto, M.	349
	468	Tsukui, T.	112
	472		250
Troupel, M.	067	Tsumiyama, T.	381
Trova, M.P.	045	Tsuruta, T.	340
Trovarelli, A.	043	Tsuruya, S.	460
Truchet, F.	350	Tsutsumi, O.	042
Trujillo, D.A.	415	Tubul, A.	233
Tsai, C.-Y.	267	Tueting, D.R.	034
Tsanaktsidis, J.	207	Tufariello, J.J.	276
Tscheschlok, K.	186	Tuladhar, S.M.	412
Tse, C.-W.	289	Turnbull, K.	163
Tso, H.-H.	462	Turner, J.A.	400
	481	Turner, S.U.	231
Tsubaki, K.	175	Tykwinski, R.	456
	359		
Tsuboi, S.	204	**U**	
	330		
	405	Uccella, N.	261
Tsuboyama, K.	167	Uchida, M.	246
Tsubuki, T.	142	Uchida, T.	155
Tsuchihashi, G.	034		324
	194	Uchida, Y.	174
	198		317
	317	Uchimura, J.	320
	341	Ucida, Y.	089
	384	Uda, H.	290
Tsuchima, K.	195	Ueda, Y.	478
Tsuchiya, T.	206	Uemura, S.	125
Tsuda, T.	165		249
	188		405
	248	Ueno, Y.	379
	408	Ugajin, S.	101
Tsuge, O.	146		111
Tsuji, J.	163	Uguen, D.	454
	170	Ujikawa, O.	047
	222	Ukai, J.	465

Ukaji, Y.	257	Utimoto. K.	407
	278		428
Ukita, T.	187		429
Umana-Ronchi, A.	366		434
Umani-Ronchi, A.	020		441
Umano, S.	450		442
Umezawa, J.	203		464
Underwood, J.M.	435	Uyehara. T.	129
Uno, M.	092		
Uozumi, Y.	136	**V**	
Upadhye, B.K.	166		
Ura, T.	013	Vågberg. J.O.	114
	201		390
	235		433
Uragaki, T.	253	Vader, J.	420
Urata, H.	013	Vaid, B.K.	006
	180	Vaid, R.K.	006
Urbanowicz, J.H.	009		090
	012		132
Uribe, J.M.	087		418
Urogdi, L.	137		420
	155	Valdes, C.	458
Urpi, F.	123	Valle, G.	237
	173	Valnot, J.-Y.	309
Urso, F.	133	Van der Steen, F.H.	358
Uruma, T.	177	Van Ende, D.	204
Uryu, T.	135	Van Hijfte, L.	110
Usami, Y.	341	Van Horn, D.E.	092
Ushio, H.	022	van der Baan, J.L.	434
Ushio, K.	331	van der Heide, F.R.	139
Ushio, Y.	145	van der Louw, J.	434
	196	van der Werf, A.	376
Utaka, M.	204	van Elburg, P.A.	409
	219	van Hijfte. L.	002
	330	van Koten. G.	358
Utimoto. K.	021	van Leusen. A.M.	067
	026	van Niel, M.B.	257
	082	Van, T.T.	380
	127	Vanderesse, R.	016
	221		041
	226	Vandevelde, O.	380
	230	Vanemon, P.	172
	238	Vankar, P.S.	254
	245	Vankar, Y.D.	447
	270	Vannoorenberghe, Y.	020
	271	Vannoorenberghe, Y.	040
	287	Vara Prasad, J.V.N.	049
	301	Varma, R.S.	297
	343	Vasantha, G.	267
Utimoto. K.	346	Vasapollo, G.	011
	388		?
	406	Vaultier. M.	

Vaultier, M. 350
Veal, W.R. 139
Veale, C.A. 413
Vedejs, E. 197
271
365
Veenstra, S.J. 233
Veeraiah, T. 007
Vel'der, Ya.L. 468
471
Velde, D.V. 250
Venanzi, L.M. 071
Venkatachalam, C.S. 225
Venkataramani, P.S. 008
Venturello, C. 320
Venturini, I. 138
Verhé, R. 142
Vermeer, P. 267
Vernhet, C. 105
Vernon, P. 384
Vervini, L.A. 276
Veschambre, H. 038
Veselovskii, A.B. 237
Veselovsky, V.V. 454
Vessiere, R. 140
Vest, G. 123
Vidari, G. 269
Viehe, H.G. 461
Vieira, P.C. 313
Vilarrasa, J. 123
Vilarrasa, J. 173
Vile, S. 190
Villalobos, A.C. 208
Villemin, D. 435
Villieras, J. 079
369
Vilsmaier, E. 277
Vincze, I. 058
Vinczer, P. 003
Vinković, V. 251
Vinogradov, M.G. 423
Virard, C. 043
Visentii, G. 012
Vo-Quang, L. 012
062
273
Vo-Quang, Y. 012
Vo-Quang, Y. 062
273
Vogiazoglou, D. 081
Volante, R.P. 356

Vottero, C. 296
Vougioukas, A.E. 024
335
Vovk, M.V. 121
Voyle, M. 191
Vukićević, R. 392
412

W

Wada, H. 405
Wada, M. 026
134
148
349
Wadgaonkar, P.P. 102
Wadman, S. 349
351
Waegell, B. 038
Wagner, A. 058
209
417
Wakabayashi, S. 170
342
Wakamatsu, K. 287
Wakamatsu, T. 164
Wakharkar, R.D. 166
Walchli, R. 279
Walker, B.J. 272
Walkup, R.D. 391
Wallace, P. 473
Waller Jr., J. 183
Walsh, C.T. 185
Walsh, R. 199
Walter, D.S. 427
Walters, M.A. 479
Walters, T.R. 293
Wamsley, E.J. 103
Wang, B.S.L. 223
Wang, C.-J. 267
Wang, D. 021
415
418
Wang, E.C. 301
Wang, G. 071
Wang, K.K. 265
Wang, M.-D. 179
Wang, W. 093
Wang, W. 193
274
Wang, X. 203
Wang, Y. 001

Wang, Y.	007	Welch, M.C.	252
	139	Welch, M.J.	436
	274	Welch, S.C.	450
Wang, Z.	029	Welta, M.	341
Ward, D.E.	042	Wen, X.	124
Ward, T.R.	071	Wen, X.-Q.	274
Ware, A.C.	339	Wender, P.A.	107
Wariishi, K.	412		465
	434	Weng, L.	418
Warren, S.	473	Wenkert, E.	078
Warshawsky, A.M.	438		133
Warwel, S.	465	Wermuth, C.-G.	164
Wasserman, H.H.	377	Westling, M.	380
Watanabe, M.	018	Whitby, R.	349
	114	.	430
	122	White, A.W.	107
	249	Whitehead, J.F.	441
Watanabe, N.	308	Whitesell, J.K.	348
Watanabe, T.	112	Whitesides, G.M.	321
	249	Whiting, D.A.	092
	250	Widdowson, D.A	258
Watanabe, Y.	077	Widener, R.K.	104
	135	Widmer, U.	054
	146	Wiemer, D.F.	477
	151		480
	180		483
	241	Wienand, A.	115
	287	Wiggins, J.M.	220
	368	Wiglesworth, C.	037
	379	Wiley, M.R.	334
	402	Williams, A.D.	222
	402	Williams, D.L.	297
	451	Williams, D.R.	172
Watt, D.S.	037		379
	182	Williams, G.M.	477
	240	Williams, R.M.	003
	310		129
Weavers, R.T.	406	Williams, R.V.	045
Weber, A.E.	337		300
Weber, E.J.	052		460
Webster, F.X.	312	Williard, P.G.	106
Weeks, J.	011	Willis, C.R.	226
Weetman, J.	410	Wilson, K.D.	361
Weglarz, M.A.	127	Wilson, L.J.	139
Wei, J.	432	Wilson, R.M.	275
Wei, Z.Y.	415	Wilson, S.R.	191
Wei, Z.Y.	418	Wingard, A.K.	281
Weinreb, S.M.	123	Winkler, J.D.	085
Weinreb, S.M.	124		095
	140		106
	176		396
Welch, J.S.	235		

Y

Winkler, T.	209
Winotai, C.	307
Wirth, D.D.	288
Wisniewski, V.	214
Wiszniewski, V.	256
Witz, P.	002
Wolfe, J.F.	364
	369
Wolin, R.L.	110
Wong, H.N.C.	081
	099
	225
	279
	289
Wong, M.O.	225
Wong, T.	279
Woodard, R.W.	308
Woodgate, P.D.	394
Woodward, P.R.	363
	401
Wölfel, G.	155
Wright, B.T.	094
Wu, A.	184
Wu, A.-h.	074
	235
Wu, C.	051
Wu, G.	457
Wu, H.	272
Wu, J.-P.	053
Wu, R.	284
Wu, S.-W.	085
Wu, T.-C.	035
	408
Wu, Y.-Y.	061
Wulff, W.D.	015
Wuts, P.G.M.	160
Wynberg, H.	023

X

Xiang,Y.B.	369
Xiao, C.	357
Xie, G.	202
Xie, L.	272
Xin, Y.	193
Xiong, H.	288
Xu, L.	202
Xu, Y.	297
Xu, Y.	402
Xu, Y.-C.	015
Xy, S.L.	460

Yadav, J.S.	269
	300
Yadav, V.	032
Yadav-Bhatnagar, N.	152
Yadla, R.	304
Yadov, V.K.	202
Yagi, M.	043
	184
Yakahashi, K.	301
Yakura, T.	453
Yam, T.M.	223
Yamada, H.	334
	405
	424
Yamada, J.	080
	119
	232
	328
	344
	349
Yamada, K.	177
	179
	401
Yamada, S.	092
	463
Yamada, T.	146
	163
	245
	369
Yamada, Y.	018
	152
	175
	198
Yamaguchi, M.	047
	053
	166
	220
	225
	299
	322
	330
	338
Yamaguchi, R.	301
Yamaguchi, Y.	334
	397
Yamakawa, K.	150
	200
	248
	419
	422

Yamakawa, K.	452
	478
Yamakawa, Y.	368
	402
Yamamoto, A.	194
	326
	367
Yamamoto, E.	269
Yamamoto, H.	021
	030
	049
	055
	065
	096
	106
	196
	215
	352
	358
	392
	421
	425
	465
Yamamoto, K.	044
	204
.	311
.	311
	444
Yamamoto, M.	177
	179
Yamamoto, N.	204
	444
Yamamoto, T.	075
	112
	194
	338
Yamamoto, Y.	030
	080
	119
	129
	232
	311
	328
	344
	349
Yamaoka, S.	316
Yamashina, N.	410
Yamashita, A.	195
Yamashita, D.S.	203
Yamashita, M.	246
Yamashita, S.	024
	148

Yamashita, S.	431
	470
Yamashita, T.	143
Yamawaki, K.	170
	201
	235
Yamazaki, S.	062
Yamazaki, T.	301
	338
	362
Yamazaki, Y.	062
Yanagi, T.	472
Yanagihara, N.	428
Yanagisawa, A.	096
Yanagiya, M.	055
Yanaguchi,.Y.	048
Yanase, M.	357
Yang, B.	274
Yang, C.-P.	151
Yang, J.	124
	465
Yang, K.E.	265
Yang, L.	328
Yang, N.C.	268
Yang, P.-F.	087
Yang, S.	190
Yang, S.-M.	280
Yang, S.H.	214
Yankep. E.	214
Yannakopoulou, K.	301
Yasuda, M.	143
Yeates, C.	349
Yefsah, R.	255
Yeh, M.C.P.	001
	025
	109
	234
	357
	447
Yi, P.	084
Yip, Y.-C.	289
Yoakim, C.	057
	337
Yoihioka, H.	436
Yokoyama, S.	111
Yona, I.	211
Yonashiro, M.	313
Yoneda, R.	262
Yoneda, T.	331
Yonemitsu, O.	055
Yoneyoshi, Y.	156
Yoon, D.C.	294

Yoon, K.B.	216	Yus, M.	132
Yoon, M.S.	060		171
	068		268
Yoon, N.M.	059		273
	100		284
Yoshida, J.	184		321
	356		351
	361		440
	426		467
Yoshida, K.	164	Yuste, F.	241
Yoshida, T.	017	Yusubov, M.S.	253
	170		
	201	**Z**	
	235		
Yoshida, Z.	027	Zablocka, M.	297
	112	Zabrowski, D.L.	292
	175	Zabłocka, M.	485
	233	Zahalka, H.A.	031
	389	Zahra, J.P.	430
	404	Zaidlewicz, M.	049
	415	Zajac Jr., W.W.	293
Yoshikawa, S.	089	Zamarlik, H.	189
Yoshikoshi, A.	035	Zamboni, R.	182
	343	Zang, X.	238
	420	Zanirato, P.	159
Yoshiltomi, S.	212	Zard, S.Z.	051
Yoshimura, J.	214		161
Yoshimura, N.	196		290
Yoshioka, H.	207	Zawadzki, S.	155
	338	Zecchini, G.P.	294
	362	Zefirov, N.S.	456
	363	Zercher, C.	114
	440	Zetta, L.	019
	441	Zezza, C.A.	157
Yoshioka, K.	365		360
Yoshioka, M.	021		368
Yoshitomi, S.	216		370
You, M.-L.	464	Zhai, D.	003
Youn, I.K.	040	Zhai, W.	003
Youn, J.-H	007	Zhaishibekov, B.S.	423
Young, J.-J.	437	Zhang, H.-Z.	062
Young, R.N.	182		102
Young, S.M.	235	Zhang, J.	297
Yu, S.-G.	280		465
Yu, Y.	455	Zhang, X.	014
	471	Zhang, Y.	082
Yugari, H.	180		094
Yuhara, M.	245		138
Yuhara, M.	405	Zhang, Y.	158
Yumoto, M.	328		202
Yus, M.	027		318
	115		456